Whiz-Kid Mathematics 2

13 - 15 Years

Teach Yourself

Arinze Oranye

COPYRIGHT NOTICE

All rights reserved. No part of this publication may be stored, reproduced in a retrieval system or transmitted, in any form or by any means, electronic, mechanical, photocopying, recording or otherwise, without the prior permission of the copyright holder.

This book may not be lent, resold, hired out or otherwise disposed of by way of trade in any form other than that in which it is published, without the written permission of the author.

Copyright © Arinze Edward Oranye 2017

ACKNOWLEDGEMENTS

The author would like to thank the following organisations for permission to reproduce photographs and table.

1) www.123rf.com for most of the pictures used in this book.
2) Pictures of students in chapter 9 algebra section, chapter 8 3d hemisphere, exercise 14a – cubes ……..Designed by Creativeart/Freepik
3) www.surveyanalysis.org …………. Statistical table on soft drinks - Chapter 18
4) Wikipedia – Picture of Fibonacci

Every effort was made to contact copyright owners/holders of materials reproduced in this book. If through oversight, any omissions will be rectified in future printings of this book if notice is given to the copyright holder.

I humbly submit to The Almighty God, for his Mercies, Guidance, Love and Grace from inception to the completion of this book.

CONTENTS

1) NUMBERS (1) REVIEW — 6
- Significant figures — 7
- Estimation and Approximation — 11
- Multiplying whole numbers — 14
- Dividing whole numbers — 20
- Multiplying by 10, 100,… — 23
- Dividing by 10, 100, 1000,.. — 25
- Functional maths — 29
- Chapter 1 review section — 30

2) NUMBERS (2) & INDICES — 37
- Multiples — 38
- LCM — 40
- Product of prime factors for LCM — 41
- Venn Diagram for LCM — 42
- Factors — 45
- Prime numbers — 47
- Prime factors — 48
- Product of prime factors — 50
- HCF — 54
- Venn Diagram for HCF — 57
- Divisibility test — 60
- Powers, squares and roots — 67
- Cubes — 68
- Cube roots — 69
- Laws of Indices — 70
- Standard form — 72
- Chapter 2 review section — 77

3) FRACTIONS — 82
- Improper fractions — 84
- Mixed numbers — 84
- Equivalent fractions — 86
- Simplifying fractions — 91
- Adding and subtracting fractions — 93
- Adding and subtracting mixed numbers — 99
- Multiplying fractions — 102
- Fraction of an amount — 106
- Multiplying mixed numbers — 109
- Dividing fractions — 112
- Chapter 3 review section — 115

4) PERCENTAGES 1 — 118
- Fractions to decimals — 121
- Percentages to fractions — 122
- Fractions to percentages — 124
- Chapter 4 Review section — 131

5) DIRECTED NUMBERS — 133
- Negative numbers — 134
- Adding and subtracting negative numbers — 139
- Multiplying and dividing negative numbers — 142
- Chapter 5 review section — 144

6) PERCENTAGES 2 — 145
- Percentages of a quantity — 146
- Percentage increase and decrease — 148
- Percentage change — 149
- Simple interest — 150
- Compound interest — 153
- Depreciation — 153
- Reverse percentages — 154
- Chapter 6 review section — 156

7) ALGEBRA 1 — 158
- Order of operations (BIDMAS) — 159
- Substitution — 160
- Formula and substitution — 163
- Equations — 164
- Fractional equations — 167
- Equations with brackets — 169
- Equations with unknown on both sides — 170
- Forming and solving equations — 172

Inequalities	174
Solving inequalities	176
Inequalities and regions	177
Chapter 7 review section	179
8) POLYGONS AND CIRCLES	182
Circles	182
Lines and angles	183
Angles at a point	183
Vertically opposite angles	183
Triangles and properties	185
Isosceles triangle	185
Equilateral triangle	185
Quadrilaterals	189
Rhombus	189
Parallelogram	190
Trapezium	190
Angles in quadrilaterals	192
Polygons	194
Circles	194
Angles and parallel lines	196
Alternate angles	196
Corresponding angles	197
Chapter 8 review section	198
9) AREA AND PERIMETER	199
Circles and perimeter	200
Area of 2-d shapes	203
Area of compound shapes	210
Shaded areas	212
Area of a circle	213
10) SEQUENCES	220
Linear sequence	221
Triangle numbers	221
Fibonacci sequence	221
Term numbers	223
Graphs and sequences	224
Nth term	225
Quadratic sequences	227
Chapter 10 review section	230
11) RATIO & PROPORTION	231
Simplifying ratios	232
Equivalent ratios	232
Dividing/sharing ratios	234
Direct proportion	235
Inverse proportion	237
Map, scale and ratio	238
12) BRACKETS & FACTORISATION	239
Expanding brackets	240
Double brackets	243
Factorisation	245
LCM of algebraic expressions	247
Venn diagrams	247
Algebraic fractions	248
Chapters 11 and 12 review sections	251
13) PROBABILITY	252
Probability scale	253
Mutually exclusive events	256
Two-way tables	257
Tree diagrams	258
Experimental probability	259
Relative frequency	259
Chapter 13 review section	261
14) RECIPROCAL & INVERSE	262
15) GRAPHS AND GRADIENTS	264
Straight line graphs	265
Horizontal line	268
Vertical line	268
Equations of straight lines	269
Gradients	270
Mid-point of a line segment	273
$Y = MX + C$	274

Intercept	274
Parallel lines	276
Perpendicular lines	276
Quadratic graphs	277
Chapters 14 and 15 review sections	279
16) DIAGRAMS	280
Scatter diagram	281
Correlation	281
Line of best fit	282
Stem and leaf diagram	283
17) AVERAGES AND RANGE	285
Mean	286
Median	287
Mode	288
Range	289
Averages from diagrams	292
Averages from grouped data	295
18) PYTHAGORAS' THEOREM	297
Length of a line segment	301
19) BEARINGS	302
20) SIMULTANEOUS EQUATION	305

1 Numbers (1) Review

This section covers the following topics:

- Rounding to significant figures
- Estimation & Approximation
- Multiplying and Dividing Whole Numbers
- Multiplying by 10, 100, 1000…
- Functional mathematics

LEARNING OBJECTIVES

By the end of this unit, you should be able to:

a) Round numbers to significant figures
b) Estimate and approximate numbers
c) Multiply and divide whole numbers
d) Multiply by 10, 100, 1000…

KEYWORDS

- Round
- Estimate
- Approximate
- Significant figure

1.1 ROUNDING TO SIGNIFICANT FIGURES

Writing a number to a given number of significant figures is almost the same as writing a number to a given number of decimal places. The **only exception** is that we must count all the figures in the number, not just the decimal parts.

The first significant figure is the first **non-zero** digit of the number. Significant figures are often abbreviated to **s.f.**

Example 1:
Consider the number   4   2   5   9

- 1st s.f → 4
- 2nd s.f. → 2
- 3rd s.f. → 5
- 4th s.f. → 9

Example 2:
Consider the number

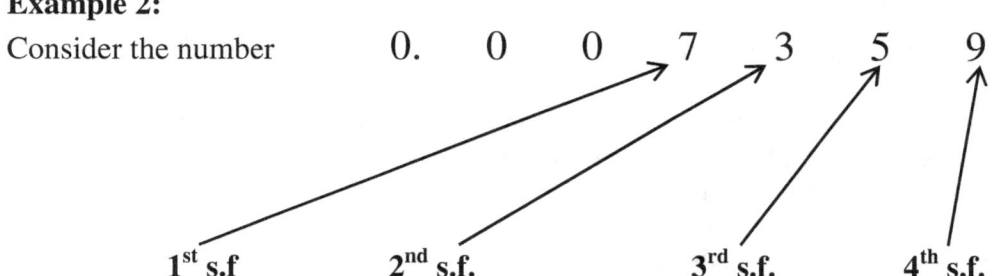

Example 3:
Consider the number

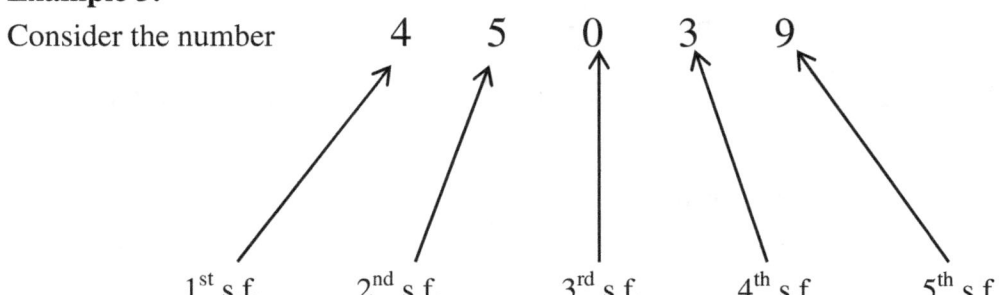

Example 4:

Write 423 to one significant figure

Solution: The first significant figure from 423 is 4. The number 2, which is after the number 4 is not up to 5, so we round down the remaining numbers to zero.

Therefore, 423 to 1 s.f. = **400**

Example 5:

Write 423 to 2 significant figures

Solution: The second significant figure from 423 is 2. The number 3, which is after the number 2 is not up to 5, so we round down to zero.

Therefore, 423 to 2 s.f. = **420**

Example 6:

Write 0.003756 to 2 significant figures

Solution: Remember that the first significant figure is the first non-zero digit. The number 7 is the second significant figure in the number 0.003756. Since the next number after 7 is 5, which is up to 5 (rounding rules), we round up.

Therefore, 0.003756 to 2 s.f. = **0.0038**

Example 7:

Write 84.375 to 3 significant figures

Solution: The 3rd significant figure from 84.375 is 3. However, 7 is up to 5 (rounding rules), so we round up.

Therefore, 84.375 to 3 significant figures = **84.4**

EXERCISE 1A

1) Write the first significant figure for the numbers given below.

a) 237

b) 34

c) 0.05

d) 0.954

e) 45876

f) 306

g) 5.67

h) 54.908

2) Write the second significant figure for the numbers given below.

a) 347

b) 0.187

c) 56.897

d) 0.107

e) 90.789

f) 78949

g) 106.9

h) 2.93

3) Round each of the numbers below to one significant figure.

a) 68

b) 786

c) 2.56

d) 1.987

e) 0.378

f) 72.99

g) 36778

h) 0.083

4) Write each of the numbers below to the given number of significant figures.

a) 4035   1 s.f.

b) 8.63   1 s.f.

c) 5.654 2 s.f.

d) 3.167 2 s.f.

e) 8.907 3 s.f.

f) 12.599 3 s.f.

5) Some lengths of leisure parks in Nigeria are given below with their lengths (not to scale).

Amusement Parks	Lengths (m)
Dreamland Africana (Lekki)	1234
Hi-Impact Planet (Lagos)	867
The Wild Bunch (Jos)	1234
Klubdelag (Lagos)	790
Fun Factory (Ibadan)	1041
Omu Resort (Lekki)	498

a) Round each length to one significant figure.

b) Round the length of *The Wild Bunch* park to 3 significant figures.

c) If the lengths of the Dreamland Africana and Fun Factory are rounded to 2 significant figures, what is the difference in their lengths?

6) A container of chocolates contains 70 chocolates to one significant figure. What is the smallest number of chocolates that could be in the container?

7) Write each number to the given number of significant figures.

a) 39   1 s.f.

b) 54.9   2 s.f.

c) 0.87   1 s.f

d) 89.54   2 s.f.

e) 6788   3 s.f

f) 56.23   1 s.f.

g) 812   1 s.f.

h) 0.8764   2 s.f.

i) 543.2   3 s.f.

j) 1.098   2 s.f.

k) 367   1 s.f.

l) 4.903   2 s.f

1.2 ESTIMATION AND APPROXIMATION

Estimation only means to make a guess very close to the correct answer. However, it might be impossible to estimate accurately. Estimating an answer to a given calculation gives an approximate answer which is close to the real answer.

We use estimation to check that the answer is about right. It is useful when shopping on a tight budget. Mental estimation of selected goods is vital so that we do not get embarrassed when trying to pay at the counter.

Rounding all the numbers to **one significant figure** is the most convenient way to estimate calculations.

Example 1: Estimate the answers to the problems below.

a) 4.2×5.3

b) 0.9×0.9

c) 48×9.6

d) $\sqrt{99} \times \sqrt{50}$

e) 30.2×8.7

f) $27.8 \div 0.98$

g) 9% of ₦89.95

Solutions: We use approximately equal to sign ($\approx$) when estimating calculations. Therefore, we convert all to one significant figure.

a) $4.2 \times 5.3 \approx 4 \times 5 = \mathbf{20}$

b) $0.9 \times 0.9 \approx 1 \times 1 = \mathbf{1}$

c) $48 \times 9.6 \approx 50 \times 10 = \mathbf{500}$

d) $\sqrt{99} \times \sqrt{50} \approx \sqrt{100} \times \sqrt{49} = 10 \times 7 = \mathbf{70}$

e) $30.2 \times 8.7 \approx 30 \times 9 = \mathbf{270}$

f) $27.8 \div 0.98 \approx 30 \div 1 = \mathbf{30}$

g) 9% of ₦89.95 $\approx$ 10% of ₦90 = **₦9**

EXERCISE 1B

1) Give an estimate for each of the following calculations.

   a) 6.83 × 9.53

   b) 18.99 × 10.09

   c) 42.97 × 9.87

   d) 48% of ₦88.90

   e) 110 + 95 + 131

2) The rent for a house is ₦995 per week. Estimate the total amount spent on rent in one year.

3) A size seven men's shoes cost ₦924. Estimate the cost of 47 shoes.

In questions 4 and 5, there are four calculations and four answers. Write down the correct answer from the list given for each calculation. The answers are estimates.

4) a) 3.2 × 9.3

   b) 20.25 ÷ 4.89

   c) 42 × 1.97

   d) 3.5 × 4.8

Answers: 4   20   30   80

5) a) 202.5 ÷ 99.73

   b) 19 × 1.83

   c) 51% of 499.80

   d) 3 × 0.98

Answers: 3   250   2   40

6) Estimate the answer/value the arrow is pointing at in the diagrams below.

a)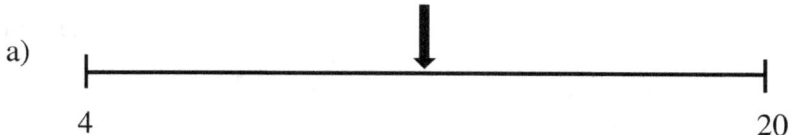

b)

c)

7) Estimate the answers to the following calculations.

a) $\dfrac{19 \times 1.8}{0.89 + 3.9}$

b) 8.93^2

c) $\sqrt{48.88} \times 1.7$

d) A tenth of 693

e) 13% of 517

f) 0.89% of 1083.45

g) In a grocery shop in London, Chukwudi bought eight tins of sardines at 95 pence each. He paid with a ten-pound note. Describe a quick way for Chukwudi to know how much he should expect as change.

1.3 MULTIPLYING WHOLE NUMBERS

Some multiplications can be done straight away in your head. However, some cannot.

$$\left.\begin{array}{l} 5 \times 3 = 15 \\ 8 \times 2 = 16 \\ 4 \times 9 = 36 \\ 7 \times 7 = 49 \\ 10 \times 10 = 100 \end{array}\right\} \text{can be done mentally}$$

257×389 needs to be worked out on paper unless you are a genius or a walking computer. Multiplication table can help when multiplying whole numbers.

TABLE 2A

×	1	2	3	4	5	6	7	8	9	10
1	1	2	3	4	5	6	7	8	9	10
2	2	4	6	8	10	12	14	16	18	20
3	3	6	9	12	15	18	21	24	27	30
4	4	8	12	16	20	24	28	32	36	40
5	5	10	15	20	25	30	35	40	45	50
6	6	12	18	24	30	36	42	48	54	60
7	7	14	21	28	35	42	49	56	63	70
8	8	16	24	32	40	48	56	64	72	80
9	9	18	27	36	45	54	63	72	81	90
10	10	20	30	40	50	60	70	80	90	100

$2 \times 8 = 16$ $7 \times 9 = 63$

Three numbers can also be multiplied together.

$1 \times 3 \times 5 = 15$

$2 \times 4 \times 8 = 64$

$3 \times 7 \times 10 = 210$

Example 1: 13 × 4

```
    13
 ×   4
   ---
    52
    ①
```

Always start your multiplication from the ones column.
3 × 4 = 12
Write down the digit **2** in the ones column and move the **1** over to the tens column. 4 × 1 = 1 plus the 1 (moved over) = 5.

Example 2: 24 × 12

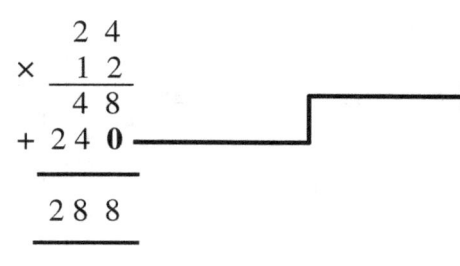

2 × 4 = 8
2 × 2 = 4

Add zero before multiplying by 1

1 × 4 = 4
1 × 2 = 2
Therefore, 24 × 12 = **288**

Example 3: 135 × 45

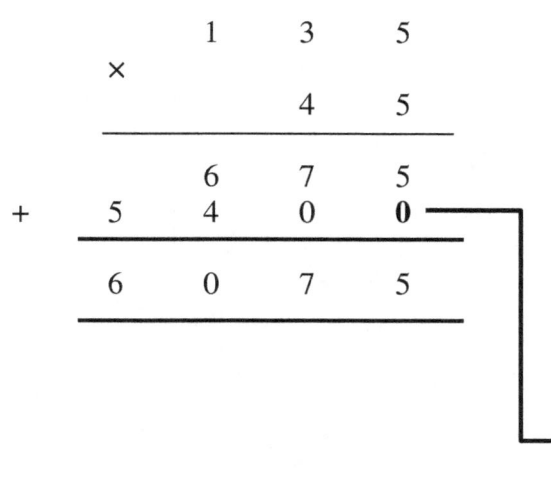

We first multiply 135 by 5 which gives 675.

Before multiplying by 4, add zero to the ones column.

Then, 135 x 4 = 540

Adding 675 and 5400 gives **6075**

THE GRID METHOD

EXAMPLE 4: Multiply 152 by 17

Split 152 into hundreds, tens and ones. That gives 100 + 50 + 2

Split 17 into tens and ones. That gives 10 + 7

Put all the numbers in a table form.

×	100	50	2
10	**1000**	**500**	**20**
7	**700**	**350**	**14**

100 × 10 = **1000**,　　100 × 7 = **700**,　　50 × 10 = **500**,　　50 × 7 = **350**

2 × 10 = **20**　　and 2 × 7 = **14**

Add up all the multiplications (blue numbers) to give

1000 + 500 + 20 + 700 + 350 + 14 = 2584

Therefore, 152 × 17 = **2584**

EXERCISE 1C

1) Without using a calculator, work out

   a) 13×8
   b) 14×9
   c) 22×8
   d) 61×7
   e) 95×4
   f) 12×14
   g) 5×32
   h) 76×3
   i) 12×7
   j) 98×5
   k) 11×7
   l) 67×9
   m) 34×12
   n) 57×13
   o) 93×76

2)

   a) $6 \times 5 \times 10$
   b) $8 \times 3 \times 10$
   c) $3 \times 4 \times 5$
   d) $3 \times 8 \times 5$
   e) $9 \times 2 \times 10$
   f) $10 \times 3 \times 5$
   g) $5 \times 5 \times 5$
   h) $3 \times 7 \times 10$
   i) $6 \times 7 \times 5$
   j) $2 \times 4 \times 10$
   k) $9 \times 2 \times 10$
   l) $7 \times 8 \times 10$
   m) $10 \times 2 \times 3$
   n) $9 \times 2 \times 4$
   o) $11 \times 10 \times 2$

3) What is 5 times 78?

4) Multiply 14 by 237

5) Tochukwu has five pens. Ikechi has fifteen times as many. How many pens does Ikechi have?

6) There are 19 cubes in a block. There are 120 blocks. How many cubes are there?

7) Answer all the questions without using a calculator.

   a) 15×16
   b) 67×57
   c) 65×765
   d) 786×12
   e) 890×12
   f) 34×198

8) Work out the product of 130 and 14

9) A fridge costs ₦25,500. How much would 12 fridges cost?

10) Copy and complete the grid below.

×	3	6	7	9	10
2					
			56		
7		42			
					110

11) The area of a flower garden is rectangular as shown below.

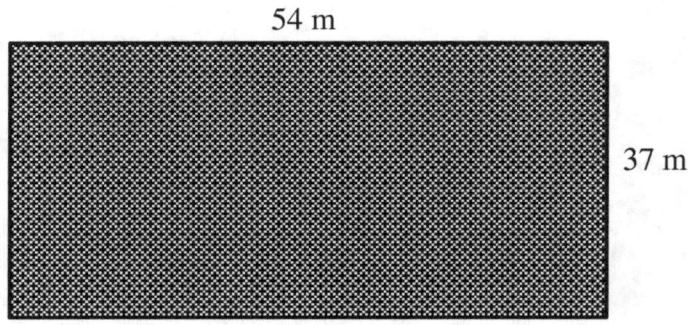

54 m

37 m

What is the area of the flower garden?

12) When you multiply 368 by 17, what is the value of the digit **2** in your answer?

13) Chinyere paid ₦23,000 into her current account which already had ₦13,400. A week later, her employer paid in three times the amount she had in her bank account. How much is in Chinyere's bank account now?

14) Copy and complete the multiplications below.

a) 13 × ☐ = 91   b) ☐ × 30 = 300   c) 45 × 67 = ☐

15) Jake saves ₦1,350 every week from his salary. How much will he save after

    a) 2 weeks   b) 5 weeks   c) 47 weeks   d) 50 weeks?

16) There are 12 eggs in a box. There are 25 boxes. How many eggs are there altogether?

17) Look at the numbers in the box below.

    8   15   16   32   37   128   171   899   4096

    a) What is the largest number multiply by 12?

    b) Multiply the third smallest number by the smallest number

    c) The product of two numbers in the box above is equal to another number in the box. Find the three numbers.

    d) Identify a square number from the box above

18) To watch Super Eagles of Nigeria play football at the National Stadium, Aminu took his two brothers, a cousin, and three friends. The average ticket costs ₦2500. How much would Aminu pay to watch the game with his entourage?

19) An Arts Theatre in Ebonyi State has 47 rows and 35 seats. How many seats are there in the Arts theatre altogether?

20) Copy and complete the multiplication table below.

×	4			9
2				
		35	40	45
7		49		
	32			

1.4 DIVIDING WHOLE NUMBERS

Division is the opposite of multiplication.

Example 1: $6 \div 3$

$6 \div 3$ simply means how many times 3 go into 6. In this case, the answer is 2

Example 2: $48 \div 12$

Knowing your times table helps in division. $48 \div 12$ means how many times 12 go into 48. In this case, the answer is 4.

Example 3: $6290 \div 5$

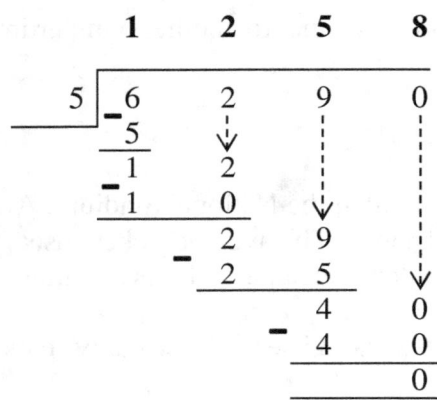

Using long multiplication, 5 goes into 6 once, so we write the **1** on top of 6 as shown.

$1 \times 5 = 5$, so we write the 5 underneath the 6 and take it away. $6 - 5 = 1$.

5 cannot go into 1, so we carry down the 2 from the next column to make 12. Now, 5 goes into 12, twice, so we write the 2 on top.

$2 \times 5 = 10$, we take away 10 from 12 to be left with 2.

Again, 5 cannot go into 2, so we carry down the 9 from the next column to make 29. 5 can go into 29 five times, so we write 5 on to and multiply by 5 to give 25.
$29 - 25 = 4$. Also, 5 cannot go into 4 so we carry down the 0 from the next column to make 40. At this point, 5 can go into 40 exactly to give 8.
We therefore write the 8 on top.
$6290 \div 5 = \mathbf{1258}$

Alternatively, there is a shorter method.

```
         1   2   5   8
             1   2   4
    5 )  6   2   9   0
```

6 divide by 5 = 1, remainder 1. We write the **1** on top and move the remainder (1) to the top of the next number to make 12.

12 ÷ 5 = 2 remainder 2. We then write the **2** on top and move the remainder (2) to the top of the next number to make 29.

29 ÷ 5 = 5 remainder 4. We then write the **5** on top and move the remainder (4) to the top of the next number to make 40.

40 ÷ 5 = **8** exactly, we then write 8 on top.

Therefore, 6290 ÷ 5 = **1258**

Example 4: 25 ÷ 4

Some division are not whole numbers.

```
         0   6.  2   5
    4 )  2   5
       - 0   ↓
             2   5
         -   2   4
                 1   0
              -      8
                     2   0
                  -  2   0
                         0
```

4 cannot go into 2 (using the system of long multiplication), so we write zero (0) on top.

0 × 4 = 0 and then take zero away from 2 to give 2.

4 cannot go into 2, so we bring down the 5 from the next column to make 25. Four (4) then goes into 25, six (6) times. We write 6 on top and multiply by 4 to give 24. 25 − 24 = 1.

We must now put a decimal point as there are no numbers left to bring down. Also, 4 cannot go into 1, we therefore bring down an imaginary zero (0) to make it 10. 4 then goes into 10 twice. We then write 2 on top. 2 × 4 = 8. 10 − 8 = 2. Again, we bring down an imaginary zero to make 20. At that point, 4 will go into 20, 5 times. We then write the 5 on top. 5 × 4 = 20, and 20 − 20 = 0

Therefore, 25 ÷ 4 = **6.25**

EXERCISE 1D

1) Work out

   a) 40 ÷ 2

   b) 9 ÷ 3

   c) 44 ÷ 11

   d) 100 ÷ 5

   e) 72 ÷ 6

   f) 182 ÷ 14

   g) 200 ÷ 20

   h) 68 ÷ 2

   i) 645 ÷ 5

   j) 144 ÷ 9

   k) 225 ÷ 3

   l) 2000 ÷ 5

   m) 3500 ÷ 5

   n) 9028 ÷ 122

   o) 15554 ÷ 7

2) Share ₦4800 equally between two people. How much will each receive?

3) Divide 144 kg by 6

4) How many 5's make 600?

5) Three people win a prize of ₦1950 between them equally. How much does each person receive?

6) A calculator factory has nine workers to produce 6741 calculators every week. How many calculators does each worker produce per week if they all work at the same rate?

7) Twenty people won a competition. They shared the ₦19,740 prize equally between them. How much did each person receive?

8) A group of 140 tourists travel by coach. Each coach holds 24 people.

a) How many coaches are needed for the tourists?

b) If 60 more tourists joined the group of travellers, how many coaches are needed?

9) Three friends, Tokunbo, Chibogu and Halima, received ₦9,250 from their Principal for hard work in school.

   How much could each person receive?

1.5 MULTIPLYING WHOLE NUMBERS BY 10, 100, 1000..

To multiply a number by 10, move the entire digit(s) one place to the left and add a zero (0). Simply put, for whole numbers; just add a zero (because there is only one zero from 10) at the end of the number.

Example 1: 3×10

Add a zero at the end of 3 to make 30.
Therefore, **$3 \times 10 = 30$** ✓

OR

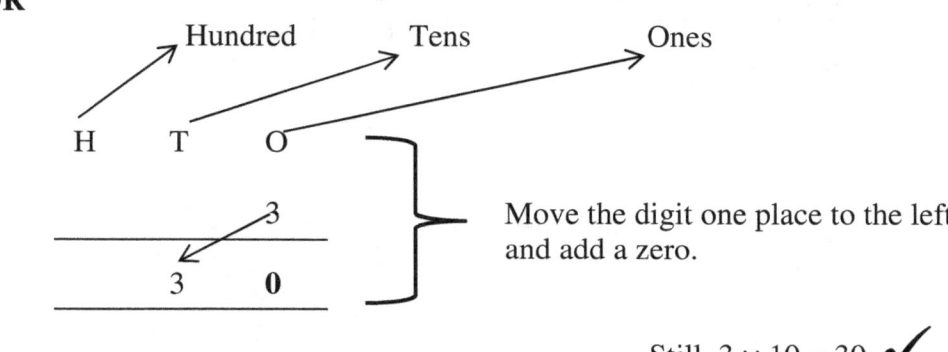

Move the digit one place to the left and add a zero.

Still, $3 \times 10 = 30$ ✓

Example 2: 12×10

Just add a zero at the end of 12 to make 120.
Therefore, **$12 \times 10 = 120$**

OR

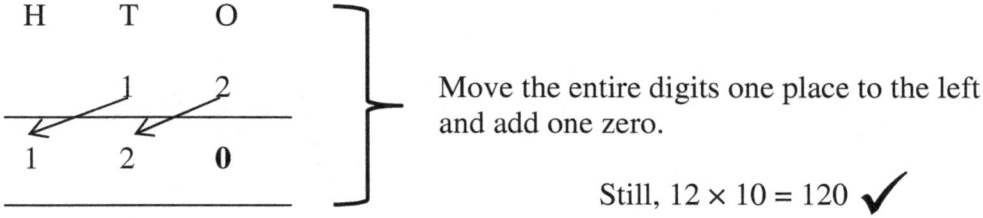

Move the entire digits one place to the left and add one zero.

Still, $12 \times 10 = 120$ ✓

Example 3: 14 × 100

Add two zeros (because a hundred has two zeros) at the end of 14 to make 1400. Therefore, **14 x 100 = 1400**

OR

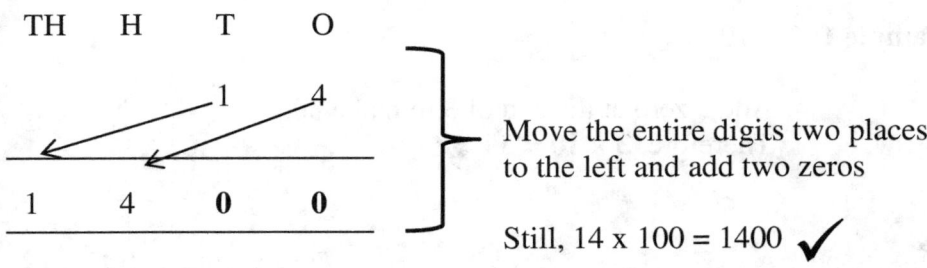

Move the entire digits two places to the left and add two zeros

Still, 14 x 100 = 1400 ✓

1.6 DIVIDING WHOLE NUMBERS BY 10, 100, 1000....

When dividing numbers by 10, move the entire digits one place to the right
When dividing numbers by 100, move the entire digits two places to the right
When dividing numbers by 1000, move the entire digits three places to the right

Likewise, to divide by 10,000, 100,000... Move all the digits to the right, according to the number of zeros.

Example 1: 70 ÷ 10

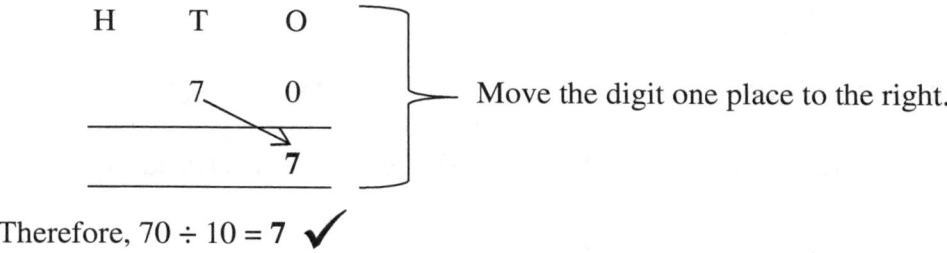

Therefore, 70 ÷ 10 = **7** ✓

Alternatively, you may cancel out the two zeros to leave you with 7 since you are dividing both sides by 10.

Example 2: 679 ÷ 10

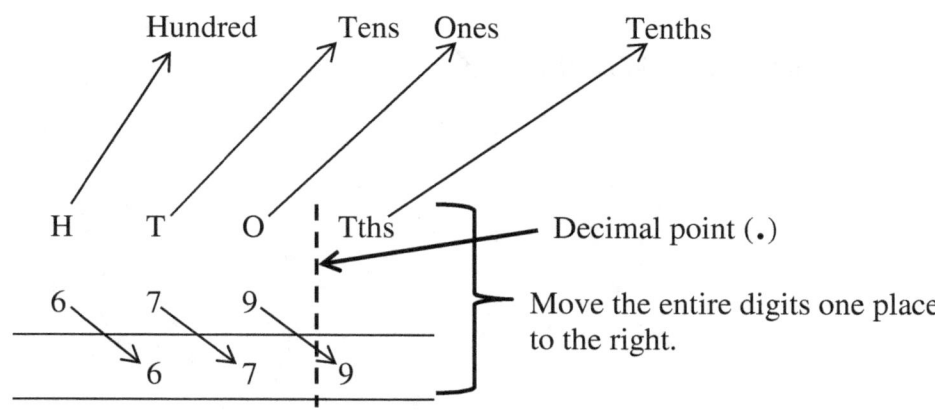

Therefore, 679 ÷ 10 = **67.9** ✓

Example 3: 546 ÷ 100

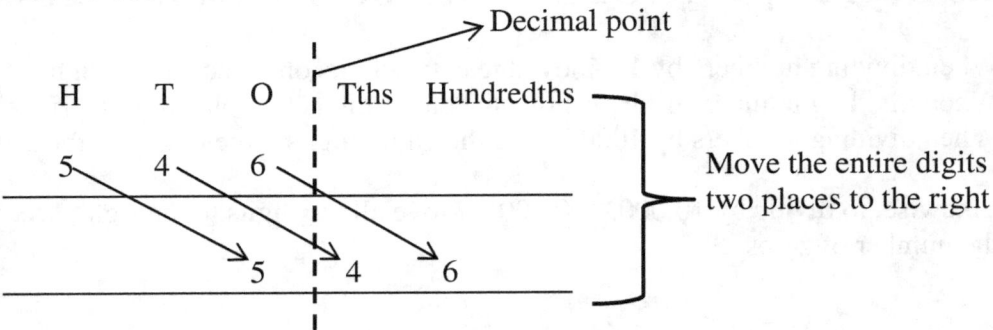

Therefore, 546 ÷ 100 = **5.46** ✓

Example 4: 762 ÷ 1000

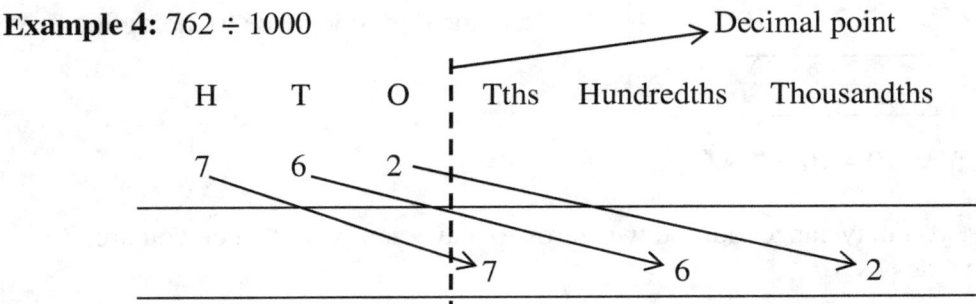

Move the entire digits three places (because there are 3 zeros in a thousand) to the right.

Therefore, 762 ÷ 1000 = **0.762** ✓

EXERCISE 1E

1) Work out without a calculator.

a) 3×10
b) 43×10
c) 77×10
d) 319×10
e) 400×10
f) 519×10
g) 76×10
h) 431×10
i) 780×10
j) 234×10
k) 980×10
l) 134×10
m) 555×10
n) 1111×10
o) 78965×10

2) Work out without using a calculator.

a) 30×100
b) 42×100
c) 6×100
d) 75×100
e) 319×100
f) 50×100
g) 419×100
h) 917×100
i) 7×100
j) 47×100
k) 9288×100
l) 1112×100
m) 976×100
n) 59×100
o) 86697×100

3) Work out without using a calculator.

a) 3×1000
b) 5×1000
c) 77×1000
d) 318×1000
e) 96×1000
f) 7×1000
g) 431×1000
h) 96×1000
i) 419×1000
j) 60×1000
k) 8776×1000
l) 92×1000
m) 1112×1000
n) 59×1000
o) 486×1000

4) Work out without a calculator.

a) 30 ÷ 10

b) 70 ÷ 10

c) 460 ÷ 100

d) 96 ÷ 10

e) 100 ÷ 100

f) 465 ÷ 10

g) 54321 ÷ 100

h) 678 ÷ 10

i) 12800 ÷ 100

j) 4000 ÷ 100

k) 83100 ÷ 100

l) 57000 ÷ 10

m) 1100 ÷ 100

n) 500 ÷ 100

o) 5000 ÷ 100

5) Work out without a calculator.

a) 9000 ÷ 1000

b) 86000 ÷ 1000

c) 32450 ÷ 1000

d) 10000 ÷ 100

e) 6540 ÷ 10

f) 4590000 ÷ 1000

g) 700000 ÷ 100

h) 2340 ÷ 10

i) 87000 ÷ 100

j) 318 ÷ 10

k) 83 ÷ 100

l) 89 ÷ 10

m) 452 ÷ 1000

n) 9874 ÷ 100

o) 408 ÷ 10

FUNCTIONAL MATHEMATICS

Some promotional items are on display as shown below.

1) What is the cost of three bottles of Nescafe?

2) How much will Blake pay for 3 cartons of Diet Coke?

3) Blake picked a packet of Tetley tea, a carton of Diet Coke and 5 packs of Pepsi Max.
   a) How much will he pay for the items?
   b) He paid with a £20 note. How much change would he receive?

4) Original Tetley tea contains 160 tea bags. In the promotion, 50% extra tea bags were added. How many tea bags are in the promotional pack as shown above?

5) Helen picked 4 bottles of Nescafe. a) How many grams altogether would that be?
   b) She gave the attendant £14. How much change would she receive?

6) Estimate the cost of 15 bottles of Nescafe.

Chapter 1 Review Section
Assessment

1) To the nearest 10, underline the numbers that would **round up**.

    12   35   43   57   76   81   89   134   789

    .... **2 marks**

2) To the nearest 100, underline the numbers that would **round down**.

    123   245   353   397   578   621   705   892   947

    .... **2 marks**

3) Round these number to the nearest 10

    a) 43   g) 456

    b) 51   h) 791

    c) 79   i) 837

    d) 135   j) 1456

    e) 278   k) 4567

    f) 312   l) 5555

    .... **12 marks**

4) Below are some mountain heights in metres.

Mountain	Height (m)
Mount Elbrus	5642
Lhotse I	8501
Manaslu	8156
Aconcagua	6959

a) Round the height of **Lhotse I** mountain to the nearest 1000 m .... **1 mark**

b) Round the height of Aconcagua mountain to the nearest 100 m .... **1 mark**

c) Round the height of Mount Elbrus to one significant figure .... **1 mark**

d) Round the height of Manaslu mountain to two significant figures .... **1 mark**

5) A number rounded to one significant figure is 4000.
Mbakwe thinks the number could be 3490.
Is Mbakwe correct? Explain fully. .... **2 marks**

6) Copy and complete.

a) 30 + ☐ = 80 .... **1 mark**

b) 47 + ☐ = 134 .... **1 mark**

c) 54 + 78 + ☐ = 564 .... **1 mark**

d) ☐ − 78 = 655 .... **1 mark**

7) Eastern shop sold 15 paintings on Friday, 25 on Sunday and 10 the following Monday.

   a) How many paintings will the shop need to order to replace the sold paintings?

   b) If the cost of one painting is ₦ 5 450, what is the cost of the paintings sold on Friday?

   c) How much was extra money realised on Sunday than on Monday?

………… **6 marks**

8)

One Tuesday, Asda supermarket had 370 tins of baked beans in stock. However, they sold 49 tins on the same day and took delivery of another 670 tins.

   a) How many tins of baked beans were still in the shop at the end of the day?

   …...... **2 marks**

   b) The cost of a tin of beans is £1.05. How much baked beans was sold on Tuesday?

   ……. **2 marks**

9) Work out

   a) 23 × 100   ……… 1 mark   d) 12 × 45   …….... **1 mark**

   b) 768 × 1000   ……… 1 mark   e) 763 × 9   ……… **2 marks**

   c) 90089 × 10   ……… 1 mark   f) 421 × 56   ……… **2 marks**

10) The cost of an annual ticket to watch Enugu Rangers football club is ₦55 120.

How much would a ten-year ticket cost? ……….. **1 mark**

11) Complete the multiplication grid.

×	2	5		7	9
3			18		
					18
7					

……….. **3 marks**

12) Below is a magic square. The sum of the entries of any row, any column or any diagonals is the same.

Complete the magic square

11		
12	10	
		9

…………. **3 marks**

13) College of the Immaculate Conception Alumni Association has ₦12 500 000 in the bank. ₦345 786 was spent on painting the walls of the school. How much money does it have left? …………. **2 marks**

14) Udoka earns ₦135 456 per month. How much does she earn in 9 months?
…………. **2 marks**

15) This ceiling light casing holds three bulbs. The cost without the bulbs is ₦3 450.

Amaka wants to buy 9 of the ceiling light casings for her new house project.

a) How much will Amaka pay for the ceiling light casings?   .... **2 marks**

The cost of a bulb is ₦430.

b) How much will she pay for bulbs **only** to fit into all the nine casings?
.... **2 marks**

c) How much will she pay altogether to fit all the bulbs and casings in her new house if the cost of labour is ₦20 000?   .... **2 marks**

16) A container is worth **£25** and can take up to 30 indoor footballs.

a) Find the cost of 5 indoor footballs. .... **1 mark**

b) If the container is full of footballs, how much is the container worth?   ....….. **2 marks**

c) To the nearest £10, what is the price of a container?   ....**1 mark**

d) Round your answer to **part a** to the nearest whole number.   .... **1 mark**

e) Round the price of three footballs to one decimal place.   .... **1 mark**

17) 2023 students are to be divided into seven equal groups. How many students will be in each group? ………………….. **2 marks**

18) Work out 245 ÷ 5 ……………..……. **1 mark**

19) The diagram shown journeys between some towns in Nigeria.

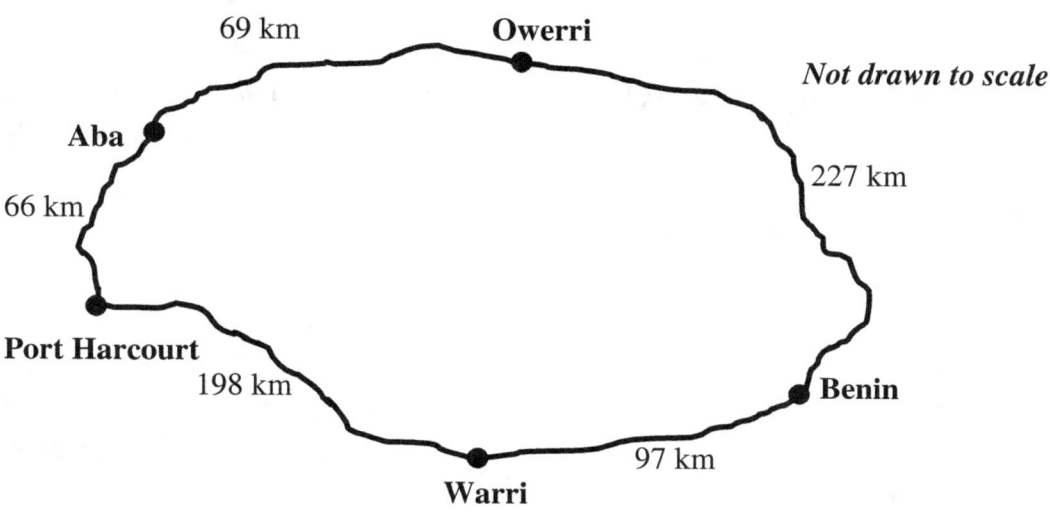

a) Which is the quicker way from Port Harcourt to Benin? Show all working out.
   ……………….. **2 marks**

b) What is the distance from Aba to Benin via Port Harcourt?
   …………………. **1 mark**

c) Wole left Warri in the morning and returned in the evening. What was his total distance for the entire journey?
   ………………… **2 marks**

d) George's residence is in Aba, and he wants to travel to Benin. His friend, Achike advised that his shortest route would be to pass through Owerri.
   Is Achike correct? Explain fully.
   ………………… **3 marks**

20) Round the answers to these calculations to **one** significant figure.

a) 34 + 87

b) 123 + 456

c) 987 – 453

d) 452 - 98

e) 1000 - 675

f) 3590 – 1999

……… **12 marks**

21) Round the answers to the calculations in question 20 to **two** significant figures
……… **12 marks**

22) Round to one decimal place.

a) 1.26

b) 23.53

c) 0.95

d) 897.345

e) 3.111

f) 98.567

……… **6 marks**

23) Estimate the answers to these calculations.

a) 23 + 45.7 ……… **2 marks**

b) 603 + 789.9 ……… **2 marks**

c) $\dfrac{6.789 \times 4.8}{5.06 + 1.6}$ ……… **2 marks**

24) Work out the following **without** using a calculator. Show all working out.

a) 270 ÷ 6 ……… **2 marks**

b) 120 ÷ 5 ……… **2 marks**

c) 3222 ÷ 9 ……… **2 marks**

d) 7896 ÷ 12 ……… **2 marks**

2 Numbers (2) & Indices

This section covers the following topics:

- Multiples Factors & Primes
- LCM and HCF
- Test for divisibility
- Squares and Cubes
- Indices
- Standard forms

LEARNING OBJECTIVES

By the end of this unit, you should be able to:

a) Understand the meaning of multiples
b) Understand and be able to write factors of numbers
c) Understand and identify prime numbers
d) Find the Highest Common Factor (HCF) of two or more numbers
e) Find the Lowest Common Multiple (LCM) of two or more numbers
f) Write or express numbers as a product of its factors
g) I can test if a number is divisible by 2, 3, 4, 5, 6, 7, 8, 9 and 10
h) Understand powers and roots
i) Use index laws for calculating indices
j) Understand and work in Standard Form

KEYWORDS
- Factors & Multiples
- Prime numbers & prime factors
- Product of prime factors
- Highest common factor and lowest common multiple
- Divisibility
- Squares, square roots, Cubes, cube roots
- Standard form

2.1 MULTIPLES

By now, you must be familiar with your necessary times tables as it will be vital in setting out multiples of numbers.

Multiples of a number are all the numbers in the multiplication table of that number. The most important fact to remember is that the first multiple of a number is the **number itself.**

Also, multiples of a number goes on forever (infinite).

The multiples of 3 are the answers to
$(1 \times 3), (2 \times 3), (3 \times 3), (4 \times 3), (5 \times 3), (6 \times 3), (7 \times 3)$………

Therefore, the multiples of 3 are **3, 6, 9, 12, 15, 18, 21**………

EXERCISE 2A

1) Write down the first four multiples of:

a) 1

b) 2

c) 5

d) 7

e) 9

f) 10

g) 13

h) 17

i) 20

2) Look at the list of numbers below.

2	5	7	12	16	22	30	36	49	54	70

a) Which numbers are multiples of 2?

b) Which numbers are multiples of 3?

c) Which numbers are multiples of 4?

d) Which numbers are multiples of 5?

e) Which numbers are multiples of 7?

3) Write down a number that is odd and a multiple of 3

4) Write down a number that is odd and a multiple of 2

5) Write down a number that is even and a multiple of 5?

6) Write down a number that is a square number and a multiple of 7

7) Write down two multiples of 4 with a sum of 28

8) Look at the list of numbers below.

a) Which numbers in the list are multiples of 11?

b) Which numbers in the list are multiples of 9?

c) Which number in the list is a square number?

9) Is 18 a multiple of 6? Explain fully.

10) Sanusi says "the first five multiples of 4 are 8, 12, 16, 20, 24."
Is Sanusi correct? Explain fully.

11) Write down:

a) the first five multiples of 7

b) the first ten multiples of 3

c) a common multiple of 3 and 7

12) Find two multiples of 6 that have a difference 18 but with a product of 360.

13) Three multiples of 9 add up to 117. Find the three numbers.

2.1.1 LOWEST COMMON MULTIPLE (LCM)

The lowest common multiple of two or more numbers is the smallest (first) of their common multiples.

Example 1: Find the LCM of 3 and 5

List some multiples of 3 and 5 as follows:

3:   3   6   9   12   15   18   21.......

5:   5   10   15   20   25   30   35.......

The lowest number which is in both lists is 15.

Therefore, **15** is the LCM of 3 and 5.

Example 2: Find the LCM of 4, 5 and 6

List some multiples of 4, 5 and 6 as follows:

4:   4   8   12   16   20   24   28   32   36   40   44   48   52   56   60   64......

5:   5   10   15   20   25   30   35   40   45   50   55   60   65.......

6:   6   12   18   24   30   36   42   48   54   60   66........

The lowest number which is on all the lists is 60.

Therefore, **60** is the LCM of 4, 5 and 6.

USING PRODUCT OF PRIME FACTORS TO WORK OUT THE LOWEST COMMON MULTIPLE (LCM)

When listing multiples of numbers to find LCM and it becomes cumbersome, it is advisable to use the prime factor method.

Example 1: Find the LCM of 16 and 20

Find all the prime factors of 16 and 20 (refer to section 4.5)

$$16: 2 \times 2 \times 2 \times 2 \longrightarrow 2^4$$
$$20: 2 \times 2 \times 5 \longrightarrow 2^2 \times 5$$

In this system, the LCM contains the highest powers of **each** factor/number.

The highest power of 2 is 2^4
The highest power of 5 is $5^1 = $ **5**

The LCM is the product of all the highest powers

$2^4 \times 5 = 2 \times 2 \times 2 \times 2 \times 5 = $ **80** ✓

Example 2: Find the LCM of 20, 24 and 30 using the product of prime method.

Find all the prime factors of 20, 24 and 30 using the factor tree method (Refer to section 4.6)

$$20: 2 \times 2 \times 5 \longrightarrow 2^2 \times 5$$
$$24: 2 \times 2 \times 2 \times 3 \longrightarrow 2^3 \times 3$$
$$30: 2 \times 3 \times 5 \longrightarrow 2 \times 3 \times 5$$

The LCM contains the highest powers of **each** number/factor.

The highest power of 2 is 2^3
The highest power of 3 is $3^1 = $ **3**
The highest power of 5 is **5**

Therefore, the LCM is the product of all the highest factors

$2^3 \times 3 \times 5 = 2 \times 2 \times 2 \times 3 \times 5 = $ **120** ✓

USING THE VENN DIAGRAM TO WORK OUT THE LOWEST COMMON MULTIPLE (LCM)

Example 1: Find the LCM of 16 and 20 using the Venn diagram.

Find the prime factors of 16 and 20 (refer to section 4.5)

$$16: 2 \times 2 \times 2 \times 2$$

$$20: 2 \times 2 \times 5$$

Draw a Venn diagram as shown below. The common prime factors are listed in the intersection part.

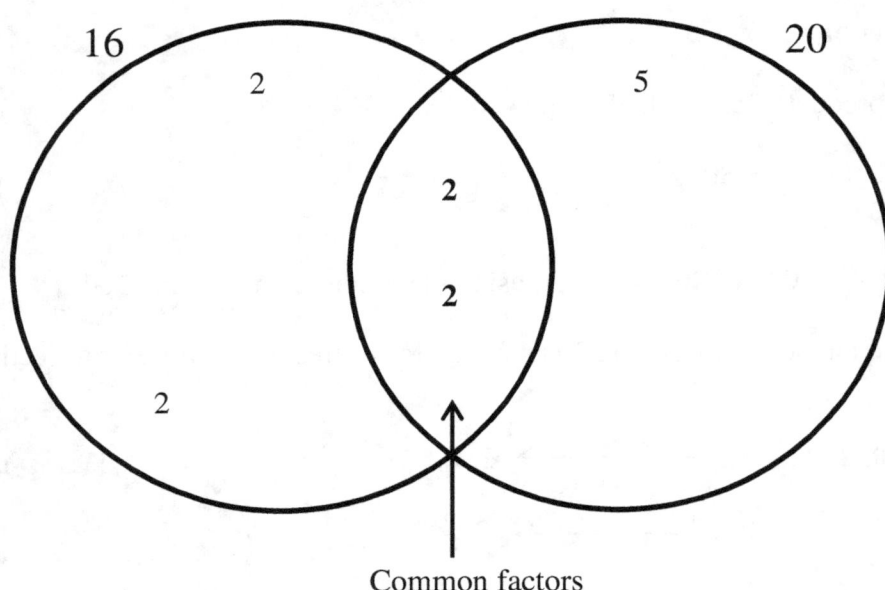

Common factors

The lowest common multiple is the product of **all** the prime factors in the diagram.

LCM = $2 \times 2 \times 2 \times 2 \times 5$ = **80** ✓

In chapter 4.6, we shall learn how to use the Venn diagram to find the highest common factor (HCF) of two or more numbers. It is simply performing the steps above and multiplying only the common factors in the intersection part above which is $2 \times 2 = 4$. Therefore, the HCF of 16 and 20 is **4**.

EXERCISE 2B

1) Find the LCM of:

   a) 2 and 4

   b) 3 and 6

   c) 5 and 7

   d) 7 and 10

   e) 6 and 8

   f) 10 and 15

   g) 15 and 20

   h) 4 and 5

2) Find the LCM of these sets of numbers.

   a) 2, 3 and 4

   b) 5, 6 and 7

   c) 10, 12 and 16

   d) 2, 7 and 8

   e) 5, 7 and 9

   f) 4, 6 and 7

   g) 30 and 45

   h) 50 and 60

3) Tony and Emma took part in a jogging experiment. Tony took 40 seconds to jog round the athletics track once and Emma 45 seconds. If they both start from the same point at the same time, how long will it take in minutes before they cross the start line together?

4) By leaving your answers in prime factors, find the LCM of the following:

a) $3 \times 3 \times 3 \times 2 \times 2$ and $2 \times 3 \times 3$

b) $2 \times 2 \times 3 \times 3 \times 5$ and $2 \times 2 \times 5 \times 5$

c) $2 \times 2 \times 3 \times 3 \times 3$ and $5 \times 5 \times 2 \times 2 \times 2 \times 3$

d) $2 \times 3 \times 7 \times 7$ and $2 \times 2 \times 2 \times 7 \times 7 \times 7$

e) $5 \times 5 \times 7 \times 11 \times 11$ and $7 \times 7 \times 7 \times 11$

f) $2 \times 3 \times 3 \times 3 \times 3 \times 4 \times 4 \times 5 \times 5 \times 5$ and $2 \times 2 \times 3 \times 3 \times 5 \times 5$

5) Jude says "The LCM of two numbers is always the product of the numbers."
Is Jude correct?
Explain fully with an example.

6) Find the LCM of 70 and 80

7) Find the LCM of 10, 20 and 40

8) Okoye and Chidi have the same number of drinking cups. Okoye placed his cups into 4 equal parts while Chidi arranged his own in 2 equal parts. Find the least number of cups they could each have.

2.2 FACTORS

Like in multiples, the knowledge of multiplications is important.

A factor of a number will divide **exactly** into that number. *Exactly* means there will be no remainder(s).

A fact about factors is that every number has at least two factors: **1 and itself**.

Any number that divides exactly into 6 is called a factor of 6.

The factors of 6 are: **1, 2, 3** and **6**

We can also find factor as pairs. Factor pairs of a number are the factors of that number.

What are the factors of 20?

Using factor pairs: (1 and 20) because $1 \times 20 = 20$
(2 and 10) because $2 \times 10 = 20$
(4 and 5) because $4 \times 5 = 20$

Putting them in order   1   2   4   5   10   20

Therefore, the factors of 20 are 1   2   4   5   10 and 20

Example 1: Find all the factors of 12.

To start with, 1 and the number itself are factors. So, 1 and 12 are factors of 12.

2 and 6 are also factors because $2 \times 6 = 12$

3 and 4 are also factors because $3 \times 4 = 12$

Putting all in order, the factors of 12 are **1   2   3   4   6 and 12**

Note: Factors of a number start with **1** and end with the number itself.

EXERCISE 2C

1) Find all the factors of:

   a) 3
   b) 4
   c) 14
   d) 24
   e) 30
   f) 35
   g) 36
   h) 48
   i) 56
   j) 84
   k) 100
   l) 120

2) Look at the following list of numbers:

   3   5   6   7   9

   From the list above, choose a number or numbers that are factors of the following:

   a) 48
   b) 21
   c) 18
   d) 27
   e) 30
   f) 108

3) Write down a number that is odd and a factor of 40

4) Write down even numbers that are factors of 18

5) Write down three factors of 50 that have a sum of 40

6) Write down two factors of 35 that have a sum of 6

7) Write down two factors of 72 that have a difference of 14

8) Anthony Joshua says that 7 is a factor of 56.
   Is Anthony correct?
   Explain fully

9) Add together all the factors of 30

2.3 PRIME NUMBERS

A prime number is a number that has only **two** factors, **1** and **itself**. This statement means that once a whole number has more than two factors, the number is **not** a prime number.

Example 1: a) list all the factors of 1, 2, 3, 4, 5, 6, 7, 8 and 9

    b) Which of the numbers are prime numbers?

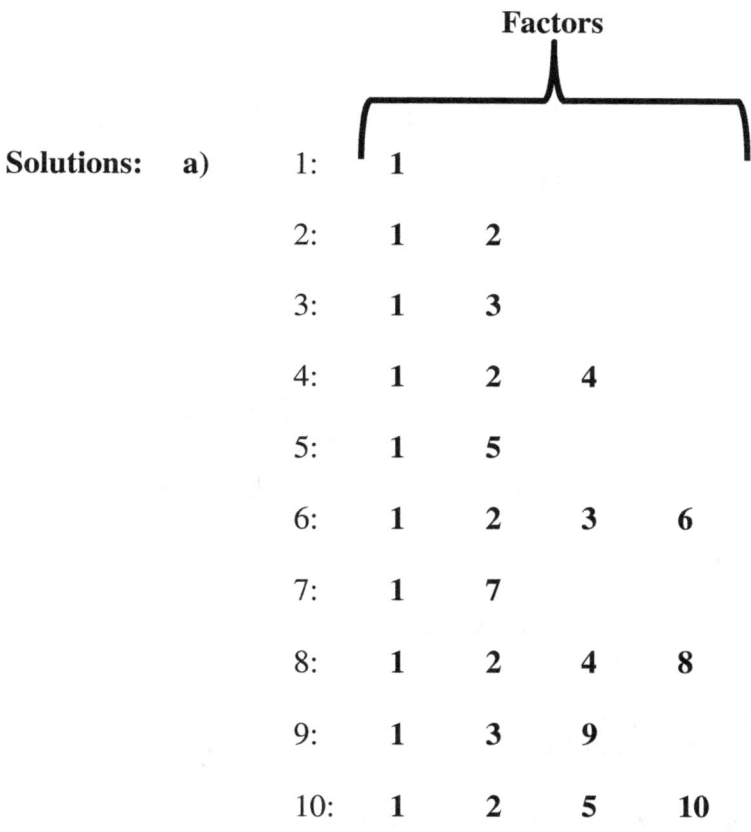

    b)   Since prime numbers have only two factors, the prime numbers from the list above are **2, 3, 5,** and **7**

Notice that the number **1** is **not** a prime number as it has only **one** factor.

Also, the number **2** is the **only even prime number.**

Example 2: Ifeanyi says: "**9** is a prime number because it is an odd number." Is he correct? Explain fully

Solution: List all the factors of 9. They are 1, 3 and 9.
However, a prime number is a number that has only two factors. 9 has three factors, and cannot be a prime number.

Therefore, **9 is not** a prime number because it has **more** than two factors, 1, 3 and 9 and a prime number must have only two factors.

The first 20 prime numbers are:

2	3	5	7	11	13	17	19	23	29
31	37	41	43	47	53	59	61	67	71

2.4 PRIME FACTORS

Prime factors of a number are the factors of that number that are prime numbers.

Example: The factors of 20 are: 1   ②   4   ⑤   10   20

From the factors, only 2 and 5 are prime numbers.

Therefore, **2** and **5** are the **prime factors** of 20.

POINTS TO NOTE

- A number is prime if the only factors are 1 and itself
- 1 **is not** a prime number since it has only one factor, 1
- 2 is the only even prime number
- Prime factors of a number are the prime numbers from factors of that number

EXERCISE 2D

1) Identify all the prime numbers from the numbers below:

   a) 1   3   7   15   19   27

   b) 2   5   8   13   18   33

   c) 6   9   14   28   31   41

   d) 10   20   25   37   59   97

   e) 4   11   17   60   78   121

2) Find the prime factors of

   a) 4   b) 12   c) 34   d) 50

3) Identify a prime number that is also a factor of 6

4) Identify a prime number that is also a factor of 24

5) a) Copy and complete the 7 by 7 square grid following the existing pattern.

1	2	3	4	5	6	7
8	9	10	11	12	13	14

b) Colour in all the prime numbers in the table above.

c) Add up all the prime numbers in the 5th column.

d) Is the number obtained in **part c** above a prime number? Explain fully.

2.5 PRODUCT OF PRIME FACTORS

In section 4.4, prime factors were discussed. Please revisit if you need extra help.

Any whole number which is not a prime number can be broken down into prime factors. It is called prime factor decomposition. The product of the prime factors obtained will always give the original number.

There are different ways to write numbers as the product of prime factors:

Method 1: Factor tree method

Example 1: Write 20 as the product of prime factors.

Step1: Find any two numbers that multiply to give 20 ⟶
If any of the factors are prime, circle it.

Step 2: For the factors **not circled, in this case, 4**, repeat step 1.

Step 3: Since all the numbers are circled, it means **no more decomposition** (breaking down). At this point, you have decomposed the number, 20 into its prime factors.

Multiply all the circled numbers $2 \times 2 \times 5$

Therefore, as a product of prime factors, $20 = 2 \times 2 \times 5$ or $2^2 \times 5$ in index forms.

Example 2: Write 320 as the product of prime factors.

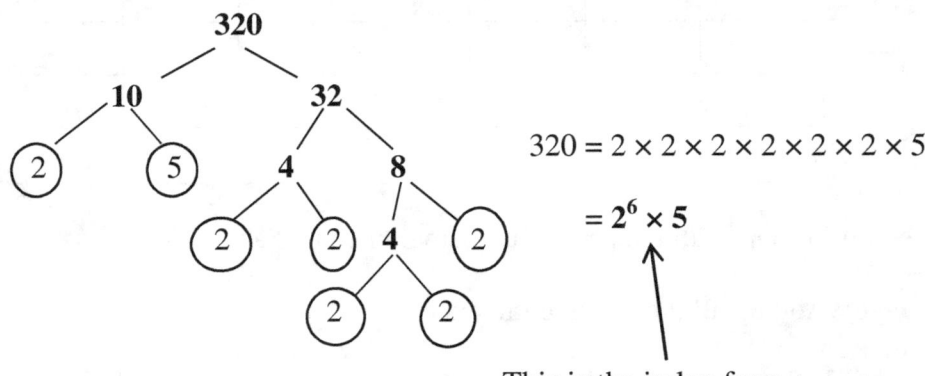

$320 = 2 \times 2 \times 2 \times 2 \times 2 \times 2 \times 5$

$= 2^6 \times 5$

↑

This is the index form

Method 2 of finding product of prime factors

This is a method of dividing the number by the smallest (first) prime number that divides into the number. The process is repeated until the answer becomes 1.

Example 3: Express 60 as the product of prime factors

The first prime number that divides into 60 is 2. Use 2 to divide. Repeat the process until the answer becomes 1.

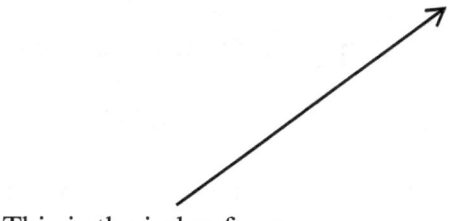

Therefore, 60 written as the product of prime factors is $2 \times 2 \times 3 \times 5 = \mathbf{2^2 \times 3 \times 5}$

This is the index form

EXERCISE 2E

1) Express each of the numbers below as a product of prime factors

   a) 30 c) 70 e) 720

   b) 55 d) 144 f) 940

2) a) Complete the factor tree.

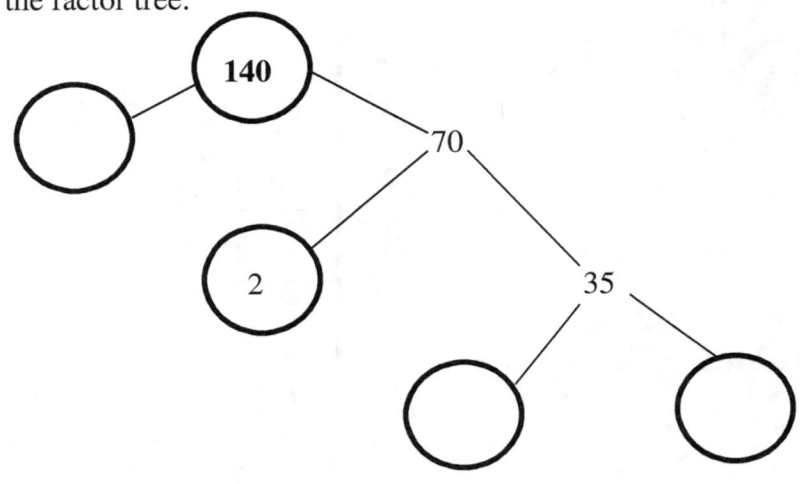

   b) Write down the prime factors of 140 in index form.

3) Write each number as a product of prime factors in index form.

   a) 150 b) 500 c) 510

4) Write the following as the product of two prime factors.

   a) 6 b) 21 143

5) Copy and complete the factor trees.

   a) 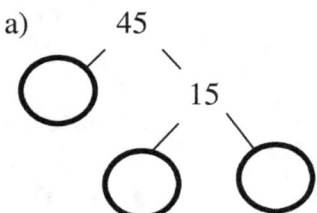   b) 210, 21, 10

6) Write in index form.

a) $2 \times 2 \times 2 \times 3 \times 3 \times 3$

b) $2 \times 2 \times 5 \times 7 \times 7 \times 7 \times 7$

c) $3 \times 3 \times 3 \times 3 \times 5 \times 5$

7) Anthony says "Prime factors will be different if you chose a different factor pair."
Is he correct?

Explain fully

8) Write each number as a product of prime factors in index form.

a) 3300

b) 1540

2.6 HIGHEST COMMON FACTOR (HCF)

The highest common factor of two or more numbers can be obtained. The first method is to find all the factors of the numbers and identify the **common factors**. The highest number of the common factors will be the HCF of the numbers.

Refer to **section 4.2** for finding factors of numbers.

Example 1: Find the HCF of 4 and 8

Identify all the factors of 4 and 8 as follows:

```
4:   1   2   (4)
8:   1   2   (4)   8
```

The common factors are 1, 2 and 4.

However, 4 is the largest number of the common factors. Therefore, **4** is the highest common factor (HCF) of 4 and 8.

It is also the highest number that can divide into both numbers.

Example 2: Find the HCF of 24 and 36

Identify all the factors of 24 and 36.

```
24:   1   2   3   4   6   8   (12)   24
36:   1   2   3   4   6   9   (12)   18   36
```

The common factors are 1, 2, 3, 4, 6 and 12.

However, the highest number common to both numbers is 12.

Therefore, **12** is the highest common factor (HCF) of 24 and 36.

It is the highest number that can divide into both numbers.

USING PRIME FACTORS TO FIND HCF

When listing factors of numbers in other to find HCF becomes cumbersome, it is advisable to use the prime factor method.

Example 1: Find the HCF of 24 and 36

Find all the prime factors of 24 and 36 (refer to section 4.5)

$$24: 2 \times 2 \times 2 \times 3 \longrightarrow 2^3 \times 3$$

$$36: 2 \times 2 \times 3 \times 3 \longrightarrow 2^2 \times 3^2$$

In this system, find the pairs of common prime factors and multiply them.

24: | 2 | 2 | 2 | 3 |
36: | 2 | 2 | 3 | 3 |
↓ ↓ ↓
2 2 3

++The HCF would be $2 \times 2 \times 3 = $ **12** ✓

ALTERNATIVELY

After finding the product of prime factors, we multiply the lowest power of **each common** prime factor to give the HCF.

$$24: 2 \times 2 \times 2 \times 3 = 2^3 \times 3$$

$$36: 2 \times 2 \times 3 \times 3 = 2^2 \times 3^2$$

LOWEST POWER OF 2 is 2^2 while the LOWEST POWER OF 3 is **3.**

Therefore, HCF of 24 and 36 = $2^2 \times 3 = 2 \times 2 \times 3 = $ **12** ✓

Example 2: Find the HCF of $3 \times 3 \times 3 \times 3 \times 5 \times 5 \times 7 \times 7 \times 7$ **and** $3 \times 3 \times 3 \times 7 \times 7$
Leave your answer in prime factors in index form.

$3 \times 3 \times 3 \times 3 \times 5 \times 5 \times 7 \times 7 \times 7 = 3^4 \times 5^2 \times 7^3$ and

$3 \times 3 \times 3 \times 7 \times 7 \quad\quad\quad\quad = 3^3 \times 7^2$

The lowest power of 3 contained in the two numbers is 3^3

The lowest power of 7 contained in the two numbers is 7^2

*Notice that the number **5** is **not** common to both numbers and as such is **not included**.*

Therefore, in prime factor index form, the HCF is $\mathbf{3^3 \times 7^2}$ ✓

Please note: The above index answer is acceptable if the question says so.
If not, work out the real answers to $3^3 \times 7^2$ as detailed below.

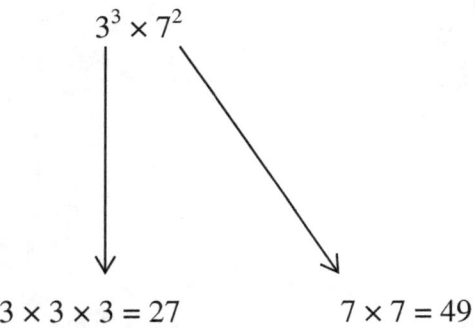

$3 \times 3 \times 3 = 27 \quad\quad\quad 7 \times 7 = 49$

Highest common factor (HCF) = 27×49

$= \mathbf{1323}$

USING THE VENN DIAGRAM TO WORK OUT THE HIGHEST COMMON FACTOR (HCF)

Example 1: Find the HCF of 16 and 20 using the Venn diagram.

Find the prime factors of 16 and 20 (refer to section 4.5)

$$16: 2 \times 2 \times 2 \times 2$$

$$20: 2 \times 2 \times 5$$

Draw a Venn diagram as shown below. The common prime factors are listed in the intersection part.

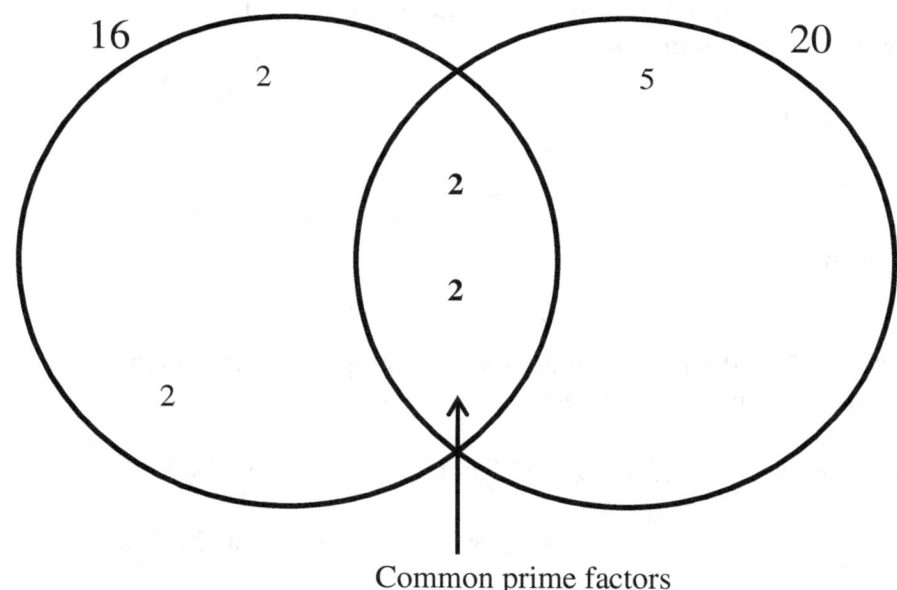

Common prime factors

The highest common factor is obtained by multiplying the numbers in the intersection part.
$$2 \times 2 = 4$$

Therefore, the HCF of 16 and 20 is **4**.

In all, you may use any method of your choice to find HCF unless told otherwise.

EXERCISE 2F

1) i) Write down all the factors of the numbers below.

   ii) Circle the common factors of each pair.

   iii) Pick out the highest common factor.

   a) 6 and 10

   b) 12 and 18

   c) 5 and 25

   d) 27 and 49

   e) 30 and 76

   f) 36 and 48

2) Find the highest common factor using the prime factor method for the following pairs of numbers.

   a) 40 and 60

   b) 120 and 150

   c) 420 and 700

3) Find the HCF of the following numbers, leaving your answers in prime factor forms using index notation.

   a) $3 \times 3 \times 5 \times 5 \times 5 \times 5 \times 7 \times 7 \times 7$ and $3 \times 5 \times 5 \times 5 \times 7 \times 7$

   b) $2 \times 2 \times 2 \times 2 \times 3 \times 3 \times 3 \times 3 \times 3 \times 3 \times 11 \times 11 \times 11$ and $2 \times 2 \times 3 \times 3 \times 3$

   c) $5 \times 5 \times 7 \times 7 \times 7 \times 13$ and $5 \times 7 \times 13$

4) Find the HCF of the following numbers using the Venn diagram method.

   a) 18 and 60

   b) 20 and 75

5) Using the index notation, find the HCF of the numbers below.

   a) 8 and 40

   b) 63 and 270

   c) 30, 60 and 90

   d) 25, 70 and 130

6) Write out the common factors of

   a) 4 and 16

   b) 12 and 20

   c) 45 and 63

   d) 8, 20 and 52

7) David needed to cut out congruent squares for making envelopes from a piece of rectangular brown cardboard paper as shown.

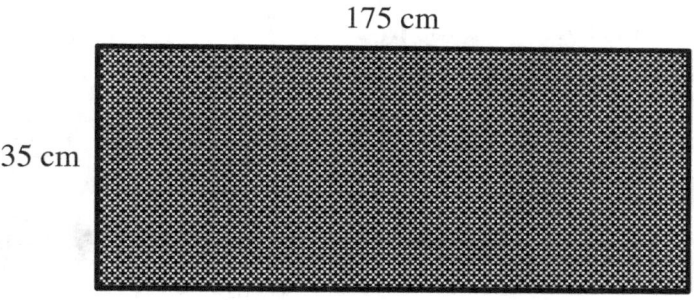

Without wasting any cardboard paper, what is the maximum square size David can use for the envelopes?

8)

Kenechukwu says "The HCF of 25 and 50 is 5."

Is she correct? Explain fully.

2.7 DIVISIBILITY TEST

To test if a whole number is divisible by 2, 3, 4, 5, 6, 7, 8, 9 or 10, use the divisibility rules which are set out below.

DIVISIBILITY BY 2

A number is divisible by 2 if the last digit is 0 or even.

Look at the numbers below.

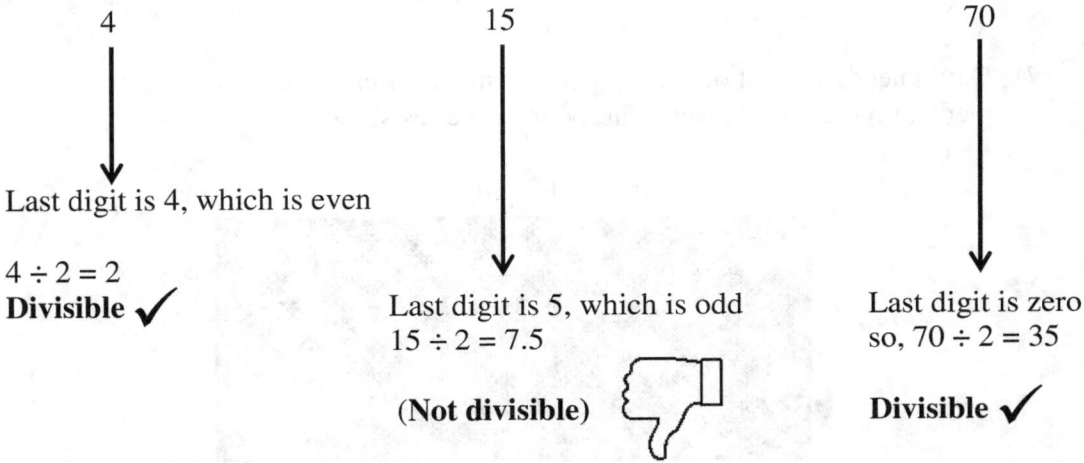

4
Last digit is 4, which is even
$4 \div 2 = 2$
Divisible ✓

15
Last digit is 5, which is odd
$15 \div 2 = 7.5$
(Not divisible)

70
Last digit is zero
so, $70 \div 2 = 35$
Divisible ✓

DIVISIBILITY BY 3

To find out if a number is divisible by 3, add up all the digits. If the sum is divisible by 3, then the number is divisible by 3.

If you cannot figure out if the added numbers are divisible by 3, add together the digits of the added numbers and divide by 3 to check.

Example 1: Check if 453 is divisible by 3.

$4 + 5 + 3 = 12$
Clearly, 12 is divisible by 3 because $12 \div 3 = 4$
Therefore, 453 **is divisible** by 3

$453 \div 3 = 151$ ✓

Example 2: Is 79796223 divisible by 3?

$7 + 9 + 7 + 9 + 6 + 2 + 2 + 3 = 45$

If you are familiar with your 3 times table, then 45 is divisible by 3 since $15 \times 3 = 45$

Yes, 79796223 **is divisible** by 3. ✓

You may use a calculator to check. $79796223 \div 3 = 26\ 598\ 741$

Example 3: Is 265 divisible by 3?

$2 + 6 + 5 = 13$

$13 \div 3 = 4.3333333.....$ (**Not** a whole number)

13 is **not** divisible by 3. Therefore 265 **is not divisible** by 3.

DIVISIBILITY BY 4

If the **last two digits** of a number are divisible by 4, then the number is divisible by 4

Example 1: Is 459 divisible by 4?

The last two digits are 59

$59 \div 4 = 14.75$ (**Not** a whole number)

No, 459 **is not divisible** by 4

Example 2: Is 2792 divisible by 4?

The last two digits are 92

$92 \div 4 = 23$ (Whole number)

Yes, 2792 **is divisible** by 4 ✓

DIVISIBILITY BY 5

A number is divisible by 5 if the last digit is 0 or 5.

Example 1: Is 345 divisible by 5?

>The last digit is 5; therefore it **is divisible** by 5. ✓
>
>345 ÷ 5 = 69 (whole number)

Example 2: Is 677 divisible by 5?

>The last digit is 7; therefore it **is not divisible** by 5
>
>677 ÷ 5 = 135.4 (**Not** a whole number)

Example 3: Is 120 divisible by 5?

>The last digit is 0; therefore it **is divisible** by 5 ✓
>
>120 ÷ 5 = 24 (whole number)

DIVISIBILITY BY 6

A number is divisible by 6 if it is divisible by both 2 and 3. Check the rules above!

Example 1: Is 156 divisible by 6?

>Step 1: Check with divisibility by 2 rules:
>   It is even; therefore it can be divided by 2 **(Yes)**
>
>Step 2: Check with divisibility by 3 rules:
>   1 + 5 + 6 = 12
>   …and 12 is divisible by 3;
>   therefore it can be divided by 3 **(Yes)**
>
>Since 156 is divisible by 2 and 3, it is divisible by 6 ✓

Example 2: Is 86 divisible by 6?

Step 1: 86 is even, so it is divisible by 2 **(Yes)**

Step 2: Check for divisibility by 3.

$$8 + 6 = 14$$
$$(14 \div 3 = 4.666666...)$$ **(X)**

14 is not divisible by 3. Therefore, 86 is not divisible by 3 as well.

Since 86 failed divisibility by 3 rules, it cannot be divided by 6.

DIVISIBILITY BY 7

For a number to be divisible by 7, double the last digit and subtract it from a number made by the other digits. If the outcome/result is divisible by 7, then the number itself is also divisible by 7.

Example 1: Is 238 divisible by 7?

Double the last digit $8 \times 2 = \mathbf{16}$

Subtract 16 from the remaining numbers $23 - 16 = \mathbf{7}$

7 is divisible by 7 ($7 \div 7 = 1$) …..(Whole number)

Therefore, 238 **is divisible** by 7 ✓

Example 2: Is 957 divisible by 7?

Double the last digit $7 \times 2 = 14$

Subtract 14 from the remaining numbers $95 - 14 = 81$

$81 \div 7 = 11.57....$ (Not a whole number)

Therefore, 957 is **not divisible** by 7

DIVISIBILITY BY 8

If the **last three** digits are divisible by 8, then the number is divisible by 8

Example 1: Is 1 547 divisible by 8?

>The last three numbers are 547.
>547 ÷ 8 = 68.375 (Not a whole number)
>
>Therefore, 1547 is **not divisible** by 8

Example 2: Is 32 984 divisible by 8?

>The last three numbers are 984
>984 ÷ 8 = 123 (Whole number)
>
>Therefore, 32 984 **is divisible** by 8 ✓

Points to note: *At times the last three numbers might be a problem to divide. Halve three times, and if the result is still a whole number, then that number is divisible by 8.*

DIVISIBILITY BY 9

A number is divisible by 9 if the sum of the digits is divisible by 9.

Example 1: Is 154 divisible by 9?

>1 + 5 + 4 = 10
>
>10 ÷ 9 = 1.1111111 (**Not** a whole number)
>
>Therefore, 154 is **not divisible** by 9

Example 2: Is 322 866 divisible by 9?

$$3 + 2 + 2 + 8 + 6 + 6 = 27$$

$$27 \div 9 = 3 \text{ (Whole number)}$$

Therefore, 322 866 **is divisible** by 9 ✓

DIVISIBILITY BY 10

A number is divisible by 10 is the last digit is zero (if the number ends in zero (0).

Example 1: Is 4 567 divisible by 10?

The last number is 7

4 567 **is not divisible** by 10 because the last number is not zero

Example 2: Is 56 430 divisible by 10?

The last number (digit) is 0

Therefore, 56 430 **is divisible** by 10 ✓

EXERCISE 2G

1) Are these numbers divisible by **2**? Give a reason.

   a) 12

   b) 26

   c) 27

   d) 69

   e) 780

   f) 7865

2) Are these numbers divisible by 3? Give a reason using divisibility tests.

   a) 12

   b) 21

   c) 34

   d) 46

   e) 180

   f) 654

3) Obinna says "287945 is divisible by 7." Is he correct? Explain using the divisibility rule.

4) Are these numbers divisible by 9? Perform divisibility tests to check.

   a) 81

   b) 108

   c) 1 553

   d) 2 345

   e) 5 166

   f) 142 911

5) Is 65432 divisible by 5? Explain fully.

6) Is 7893451 divisible by 8? Explain fully

7) Is 6754320 divisible by 2 and 10? Explain fully.

8) Using the divisibility test, check whether the following numbers are divisible by 4.

   a) 876

   b) 1258

2.8 POWERS AND ROOTS

Powers are often called **indices**. Repeated multiplications by the same number can be shown using **index notations**.

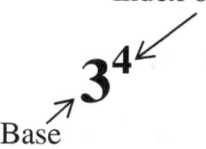

The number, 4, indicates the number of times the base number, 3, is multiplied.
So, $3^4 = 3 \times 3 \times 3 \times 3 = 81$
It is read as 3 to the power of 4.

Note: 3^4 **is not** 3×4 as this would give 12, which is not the answer.

SQUARE NUMBERS

When a number is multiplied by itself, we say the number has been **squared**.
5 squared = $5^2 = 5 \times 5 = $ **25**
7 squared = $7^2 = 7 \times 7 = $ **49**
9 squared = $9^2 = 9 \times 9 = $ **81**

25, 49 and 81 are square numbers.
Similarly,
$2.4^2 = 2.4 \times 2.4 = $ **5.76**
$1.6^2 = 1.6 \times 1.6 = $ **2.56**

Therefore, when a number is multiplied by itself, the result or outcome is called a **square number**.

A common *misconception* is thinking that to square a number is simply to multiply by 2. It is wrong.

SQUARE OF NEGATIVE NUMBERS

When two negative numbers are multiplied together, the answer is always a **positive** number.

$(-2)^2 = -2 \times -2 = 4$
$(-5)^2 = -5 \times -5 = 25$

Therefore, the square of a negative number is a positive number.

SQUARE ROOT ($\sqrt{\ }$)

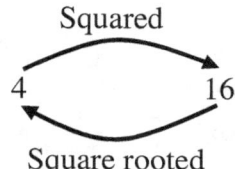

Example 1
$\sqrt{16} = $ **4** ....because $4 \times 4 = 16$
$\sqrt{49} = $ **7** ....because $7 \times 7 = 49$

Remember, the square root of a number could be positive or negative. So,

$\sqrt{16} = \pm 4$ which means **4** or **-4**
Proof: $4 \times 4 = 16$ and $-4 \times -4 = 16$

Example 2: The area of a square is $144 m^2$. Work out the length of the sides.

Solution:
$\sqrt{144} = 12$
Therefore, the length of each side $x = $ **12 m**.
Check: $12 \times 12 = 144$

SQUARE ROOT OF BIG NUMBERS

At times, we may be required to work out the square root of big numbers without a calculator.
The knowledge of prime factors is important.

Example 1: Work out the square root of 3136.

Using the knowledge of prime factor decomposition,

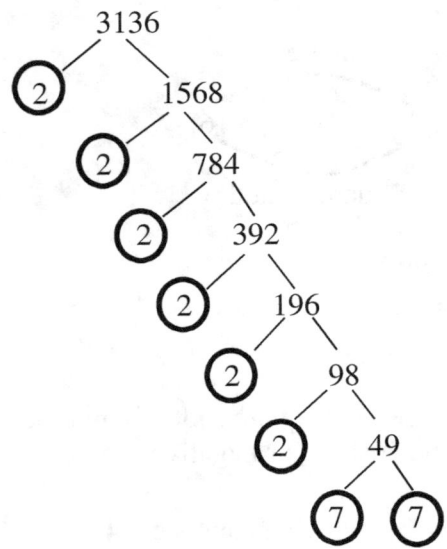

$= 2^6 \times 7^2$

Pair the prime numbers as follows:

$(2 \times 2) \times (2 \times 2) \times (2 \times 2) \times (7 \times 7)$

$\quad\downarrow\qquad\quad\downarrow\qquad\quad\downarrow\qquad\quad\downarrow$

$\quad 2 \;\times\; 2 \;\times\; 2 \;\times\; 7 \;=\; 56$

Therefore, $\sqrt{3136} = \mathbf{56}$ ✓

Example 2: Work out $\sqrt{4761}$

As a product of prime factors,
$4761 = 3^2 \times 23^2$.
Writing in pairs gives $(3 \times 3) \times (23 \times 23)$

Therefore, $\sqrt{4761} = 3 \times 23 = \mathbf{69}$ ✓

Check: $69 \times 69 = 4761$

Example 3: Work out the values of

a) $\sqrt{\frac{49}{81}}$ b) $\sqrt{1\frac{9}{16}}$ c) $\sqrt{\frac{128}{162}}$

Solutions:

a) $\sqrt{\frac{49}{81}} = \frac{\sqrt{49}}{\sqrt{81}} = \frac{7}{9}$

From converting $1\frac{9}{16}$ to improper fraction

b) $\sqrt{1\frac{9}{16}} = \sqrt{\frac{25}{16}} = \frac{\sqrt{25}}{\sqrt{16}} = \frac{5}{4} = \mathbf{1\frac{1}{4}}$

c) Divide both numbers by 2 to get square numbers.

$\sqrt{\frac{128}{162}} = \sqrt{\frac{64}{81}} = \frac{\sqrt{64}}{\sqrt{81}} = \frac{\mathbf{8}}{\mathbf{9}}$

CUBE NUMBERS

The number 27 can be written as $3 \times 3 \times 3$ or 3^3. We say *three cubed.*

Examples
$1^3 = 1 \times 1 \times 1 = \mathbf{1}$
$2^3 = 2 \times 2 \times 2 = \mathbf{8}$
$3^3 = 3 \times 3 \times 3 = \mathbf{27}$

The numbers 1, 8 and 27 are all **cube numbers**.

Cube numbers are formed when a number is multiplied by itself three times.

Note: 4^3 is not $4 \times 3 = 12$

It is $4 \times 4 \times 4 = 64$ ✓

CUBE ROOTS ($\sqrt[3]{}$)

Finding the cube root of a number is the opposite of finding the cube of that number.

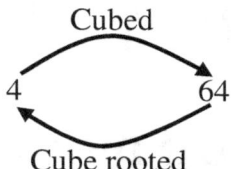

Hence, the cube root of 64 is ($\sqrt[3]{64}$) = **4**, because $4 \times 4 \times 4 = 64$

$\sqrt[3]{8} = 2$, because $2 \times 2 \times 2 = 8$

Likewise,
$\sqrt[3]{-8} = $ **-2,** because $-2 \times -2 \times -2 = -8$

EXERCISE 2H

1) From the list of numbers below, write down
a) a square number   b) a cube number
c) the square root of 169
d) the cube root of 27.

| 1 | 3 | 13 | 18 | 64 |
| 125 | 338 | 500 | | |

2) Write down the sum of the first five square numbers.

3) Calculate
a) 9^2 b) 1.5^2 c) 13^2 d) $(-11)^2$

4) Work out
a) $\sqrt{225}$ b) $\sqrt{289}$ c) $\sqrt{441}$ d) $\sqrt{6.25}$

5) Is it possible to work out $\sqrt{-49}$? Explain fully.

6) Work out the value of $(-3)^2 + (-10)^2$

7) Find the difference between $(-6)^2$ and 11^2

8) Work out
a) 5^3 b) $(-4)^3$ c) $(\frac{1}{2})^3$ d) $(\frac{2}{5})^3$

9) Work out
a) $\sqrt[3]{64}$ b) $\sqrt[3]{512}$ c) 7^3 d) 1.2^3

10) Copy and complete.
a) $\sqrt{?} = 11$
b) $8 - \sqrt{1} = ?$
c) $25 + \sqrt{?} = 30$
d) $1002 + \sqrt{400} = ?$

11) Work out
a) $6^2 + 3^3 + 1^2$
b) $\sqrt{484}$
c) $\sqrt{9604}$
d) $\sqrt{4225}$
e) $\sqrt{\frac{400}{100}}$
f) $\sqrt{2\frac{14}{25}}$
g) $\sqrt{\frac{121}{625}}$
h) $\sqrt[3]{343}$
i) $4.5^2 - 2.3^2$
j) $\sqrt{3^2 + 4^2}$
k) Work out the negative square root of   a) 4
b) 196   c) 900

2.9 LAWS OF INDICES

We already know that $5^2 = 5 \times 5 = 25$. In index notation, $3 \times 3 \times 3 = 3^3$ but the value is 27.

1) **Multiplication Rule**: To multiply powers of the same numbers or letters, add the index numbers.

$$C^x \times C^y = C^{x+y}$$

The base numbers or letters must be the same for multiplication rule to work.

Example 1: Simplify $4^2 \times 4^3$ and write your answer in index form.

$4^2 \times 4^3 = 4^{2+3} = \mathbf{4^5}$

Example 2: Simplify $t^{10} \times t^{12}$

$t^{10} \times t^{12} = t^{10+12} = \mathbf{t^{22}}$

Example 3: Simplify $6^5 \times 6^{-2}$

$6^5 \times 6^{-2} = 6^{5+-2} = \mathbf{6^3}$

Example 4: Simplify $5d^6 \times 2d^3$
First multiply the numbers: $5 \times 2 = 10$
Then, $d^6 \times d^3 = d^{6+3} = d^9$
Therefore, $5d^6 \times 2d^3 = \mathbf{10d^9}$

2) **Division rule:** To divide powers of the same numbers or letters, subtract the index numbers.

$$C^x \div C^y = C^{x-y}$$

Just like multiplication rule, the base numbers or letters must be the same before applying division rule/law.

Example 5: Simplify $12^7 \div 12^5$
$12^7 \div 12^5 = 12^{7-5} = \mathbf{12^2}$

Example 6: Simplify $x^7 \div x^3$
$x^7 \div x^3 = x^{7-3} = \mathbf{x^4}$

Example 7: Simplify $20c^7 \div 4c^2$
$= 20 \div 4 = 5, \quad c^7 \div c^2 = \mathbf{c^5}$
Therefore, $20c^7 \div 4c^2 = \mathbf{5c^5}$

Example 8: Simplify $\dfrac{5^8 \times 5^7}{5^4 \div 5^2}$

Numerator: $5^8 \times 5^7 = 5^{8+7} = 5^{15}$
Denominator: $5^4 \div 5^2 = 5^{4-2} = 5^2$
$= \dfrac{5^{15}}{5^2} = 5^{15-2} = \mathbf{5^{13}}$

NEGATIVE INDICES RULE

$x^{-n} = \dfrac{1}{x^n}$

Example 9: $4^{-2} = \dfrac{1}{4^2} = \dfrac{1}{16}$

Example 10: $c^5 \div c^7$
$= c^{5-7} = c^{-2} = \dfrac{1}{c^2}$

Example 11: $e^{-4} \div e^{-6}$
$= e^{-4--6} = \mathbf{e^2}$

POWERS OF ZERO

Any number raised to the power zero is equal to 1. $d^5 \div d^5 = d^{5-5} = d^0 = \mathbf{1}$ since any number divided by itself is **1**.

Therefore, $5^0 = 1$, $35^0 = 1$, $567^0 = 1$

RAISED POWERS

Example 1: Simplify $(3^2)^3$

This is the same as $3^2 \times 3^2 \times 3^2$
$= 3^{2+2+2} = \mathbf{3^6}$

Generally, for one power raised to another power, **multiply the indices**.
$(3^2)^3 = 3^{2 \times 3} = \mathbf{3^6}$

Example 2: Simplify $(5^{-2})^3$
This is $5^{-2 \times 3} = \mathbf{5^{-6}}$

Example 3: Simplify $(w^4)^5 \times w^7$
This is $w^{4 \times 5} \times w^7 = w20 \times w^7 = \mathbf{w^{27}}$

EXERCISE 2I

1) Write the following using index notation.

a) $5 \times 5 \times 5$
b) $6 \times 6 \times 6 \times 6 \times 6$
c) $13 \times 13 \times 13 \times 13 \times 13 \times 13 \times 13$
d) $4 \times 4 \times 4 \times 7 \times 7$

2) Evaluate these powers without using a calculator.

a) 1^2
b) 2^3
c) 17^2
d) 10^3
e) 20^0
f) 50^3
g) 22^2
h) 30^3

3) Evaluate the following.

a) $6^2 + 2^2$
b) $7^2 - 3^2$
c) $11^2 + 3^2 + 5^3$
d) $(6-2)^2$
e) $1^3 + 3^3 + 7^2$
f) $3^2 + 2^3 - 6^0$
g) $9^0 - 8^2$
h) $15^2 - 12^2$

4) Simplify these and leave your answers in index form.

a) $2^2 \times 2^6$
b) $4^2 \times 4^5$
c) $10^4 \times 10^5$
d) $17^2 \times 17^9$
e) $3^1 \times 3^0$
f) $9^8 \times 9$
g) $e^3 \times e^5$
h) $y^7 \times y^9$
i) $w \times w^4$
j) $3d^2 \times 2d^8$
k) $7h^3 \times 15h^5$
l) $y \times y \times y \times y$

5) Simplify these and leave your answers in index form.

a) $4^8 \div 4^5$
b) $3^9 \div 3^2$
c) $n^5 \div n^3$
d) $c^{13} \div c^7$
e) $y^9 \div y^3$
f) $12w^8 \div 4w^3$
g) $5^6 \div 5^2$
h) $15w^9 \div 5w$
i) $3d^2 \div d$
j) $(5^{-3})^2$

6) Write as an ordinary number.
a) 4^{-1} b) 3^{-1} c) 5^{-2} d) 13^{-3}

7) Simplify and write your answers in index form.
a) $5^3 \div 5^5$
b) $5^{-3} \div 5^{-2}$
c) $10^6 \times 10^{-4}$
d) $5^3 \times 5^4 \div 5^2$
e) $\dfrac{n^{12} \div n^7}{n^8 \div n^4}$
f) $\dfrac{8^4 \times 8^4}{8^5 \div 8^2}$
g) $\dfrac{11^{15}}{11^3 \times 11^6}$
h) $(15^{-3})^4$

8) Write **true** or **false**.
a) $5^2 \times 5^3 = 5^6$
b) $8^7 \div 8^2 = 8^5$
c) $12^{-1} = \frac{1}{12}$
d) $175^0 = 175$
e) $y^{12} \times y^0 = y^{12}$
f) $4^{-2} \div 4^{-4} = y^{-6}$

9) Simplify fully

a) $6x^2 \times 2x^4$
b) $25a^6 \div 5a^2$
c) $4xy \times 4xy$
d) $(5x)^2$
e) $\dfrac{88p^4y^2}{11p^2y}$

2.10 STANDARD FORM

Standard form is another way of writing very large or very small numbers. However, most numbers can be written using standard form.

Standard form is based on our decimal system which is based on the powers of ten. Index notations are used when writing powers of ten.

Examples:
$10^1 = 10$
$10^2 = 10 \times 10 = 100$
$10^3 = 10 \times 10 \times 10 = 1000$
$10^4 = 10000$ ….and so on

Remember, $10^0 = 1$

Negative powers of 10 can also be used when writing decimals.
Examples:

$10^{-1} = \dfrac{1}{10^1} = \dfrac{1}{10} = 0.1$

$10^{-3} = \dfrac{1}{10^3} = \dfrac{1}{1000} = 0.001$

If we use a calculator to work out the answer to 600000 × 50000, the calculator may display the answer as

$3E^{10}$ OR 3×10^{10} OR 3^{10}

Look at the number 8×10^{11}, the *11* means that 8 is followed by 11 zeros. This could be seen as 800 000 000 000.

Therefore, 8×10^{11}, 1.3×10^{-8}, 2×10^3… are all examples of numbers in standard form.

A number is in standard form if it is written in the form $a \times 10^n$ where *a* is a number between 1 and 10 and *n* is a positive or negative whole number.

In other words, numbers written in standard form must have **two parts**.

PART 1   PART 2

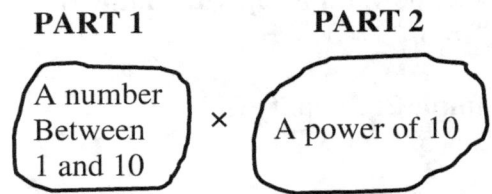

Examples
1) 2×10^3 is in standard form
2) 0.2×10^5 is **NOT** in standard form
3) $6.5^2 \times 10^{-7}$ is in standard form
4) 12×10^5 is **NOT** in standard form

Example 5: Write 235 in standard form.

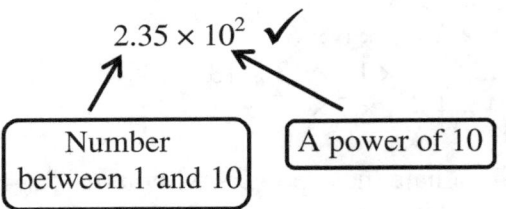

Example 6: Write the following numbers in standard form.

a) 35   b) 6000   c) 27500   d) 862.37

Answers:

a) $35 = \mathbf{3.5 \times 10^1}$
b) $6000 = \mathbf{6 \times 10^3}$
c) $27500 = \mathbf{2.75 \times 10^4}$
d) $862.37 = \mathbf{8.6237 \times 10^2}$

Example 7: Write these numbers
a) 2×10^3 b) 6.3×10^5 c) 3.261×10^9
as ordinary numbers.

Answers:
a) $2 \times 10^3 = 2 \times 1000 =$ **2000**
b) $6.3 \times 10^5 = 6.3 \times 100000 =$ **630000**
c) $3.261 \times 10^9 = 3.261 \times 1000000000$
$=$ **3261000000**

Example 8: Write these numbers in standard form. a) 0.002 b) 0.000023 c) 0.00405 d) 0.04007

Answers:
a) $0.002 = 2 \times 10^{-3}$
b) $0.000023 = 2.3 \times 10^{-5}$
c) $0.00405 = 4.05 \times 10^{-3}$
d) $0.04007 = 4.007 \times 10^{-2}$

Example 9: Write these numbers as ordinary numbers (decimal fractions)
a) 7×10^{-3} b) 9.5×10^{-5}
c) 6.203×10^{-7} d) 4.257×10^{-6}

Answers:
a) $7 \times 10^{-3} = \frac{7}{1000} = 0.007$
b) $9.5 \times 10^{-5} = \frac{9.5}{100000} = 0.000095$
c) $6.203 \times 10^{-7} = 0.0000006203$
d) $4.257 \times 10^{-6} = 0.000004257$

EXERCISE 2J

Write the following numbers in standard form.

1) 200
2) 500
3) 4000
4) 120000
5) 650
6) 2345
7) 3037
8) 10000
9) 340000
10) 39000
11) 40
12) 54
13) 989
14) 1345
15) three thousand
16) sixty thousand
17) 1 billion
18) 7893000
19) 9050
20) 9999

21) The population of Nigeria is estimated to be 210 000 000. Write this in standard form.

22) In chemistry, a hydrogen atom weighs about 0.000 000 000 000 000 000 000 00168 g. Write this number in standard form.

23) The speed of a car is 200 km/s. Write this in **m/s** and in standard form.

24) Which of the numbers below are in standard form?

a) 24×10^3 b) 3.5×10^{-2} c) 485×10^{-3}

25) Write 4 trillion in standard form.

EXERCISE 2K

Write the following numbers in standard form.

1) 0.02
2) 0.003
3) 0.00005
4) 0.000123
5) 0.0809
6) 0.000007
7) 0.00064
8) 0.0000444
9) 0.012
10) 0.045
11) 0.00323
12) 0.009
13) 0.000084
14) 0.06128

15) Write these numbers as ordinary numbers (decimal fractions).
a) 1.8×10^{-5}
b) 6.7×10^{-3}
c) 2.05×10^{-4}
d) 7.3×10^{-2}
e) 8.9×10^{-1}
f) 5.02×10^{-3}

ORDERING NUMBERS IN STANDARD FORM

To order numbers means to write them in order of size (smallest to highest or highest to smallest). However, it could be difficult when dealing with negative powers (very small numbers).

RULE 1
Start by comparing the powers of 10. Know that 10^{-7} is smaller than 10^{-6} and 10^{-5} is smaller than 10^{-4}, and so on.

Therefore, 3.5×10^{-7} is smaller than 2.4×10^{-5}.

RULE 2
If the numbers have the same powers of 10, compare the number part of the standard form.

6.7×10^{-3} is bigger than 5.2×10^{-3} and 2.3×10^{-11} is bigger than 1.2×10^{-11} since they have the same powers of 10.

Example 1: Write these numbers in order from smallest to highest.
$4.3 \times 10^{-2}, 6.2 \times 10^{-7}, 2.6 \times 10^{-5}, 3.4 \times 10^{-2}$

Solution:
10^{-7} is the smallest, so 6.2×10^{-7} is the smallest number. This is then followed by 2.6×10^{-5}, then 3.4×10^{-2} and finally 4.3×10^{-2}.

From smallest to highest, the numbers are **$6.2 \times 10^{-7}, 2.6 \times 10^{-5}, 3.4 \times 10^{-2}$ and 4.3×10^{-2}**

CALCULATIONS IN STANDARD FORM

To multiply numbers in standard form, the ***number*** parts are multiplied together and then the powers of ten after.

Example 1:
Work out $(3 \times 10^4) \times (4 \times 10^5)$ and leave your answer in standard form.

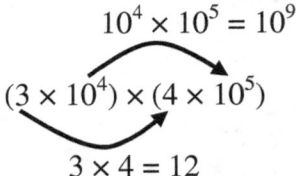

$$10^4 \times 10^5 = 10^9$$
$$(3 \times 10^4) \times (4 \times 10^5)$$
$$3 \times 4 = 12$$

This becomes 12×10^9.
However, 12×10^9 is **not** in standard form since the first part must be a number between 1 and 10 (exclusive) and 12 is more than 10.
In standard form, $12 \times 10^9 =$ **1.2×10^{10}**

Example 2:
Work out $(3.5 \times 10^3) \times (2.5 \times 10^{11})$ and leave your answer in standard form.

$(3.5 \times 2.5) \times (10^3 \times 10^{11}) =$ **(8.75×10^{14})**

Example 3:
Work out $(2.4 \times 10^{-5}) \div (4.8 \times 10^6)$

$$\left(\frac{2.4}{4.8}\right) \times \left(\frac{10^{-5}}{10^6}\right)$$

$= (2.4 \div 4.8) \times (10^{-5} \div 10^6)$
$= 0.5 \times 10^{-11}$

In standard form, $0.5 \times 10^{-11} =$ **5×10^{-12}**

Example 4: Work out $(3.5 \times 10^{-5})^2$ and leave your answer in standard form.

Remember: $(3.5 \times 10^{-5})^2$
$= (3.5 \times 10^{-5}) \times (3.5 \times 10^{-5})$
$= (3.5 \times 3.5) \times (10^{-5} \times 10^{-5})$
$= 12.25 \times 10^{-10}$
In standard form,
$12.25 \times 10^{-10} = \mathbf{1.225 \times 10^{-9}}$

Example 5: Express the answer to $2\,700 \times 3\,000\,000$ in standard form

First convert to standard form.
$2\,700 = 2.7 \times 10^3$, $3\,000\,000 = 3 \times 10^6$
It becomes $(2.7 \times 3) \times (10^3 \times 10^6)$
$= \mathbf{8.1 \times 10^9}$

ADDING AND SUBTRACTING IN STANDARD FORM

When adding or subtracting in standard form, work out the calculations in the brackets and then add or subtract.

Example 6
Work out $(3.2 \times 10^3) + (4.1 \times 10^2)$

    3200   +   410

$3200 + 410 = 3610$
In standard form, $\mathbf{3.61 \times 10^3}$

Example 7
Work out $(5.2 \times 10^{-3}) - (2.7 \times 10^{-4})$

    0.0052   -   0.00027

$0.0052 - 0.00027 = 0.00493$
In standard form $= \mathbf{4.93 \times 10^{-3}}$

EXERCISE 2L

1) Calculate the following and give your answer in standard form.

a) 10×200 f) $(40)^2$
b) 400×7 g) $(1.3 \times 10^5) \times (2.2 \times 10^3)$
c) 35×10 h) $(7.5 \times 10^{13}) \times (3 \times 10^{-4})$
d) 12×5 i) $(20)^3$
e) 500×50 j) $(2.3 \times 10^{-5}) \times (1.2 \times 10^{-6})$

2) If $c = 3 \times 10^6$ and $d = 2 \times 10^{11}$, find
a) cd   b) $c \div d$ c) $c + d$ d) $d - c$

3) These numbers are in standard form; arrange them in order of size (smallest first)

a) 2.3×10^5, 8.6×10^4, 1.5×10^8
b) 6.5×10^{-3}, 2.5×10^{-6}, 3.7×10^{-3}, 6.9×10^{-3}
c) 6.1×10^{-9}, 7.1×10^{-9}, 4.5×10^{-6}, 3.4×10^{-6}

4) Work out the following and write your answer in standard form.

a) $40000 \div 20$
b) $6000 \div 12$
c) $(4.8 \times 10^4) \div (1.2 \times 10^6)$
d) $(3 \times 10^{-5}) \div (1.5 \times 10^4)$
e) $(9 \times 10^{-2}) \div (3 \times 10^{-3})$
f) $(2 \times 10^{-5}) \times 10^{-7}$
g) $(4 \times 10^{-4}) + (3.5 \times 10^{-6})$
h) $(8.8 \times 10^{-3}) - (2.2 \times 10^{-5})$

5) The radius of the sun is 1.39×10^6 km approximately.
a) Find the diameter of the sun.
b) Find the surface area if the sun is considered a sphere. 2.43×10^{13} km^2

6) A computer part is rectangular in shape. Calculate the area of the part.

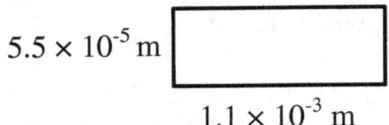

5.5×10^{-5} m

1.1×10^{-3} m

7) The mass of the Earth is 5 973 600,000 000 000 000 000 000 kg.

a) Write this is standard form

b) The mass of Venus is 4.87×10^{24} kg, work out the difference in mass between Venus and Earth.
Leave your answer in standard form.

c) If the radius of Venus is 6 052 km, work out the surface area if it is spherical.

d) If the mass of the Sun is 333000 times mass of the Earth, calculate the mass of the Sun.

8)

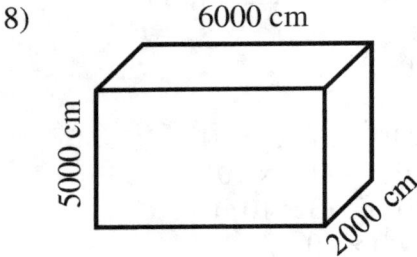

6000 cm

5000 cm

2000 cm

a) Calculate the volume of the cuboid.

b) Write your answer to part **a** in standard form.

c) Calculate the area of the base of the cuboid and leave your answer in standard form.

Chapter 2 Review Section
Assessment

1) Write down the first five multiples of

   a) 3 .... **1 mark**

   b) 4 .... **1 mark**

   c) 14 .... **1 mark**

2) Write down all the factors of

   a) 13 .... **1 mark**

   b) 36 .... **1 mark**

   c) 104 .... **1 mark**

3) From the cloud, write down the number(s) that are

   a) factors of 35

   b) factors of 63

   c) multiples of 6

   d) prime numbers

   e) multiples of 92

   Numbers in cloud: 23, 2, 72, 7, 6, 9

   .... **5 marks**

4) Write 250 as a product of prime factors in index form. .... **3 marks**

5) Work out the HCF of

   a) 27 and 63 .... **2 marks**

   b) 400 and 500 .... **3 marks**

6) Work out the LCM of

   a) 2 and 7 .... **2 marks**

   b) 35 and 50 .... **2 marks**

   c) 182 and 420 .... **3 marks**

7) Using the Venn diagram method, find the LCM and HCF of

   a) 66 and 132 .... **5 marks**

   b) 420 and 560 .... **5 marks**

8) Find the HCF of the following and leave your answers in prime factors and index notation.

   a) $2 \times 2 \times 3 \times 3 \times 3 \times 5 \times 5 \times 5$ and $2 \times 3 \times 3 \times 3 \times 3 \times 5$ .... **1 mark**

   b) $7 \times 7 \times 7 \times 7 \times 11 \times 11$, $7 \times 7 \times 13 \times 13 \times 13 \times 13$ and

   $7 \times 11 \times 13 \times 13$ .... **1 mark**

9) Azubuike says "The Lowest common multiple of two numbers cannot be one of the numbers."

   Is Azubuike correct? Explain fully.

   .... **2 marks**

10) The Lowest common multiple of 8 and another number is 72. If the other number is more than 8 but less than 15, what is the other number?

.... **2 marks**

11) Tolu, Nnamdi and Okechukwu started at the same time to ring a school bell.

Tolu rings a bell every 3 seconds.

Nnamdi rings a bell every 5 seconds.

Okechukwu rings a bell every 11 seconds.

How long will it take before they ring the bell at the same time?

.... **3 marks**

12) a) Write 54 as a product of its prime factors. .... **2 marks**

b) Express your answer in index form .... **1 mark**

c) Find the HCF of 54 and 126 .... **2 marks**

13)

Dictionaries are sold in packs of 12.

Scrabbles are sold in packs of 20.

A shop wants to buy the same number of dictionaries and scrabbles.

What is the lowest number of packs of dictionaries and scrabbles they could buy?

.... **3 marks**

14) Henry says "1 is a prime number and 2 is not."
    Is Henry correct?

   Explain fully. ………. **3 marks**

15) Write down the next **five** prime numbers larger than 36.
    ………. **2 marks**

16) Draw a factor tree for the following numbers.

   a) 16 ………. **2 marks**

   b) 144 ………. **2 marks**

17) Is the number 189357 divisible by 3?

   Explain your answer.
   ………. **3 marks**

18) Write down the prime factor of 750. ………. **2 marks**

19) From these set of numbers

   | 12 | 16 | 32 | 33 | 47 | 54 |

   Write down

   a) a prime number ………. **1 mark**

   b) a multiple of 9 ………. **1 mark**

   c) a multiple of 8 ………. **1 mark**

   d) a factor of 144 ………. **1 mark**

   e) a square number ………. **1 mark**

20) Copy and complete the table.

x	1	4	9	16	25	144
x^2						
$\sqrt{x}$						
$-\sqrt{x}$						

..........................18 marks

21) Which number in each pair is greater?

    a) 2^2 or $(-3)^2$ ....1 mark
    b) 7^2 or $(-7)^2$ ....1 mark
    c) $\sqrt{144}$ or 4^2 ....1 mark
    d) $\sqrt[3]{216}$ or 2^3 ....1 mark
    e) 0.3^2 or 0.3^3 ....1 mark

22) Simplify these and leave your answer in index form.

    a) $7^2 \times 7^8$ ....1 mark   d) $\dfrac{(4^5)^2 \div (4^2)^5}{(4^3)^2 \times 4}$ ....3 marks
    b) $3^6 \times 3^2$ ....1 mark
    c) $12^3 \div 12^7$ ....1 mark   e) $a^{-5} \times a^{-11}$ ....1 mark

23) Write the following in standard form.

    a) 345   d) 0.0012   g) 8732
    b) 28000   e) 0.0000098   h) 6 billion
    c) 8980000000   f) 0.405   i) 0.0000915

..........................9 marks

24) Write as decimal numbers.

    a) 6×10^3 c) 7×10^{-5} e) 9.4×10^{-4}
    b) 4.2×10^{-1} d) 6.6×10^{-6} f) 1.4×10^{-2}

..........................6 marks

25) Work out the following and write your answer in standard form.

    a) $(7.6 \times 10^5) \times (2 \times 10^3)$ d) $(5.5 \times 10^{-7}) \div (1.1 \times 10^{-3})$
    b) $(4.3 \times 10^{-2}) \times (3.5 \times 10^{-4})$ e) $(9.2 \times 10^8) + (4 \times 10^7)$
    c) $(8.4 \times 10^4) \div (4.2 \times 10^6)$ f) $(5.9 \times 10^4) - (2.3 \times 10^3)$

..........................12 marks

3 Fractions

In this section, we shall consider the following:

- Types of fractions
- Equivalent fractions
- Simplifying fractions
- Adding and subtraction fractions
- Multiplying and dividing fractions
- Fraction of an amount

LEARNING OBJECTIVES

By the end of this unit, you should be able to

a) Understand what fractions are

b) Work out equivalent fractions

c) Simplify fractions to their lowest form

d) Add and subtract fractions including mixed numbers

e) Multiply and divide fractions including mixed numbers

f) Work out fraction of an amount

KEYWORDS

- Fractions
- Numerator
- Denominator
- Add and Subtract
- Mixed number
- Simplify
- Divide and multiply

3.1 UNDERSTANDING FRACTIONS

Fractions usually have two parts, the **numerator** and **denominator**.

The numerator is the top number while the denominator is the bottom number in a fraction.

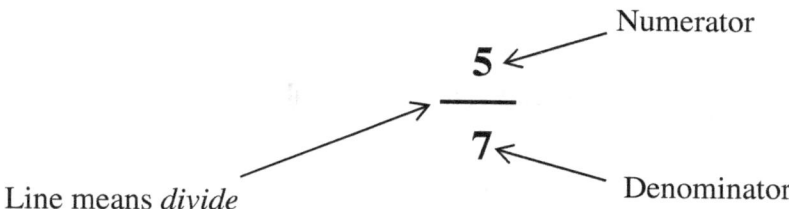

When a whole number is divided into equal parts, each of the parts is a fraction of the whole. Fractions could also be written with a slash (/) instead of a horizontal line (—) like 2/3, 4/7, 5/7.

3.2 TYPES OF FRACTIONS

There are three types of fractions at this level.

a) Proper Fractions

b) Improper fractions

c) Mixed Numbers

PROPER FRACTIONS

In a proper fraction, the numerator (top number) is less than the denominator (bottom number).

Examples of proper fractions: $\frac{2}{3}$ $\frac{4}{7}$ $\frac{12}{15}$

IMPROPER FRACTIONS

In an improper fraction, the numerator is bigger than the denominator.

Examples of improper fractions: $\frac{4}{3}$ $\frac{15}{4}$ $\frac{24}{7}$

MIXED NUMBERS

A whole number combined with a fraction is called a mixed number.

Examples of mixed numbers: $1\frac{2}{3}$ $7\frac{3}{4}$ $11\frac{1}{7}$

3.3 CONVERSION FROM MIXED NUMBERS TO IMPROPER FRACTIONS

Example 1: Convert $2\frac{3}{4}$ to improper fraction.

- Multiply the denominator (bottom number) by the whole number
- Add the outcome to the numerator
- Finally, divide the result by the denominator (**which stays the same**)

$4 \times 2 = 8$

$8 + 3 = 11$

$\frac{11}{4}$ ✓

Example 2: Change $5\frac{1}{2}$ to improper fraction.

$2 \times 5 = 10$

$10 + 1 = 11$

$\frac{11}{2}$ ✓

3.4 CONVERSION FROM IMPROPER FRACTION TO A MIXED NUMBER

Example 1: Convert $\frac{9}{4}$ to a mixed number.

This is a normal division with the remainder on top as the numerator.

Four (4) goes into 9 two (**2**) times, remainder one (**1**) ⟶ $2\frac{1}{4}$ ✓

Example 2: Convert $\frac{29}{8}$ to a mixed number.

8 goes into 29 three (**3**) times remainder five (**5**) ⟶ $3\frac{5}{8}$ ✓

3.5 SHADED FRACTIONS

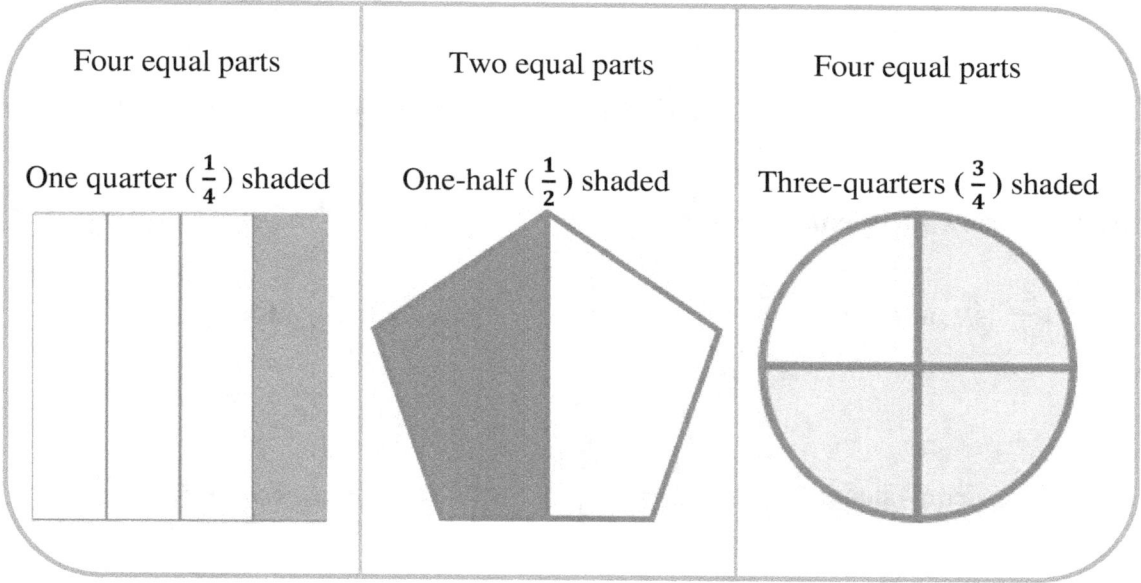

Four equal parts	Two equal parts	Four equal parts
One quarter ($\frac{1}{4}$) shaded	One-half ($\frac{1}{2}$) shaded	Three-quarters ($\frac{3}{4}$) shaded

3.6 EQUIVALENT FRACTIONS

These are fractions that have the same value though they may look different.
If **the same number multiplies both the numerators and denominators**, the value of the fraction remains the same provided that the number is not zero (0).

A

B

$\dfrac{2}{4}$

$\dfrac{1}{2}$

Shapes A and B are the same and are called equivalent fractions. They look different but are the same.

How to form equivalent fractions

$\dfrac{2}{3}$ is equivalent to $\dfrac{2}{3} \times \dfrac{4}{4} = \dfrac{8}{12}$ ……………. **Example 1**

$\dfrac{5}{11}$ is equivalent to $\dfrac{5}{11} \times \dfrac{3}{3} = \dfrac{15}{33}$ ……………. **Example 2**

Also, equivalent fractions can be obtained by **dividing** the numerator and the denominator by the same number provided that number is not zero (0).

$\dfrac{16}{20}$ is equivalent to $\dfrac{16 \div 4}{20 \div 4} = \dfrac{4}{5}$ .... Example 3

Example 4: Draw pictorial diagram to show that $\dfrac{3}{5} = \dfrac{9}{15}$

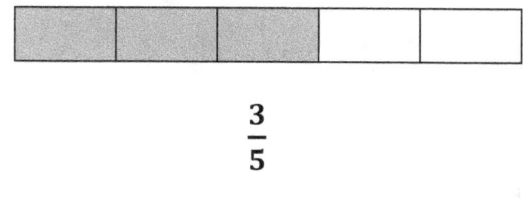

$\dfrac{3}{5}$

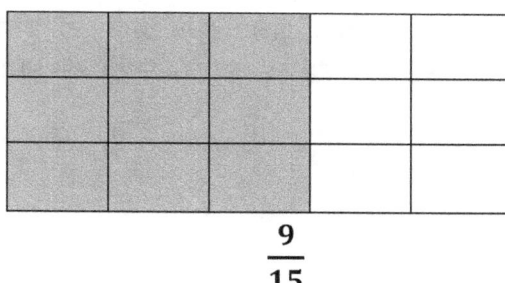

$\dfrac{9}{15}$

Example 5: Fill in the missing numbers.

a) $\dfrac{1}{3} = \dfrac{5}{\Box}$
b) $\dfrac{3}{7} = \dfrac{\Box}{21}$
c) $\dfrac{12}{18} = \dfrac{\Box}{3}$

Solutions

a) $\dfrac{1}{3} = \dfrac{5}{15}$ (× 5 top and bottom)

b) $\dfrac{3}{7} = \dfrac{9}{21}$ (× 3 top and bottom)

c) $\dfrac{12}{18} = \dfrac{2}{3}$ (÷ 6 top and bottom)

Note: Whatever you do to the top, you must do to the bottom. In equivalent fractions, × or ÷ is used to find any missing number.

Example 6: Arrange the following fractions in order of size, smallest first.

$$\frac{2}{3}, \frac{1}{5}, \frac{5}{6} \text{ and } \frac{7}{10}$$

The easiest way to compare fractions is to make the denominators the same using equivalent fractions. We also use our knowledge of the lowest common multiple.

The LCM of 3, 5 and 6 is 30.
Therefore, we find equivalent fractions with 30 as the denominator.

$$\frac{2 \times 10}{3 \times 10} = \frac{20}{30}$$

$$\frac{1 \times 6}{5 \times 6} = \frac{6}{30}$$

$$\frac{5 \times 5}{6 \times 5} = \frac{25}{30}$$

$$\frac{7 \times 3}{10 \times 3} = \frac{21}{30}$$

Since denominators of the equivalent fractions are now 30, the fraction with the lowest numerator is the lowest fraction. Likewise, the fraction with the highest numerator is the highest fraction.

Therefore, the order of size of smallest to biggest is $\frac{1}{5}, \frac{2}{3}, \frac{7}{10}$ and $\frac{5}{6}$

EXERCISE 3A

In questions **1** to **6**, write down the fractions shaded.

1)

2)

3)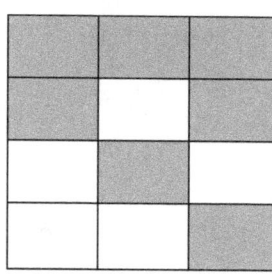

4)

5)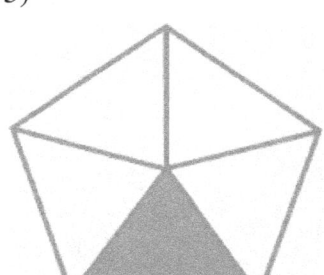

6)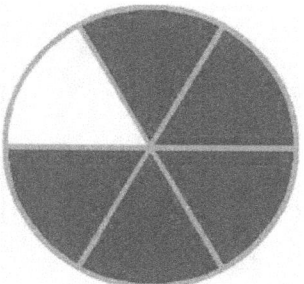

7) Copy and shade $\frac{1}{3}$ of the shape below.

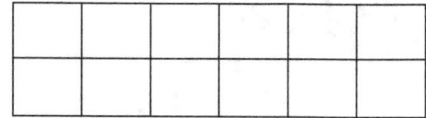

8) Copy and shade $\frac{1}{4}$ of the shape in number 7 above.

9) Copy and shade $\frac{3}{4}$ of the shape in question 7 above.

Change each of these improper fractions to mixed numbers.

10) $\dfrac{3}{2}$ 11) $\dfrac{14}{6}$ 12) $\dfrac{35}{11}$ 13) $\dfrac{122}{40}$

Change each of these mixed numbers to improper fractions.

14) $1\dfrac{3}{5}$ 15) $2\dfrac{1}{3}$ 16) $6\dfrac{2}{5}$ 17) $9\dfrac{5}{7}$

In questions 18 to 28, copy and complete the equivalent fractions.

18) $\dfrac{3}{4} = \dfrac{9}{?}$

19) $\dfrac{1}{2} = \dfrac{7}{?}$

20) $\dfrac{2}{5} = \dfrac{?}{15}$

21) $\dfrac{7}{9} = \dfrac{?}{18}$

22) $\dfrac{7}{10} = \dfrac{?}{30}$

23) $1 = \dfrac{?}{5}$

24) $\dfrac{3}{5} = \dfrac{21}{?}$

25) $\dfrac{30}{48} = \dfrac{?}{8}$

26) $\dfrac{7}{35} = \dfrac{?}{5}$

27) $\dfrac{1}{2} = \dfrac{?}{4} = \dfrac{?}{8}$

28) $\dfrac{1}{3} = \dfrac{?}{9} = \dfrac{?}{12} = \dfrac{?}{15}$

Equivalent Fractions

90

3.7 SIMPLIFYING FRACTIONS

When it is *impossible* to find a number which is not 1, that will divide exactly into the numerator (top number) and denominator (bottom number), the fraction is said to be in its **simplest form or lowest term.**

The fractions $\frac{2}{3}, \frac{7}{11}, \frac{12}{13}$ are in their simplest forms or lowest terms because no number can divide both the numerator and denominator exactly.

Advice: To simplify a fraction to its lowest form, always divide the fraction by their **highest common factor** (HCF).

Example 1:

Is the fraction $\frac{4}{8}$ in its lowest form?

NO, because 2 or 4 can divide both numbers. In its simplest form,

$$\frac{4 \div 4}{8 \div 4} = \frac{1}{2}$$

$\frac{1}{2}$ is the lowest form of $\frac{4}{8}$

91

Example 2: Reduce $\frac{25}{30}$ to its lowest form.

Divide both numbers by 5.

$$\frac{25 \div 5}{30 \div 5} = \frac{5}{6}$$

Example 3:

Write $\frac{16}{40}$ in its lowest form.

2, 4 and 8 are the common factors of 16 and 40. We may use any of the common factors to divide.

$$\frac{16 \div 2}{40 \div 2} = \frac{8}{20} \qquad \frac{8 \div 2}{20 \div 2} = \frac{4}{10} \qquad \frac{4 \div 2}{10 \div 2} = \frac{2}{5} \checkmark$$

↑ *Not in simplest form* ↑ *not in simplest form* ↑ *in simplest form*

It took three calculations to get to the fraction in its lowest form.

However, it is advisable to use the **highest common factor** when dividing both numbers. The answer will be obtained quicker in that way. Dividing by 2 or 4 will still give a correct answer, but more calculations are needed.

The highest common factor, in this case, is 8.

Use 8 to divide both numbers.

$$\frac{16 \div 8}{40 \div 8} = \frac{2}{5} \checkmark$$

In one simple calculation, $\frac{16}{40}$ to its lowest form is $\frac{2}{5}$

3.8 ADDING AND SUBTRACTING FRACTIONS

Before fractions can be added or subtracted, the **denominators** (bottom numbers) must be the same. If they are the same, simply add the numerators and divide by one of the denominators (since they are the same). Remember, do not add the denominators.

Example 1: $\quad \frac{2}{7} + \frac{4}{7} = \frac{2+4}{7} = \left(\frac{6}{7}\right) \checkmark$

(the Same denominator, so add the numerators)

Also when subtracting, $\frac{2}{5} - \frac{1}{5} = \frac{2-1}{5} = \left(\frac{1}{5}\right) \checkmark$

(the Same denominator, so subtract the numerators)

WHEN DENOMINATORS ARE NOT THE SAME

When fractions have different denominators, we **must** make them the same by finding a common denominator, preferably, the lowest common multiple (LCM). Then write each fraction as an equivalent fraction and perform the given calculation(s).

See sections 4.1.1 for LCM and 5.6 for equivalent fractions.

Example 2: $\dfrac{1}{3} + \dfrac{2}{5}$

Since the denominators 3 and 5 are not the same, we must make them the same before adding. The LCM of 3 and 5 is 15, so we make the denominators equal 15.

Find the equivalent fractions using 15 as the denominator.

$$\dfrac{1 \times 5}{3 \times 5} + \dfrac{2 \times 3}{5 \times 3}$$

$$= \dfrac{5}{15} + \dfrac{6}{15}$$

$$= \dfrac{5 + 6}{15}$$

$$= \dfrac{11}{15} \checkmark$$

Example 3: $\dfrac{3}{4} - \dfrac{3}{5}$

$$\dfrac{3 \times 5}{4 \times 5} - \dfrac{3 \times 4}{5 \times 4}$$

$$= \dfrac{15}{20} - \dfrac{12}{20}$$

$$= \dfrac{3}{20} \checkmark$$

EXERCISE 3B

1) Write each fraction in its simplest form.

a) $\dfrac{10}{20}$ f) $\dfrac{4}{18}$ k) $\dfrac{20}{75}$

b) $\dfrac{14}{21}$ g) $\dfrac{10}{48}$ l) $\dfrac{44}{80}$

c) $\dfrac{6}{18}$ h) $\dfrac{55}{88}$ m) $\dfrac{50}{250}$

d) $\dfrac{6}{33}$ i) $\dfrac{5}{15}$ n) $\dfrac{200}{500}$

e) $\dfrac{10}{40}$ j) $\dfrac{12}{40}$ o) $\dfrac{44}{44}$

2) Make both fractions below, so they have the same denominator.

a) $\dfrac{4}{5}$ and $\dfrac{2}{3}$

b) $\dfrac{1}{6}$ and $\dfrac{4}{5}$

c) $\dfrac{3}{4}$ and $\dfrac{1}{2}$

3) Write each of these fractions as a mixed number.

a) Eight fifths   e) $\dfrac{9}{5}$

b) Twenty nine sevenths   f) $\dfrac{17}{4}$

c) Fifty-five thirds   g) $\dfrac{142}{13}$

d) Eleven sixths   h) $\dfrac{200}{15}$

4) Work out each of the following calculations in its simplest form. Write as a mixed number when necessary.

a) $\frac{1}{9} + \frac{3}{9}$

b) $\frac{2}{7} + \frac{1}{7}$

c) $\frac{4}{6} + \frac{1}{6}$

d) $\frac{1}{10} - \frac{1}{10}$

e) $\frac{1}{3} + \frac{1}{4}$

f) $\frac{1}{5} + \frac{1}{9}$

g) $\frac{2}{3} + \frac{1}{5}$

h) $\frac{1}{9} - \frac{1}{9}$

i) $\frac{7}{9} - \frac{1}{5}$

j) $\frac{2}{8} + \frac{3}{5}$

k) $\frac{1}{2} + \frac{1}{7}$

l) $\frac{4}{7} + \frac{1}{10}$

m) $\frac{1}{2} + \frac{1}{3} + \frac{1}{4}$

n) $\frac{2}{3} + \frac{3}{5} - \frac{1}{6}$

o) $\frac{4}{5} - \frac{1}{4} + \frac{3}{5}$

5) Copy and complete

a) $\frac{1}{3} + \square = 1$

b) $\frac{1}{9} + \square = 1$

c) $1 - \frac{2}{3} = \square$

d) $\frac{15}{17} + \square = 1$

e) $\square + \frac{6}{13} = 1$

f) $1 - \square = \frac{15}{19}$

6) Write the fractions below in the order of size, largest first.

a) $\frac{1}{2}, \frac{2}{3}, \frac{5}{6}$

b) $\frac{7}{8}, \frac{1}{4}, \frac{1}{2}$

c) $2\frac{1}{3}, 3\frac{2}{3}, 2\frac{2}{3}$

7) What fraction is shaded

a) red

b) green

c) yellow

d) blue?

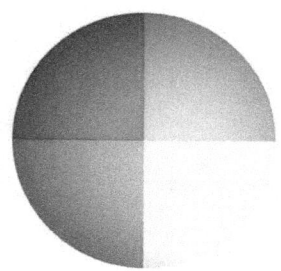

8) Funmi gave $\frac{1}{3}$ of her books to Abubakar and $\frac{1}{5}$ of her books to Chichi. What fraction of Funmi's book does she still have?

9) Five mathematics books are placed side by side and the width shown.

   a) What is the total width, **b metre** of the five books?

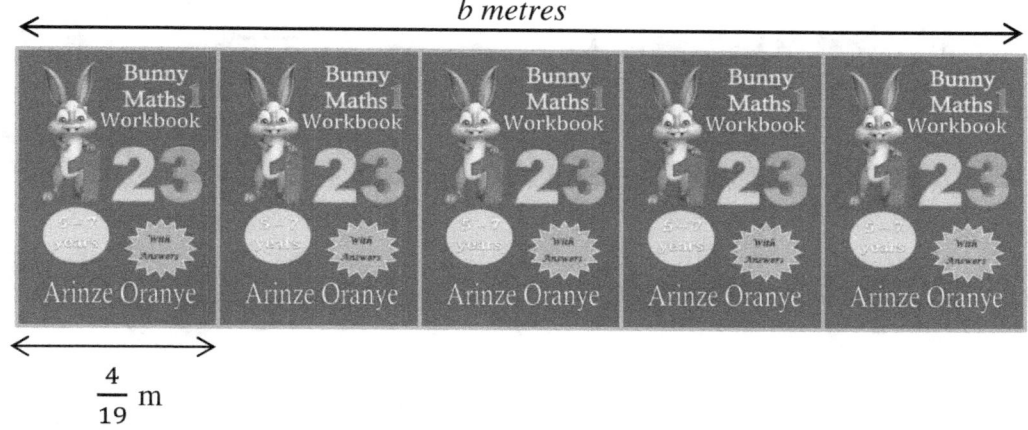

   b) A student bought two of the books. What is the combined width of the remaining books?

10) All are equivalent fractions apart from one. Pick the odd one out and give a reason.

$$\frac{4}{15} \qquad \frac{8}{30} \qquad \frac{5}{45} \qquad \frac{20}{75} \qquad \frac{40}{150}$$

11) Complete the missing numbers in the boxes.

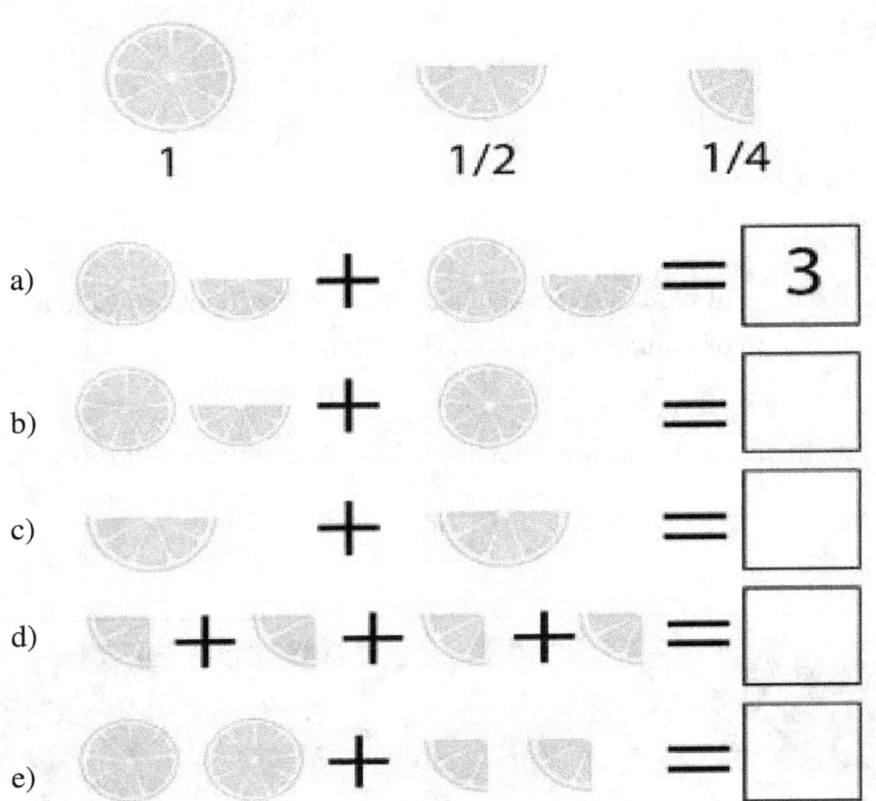

ADDING AND SUBTRACTING MIXED NUMBERS

Example 1: Work out $2\frac{1}{2} + 3\frac{1}{3}$

Solution: Change the mixed numbers to improper fractions (Refer to section 5.3).

For mixed number $2\frac{1}{2}$, change to $\frac{(2 \times 2) + 1}{2} = \frac{5}{2}$

For mixed number $3\frac{1}{3}$, change to $\frac{(3 \times 3) + 1}{3} = \frac{10}{3}$

Therefore, $\frac{5}{2} + \frac{10}{3}$

The denominators are not the same, so we make them the same.

$$= \frac{5 \times 3}{2 \times 3} + \frac{10 \times 2}{3 \times 2}$$

$$= \frac{15}{6} + \frac{20}{6} = \frac{35}{6}$$

However, the answer is top heavy (improper fraction) and should be written as a mixed number (refer to section 5.4).

Therefore, $\frac{35}{6} = 5\frac{5}{6}$

$2\frac{1}{2} + 3\frac{1}{3} = 5\frac{5}{6}$ ✓

A quicker method is to add up the whole numbers and fractions separately.

$2 + 3 = \boxed{5}$

$$\frac{1}{2} + \frac{1}{3} = \frac{1 \times 3}{2 \times 3} + \frac{1 \times 2}{3 \times 2} = \frac{3}{6} + \frac{2}{6} = \boxed{\frac{5}{6}}$$

Therefore, 5 and $\frac{5}{6}$ gives $5\frac{5}{6}$ ✓   ………..same answer.

Example 2: Work out $4\frac{1}{2} - 3\frac{1}{3}$

Solution: Change the mixed numbers to improper fractions (Refer to section 5.3).

For mixed number $4\frac{1}{2}$, change to $\frac{(2 \times 4) + 1}{2} = \frac{9}{2}$

For mixed number $3\frac{1}{3}$, change to $\frac{(3 \times 3) + 1}{3} = \frac{10}{3}$

Therefore, $\frac{9}{2} - \frac{10}{3}$

The denominators are not the same, so we make them the same.

$$= \frac{9 \times 3}{2 \times 3} - \frac{10 \times 2}{3 \times 2}$$

$$= \frac{27}{6} - \frac{20}{6}$$

$$= \frac{7}{6}$$

However, the answer is top heavy (improper fraction) and should be written as a mixed number (refer to section 5.4).

Therefore, $\frac{7}{6} = 1\frac{1}{6}$

$4\frac{1}{2} - 3\frac{1}{3} = 1\frac{1}{6}$ ✓

EXERCISE 3C

Work out the following and leave your answers in their simplest form.

1) $1\frac{2}{3} + 1\frac{2}{3}$

2) $2\frac{2}{3} + 1\frac{2}{5}$

3) $4\frac{2}{3} + 3\frac{1}{3}$

4) $5\frac{2}{3} - 2\frac{2}{3}$

5) $7\frac{1}{3} - 1\frac{1}{4}$

6) $1\frac{2}{3} + 1\frac{2}{3} + 1\frac{2}{3}$

7) $1\frac{2}{3} - \frac{3}{5} + \frac{2}{7}$

8) $3\frac{2}{7} + \frac{1}{3}$

9) $1\frac{2}{3} - 1\frac{2}{3}$

10) $8\frac{2}{3} - 3\frac{4}{5}$

11) On Tuesday, Edward cycled $9\frac{2}{3}$ kilometres to work from his house.

On his way back, he cycled $4\frac{1}{4}$ km and his bike had a puncture. Edward had a lift back to his house by a friend, Andrew.

How far did Andrew travel to get Edward back to his house?

3.9 MULTIPLYING FRACTIONS

Multiplying fractions is the easiest of all the fraction calculations. To multiply fractions, multiply the numerators (top) numbers and multiply the denominators (bottom numbers). Cancel down when possible.

Example 1: Work out $\frac{1}{4} \times \frac{2}{5}$

Multiply the numerators: $1 \times 2 = 2$

Multiply the denominators: $4 \times 5 = 20$

Therefore, $\frac{1}{4} \times \frac{2}{5} = \frac{2}{20}$

A more structured approach would be

$$\frac{1}{4} \times \frac{2}{5} = \frac{1 \times 2}{4 \times 5} = \frac{2}{20}$$

But $\frac{2}{20}$ is not in its simplest form, so we cancel down

$$\frac{2 \div 2}{20 \div 2} = \frac{1}{10} \checkmark$$

Example 2: Work out $\frac{3}{7} \times \frac{5}{8}$

$$= \frac{3 \times 5}{7 \times 8}$$

$$= \frac{15}{56} \checkmark$$

Example 3: The diagram shows a square of side **1 metre** which is divided into Q, R, S, T rectangles.

a) Work out the area of **each** rectangle Q, R, S and T.

b) Show that the total area of the square is 1m²

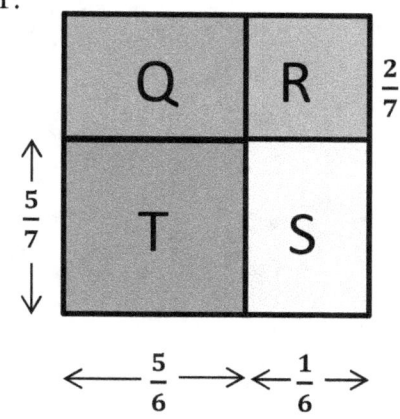

Solution:

Area of a rectangle = length × width

Area of **Q** = $\frac{2}{7} \times \frac{5}{6} = \frac{10}{42} = \frac{5}{21}$ m²

Area of **R** = $\frac{2}{7} \times \frac{1}{6} = \frac{2}{42} = \frac{1}{21}$ m²

Area of **S** = $\frac{5}{7} \times \frac{1}{6} = \frac{5}{42}$ m²

Area of **T** = $\frac{5}{7} \times \frac{5}{6} = \frac{25}{42}$ m²

For question b, total area of square is the total area of rectangles Q, R, S and T

$$\frac{5 \times 2}{21 \times 2} + \frac{1 \times 2}{21 \times 2} + \frac{5}{42} + \frac{25}{42}$$

$$= \frac{10}{42} + \frac{2}{42} + \frac{5}{42} + \frac{25}{42}$$

$$= \frac{10 + 2 + 5 + 25}{42}$$

$$= \frac{42}{42}$$

$$= 1$$

Therefore, the total area of the square is **1 m²** ✓

EXERCISE 3D

1) Work out and simplify your answers where possible.

   a) $\dfrac{1}{5} \times \dfrac{2}{3}$ e) $\dfrac{3}{5} \times \dfrac{3}{5}$ i) $\dfrac{2}{9} \times \dfrac{6}{7}$

   b) $\dfrac{2}{5} \times \dfrac{2}{7}$ f) $\dfrac{1}{4} \times \dfrac{7}{8}$ j) $\dfrac{12}{15} \times \dfrac{1}{7}$

   c) $\dfrac{3}{4} \times \dfrac{6}{7}$ g) $\dfrac{5}{7} \times \dfrac{8}{9}$ k) $\dfrac{8}{10} \times \dfrac{3}{4}$

   d) $\dfrac{5}{6} \times \dfrac{2}{9}$ h) $\dfrac{3}{12} \times \dfrac{1}{8}$ l) $\left(\dfrac{2}{20}\right)^2$

2) The height of the Nigerian flag is $\dfrac{6}{7}$ m.
   If fourteen flags are placed on top of each other, what would be the total length of all the flags?

3) Work out the area of the rectangular field.

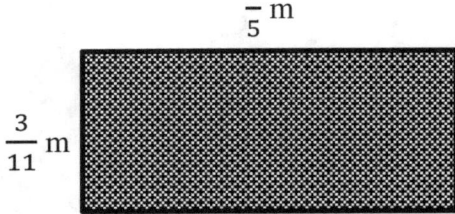

4) Work out the area of the triangle.

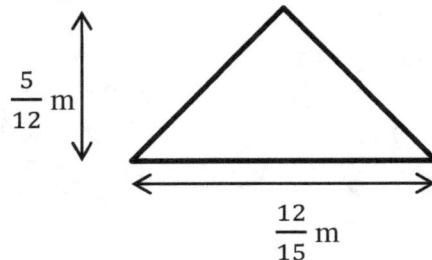

5) A rectangle is divided into four parts as shown below. Work out the area of **each** part of the rectangle. All lengths are in metres.

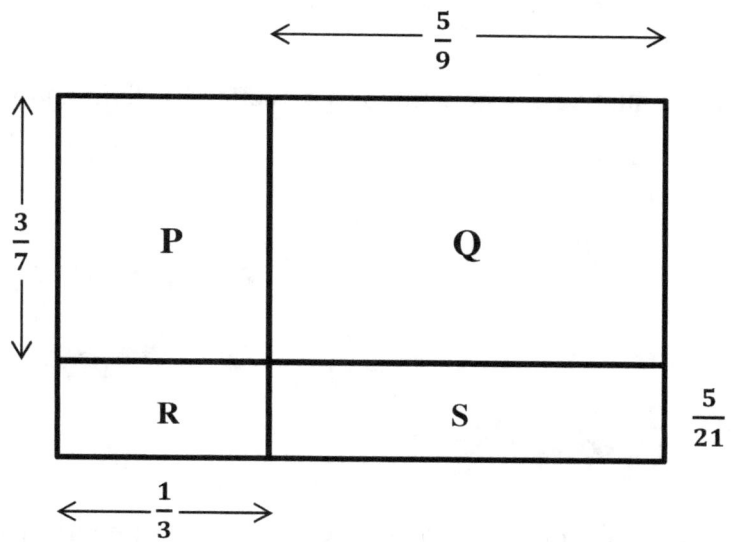

6) Work out the following.

a) $\dfrac{1}{3} \times \dfrac{1}{3} \times \dfrac{1}{3}$

b) $\dfrac{2}{3} \times \dfrac{3}{4} \times \dfrac{1}{5}$

c) $\dfrac{4}{11} \times \dfrac{9}{13}$

7) Nnaemeka says "$\dfrac{5}{6} \times \dfrac{1}{3} = \dfrac{6}{9}$"

a) Is he correct? Explain fully

b) If not, what did Nnaemeka do wrong?

FRACTION OF AN AMOUNT

Example 1: Work out $\frac{3}{5} \times 20$

Remember, every whole number can be written as a fraction by dividing by 1. 20 is the same as $\frac{20}{1}$

$$\frac{3}{5} \times \frac{20}{1}$$

$$= \frac{3 \times 20}{5 \times 1}$$

$$= \frac{60}{5} = 12 \checkmark$$

Alternatively, you may consider dividing by the denominator and multiplying by the numerator. It is the same thing as cancelling down. $20 \div 5 = 4$. Then, $4 \times 3 = 12$.

$$\frac{3}{\cancel{5}_1} \times \frac{\cancel{20}^4}{1} = 3 \times 4 = 12$$

Example 2: Work out $\frac{3}{5}$ of ₦100

In mathematics, '**Of**' in this instance is multiplication (×).

It is the same method of working out fractions as in examples 1 and 2 above.

$$\frac{3}{5} \text{ of } ₦100 = \frac{3}{5} \times ₦100$$

$$= \frac{3}{\cancel{5}_1} \times \cancel{₦100}^{₦20}$$

$$= 3 \times ₦20$$

$$= ₦60 \checkmark$$

⎱ Dividing by the denominator and multiplying by the numerator

Alternatively; $\frac{3}{5}$ of ₦100

$$= \frac{3}{5} \times \frac{100}{1}$$

$$= \frac{3 \times 100}{5 \times 1}$$

$$= \frac{300}{5}$$

$$= ₦60$$

Example 3: $450 \times \frac{3}{9}$

Method 1: Multiply 450 by 3, and then divide by 9

$$450 \times 3 = 1350$$

$$1350 \div 9 = \mathbf{150} \checkmark$$

Method 2: Divide 450 by 9, and then multiply by 3

$$450 \div 9 = 50$$

$$50 \times 3 = \mathbf{150} \checkmark$$

⇩

Cancelling down method

$$\frac{50 \times 3}{1} = \mathbf{150} \checkmark$$

EXERCISE 3E

1) Work out the following.

   a) $\frac{1}{5} \times 5$

   b) $\frac{2}{5} \times 10$

   c) $\frac{3}{4} \times 12$

   d) $\frac{7}{8} \times 24$

   e) $40 \times \frac{1}{5}$

   f) $45 \times \frac{2}{9}$

   g) $120 \times \frac{5}{12}$

   h) $300 \times \frac{3}{60}$

   i) $\frac{1}{5} \times 200$

   j) $275 \times \frac{1}{5}$

   k) $2000 \times \frac{7}{40}$

   l) $\frac{1}{13} \times 13$

2) Work out the following.

   a) $\frac{1}{5}$ of ₦200

   b) $\frac{1}{3}$ of ₦150

   c) $\frac{1}{8}$ of $400

   d) $\frac{1}{9}$ of 54 kg

   e) $\frac{5}{8}$ of 64

   f) $\frac{4}{5}$ of ₦1500

   g) $\frac{1}{5}$ of £77.50

   h) $\frac{17}{35}$ of 2100 kg

3)

Emeka paid $\frac{2}{3}$ of the cost of Obiora's jacket for a similar black jacket.

a) How much did Emeka pay for the jacket?

b) Arinze says "$\frac{4}{5}$ of what Emeka paid is more than half the cost of Obiora's jacket."

Is Arinze correct?
Explain fully.

MULTIPLYING WITH MIXED NUMBERS

To multiply mixed numbers, it is advisable to change them to improper fractions (Refer to section 5.3)

Example 1: Work out $3\frac{2}{5} \times 7\frac{1}{2}$

Change to improper fractions

$3\frac{2}{5}$ becomes $\frac{(5 \times 3) + 2}{5} = \frac{17}{5}$ and $7\frac{1}{2}$ becomes $\frac{(2 \times 7) + 1}{2} = \frac{15}{2}$

Multiplying the two improper fractions: $\frac{17}{5} \times \frac{15}{2} = \frac{17 \times 15}{5 \times 2}$

$$= \frac{255}{10} = 25.5$$

$$= 25\frac{1}{2} \checkmark$$

Example 2: Work out $5\frac{1}{4} \times \frac{2}{7}$

$5\frac{1}{4}$ becomes $\frac{(4 \times 5) + 1}{4} = \frac{21}{4}$

Multiplying both fractions gives $\frac{21}{4} \times \frac{2}{7}$

$$= \frac{21 \times 2}{4 \times 7}$$

$$= \frac{42}{28} \quad \text{(Cancel down by dividing both numbers by 2)}$$

$$= \frac{21}{14} \quad \text{(Cancel down by dividing both numbers by 7)}$$

$$= \frac{3}{2} \quad \text{(Change back to a mixed number)}$$

$$= 1\frac{1}{2} \checkmark$$

Example 3: Multiply $4 \times 2\frac{3}{5}$

Change $2\frac{3}{5}$ to an improper fraction. $\dfrac{(5 \times 2) + 3}{5} = \dfrac{13}{5}$

Multiplying gives $4 \times \dfrac{13}{5}$

$$= \dfrac{4 \times 13}{5}$$

$$= \dfrac{52}{5}$$

$$= 10\frac{2}{5} \checkmark$$

EXERCISE 3F

1) Write the following mixed numbers to improper fractions.

   a) $1\frac{3}{4}$ b) $5\frac{3}{5}$ c) $11\frac{4}{9}$ d) $17\frac{1}{3}$

2) Work out the following but give your answer as a mixed number when applicable.

   a) $2\frac{3}{5} \times \frac{3}{5}$ b) $1\frac{3}{5} \times 2\frac{3}{7}$ c) $7\frac{3}{5} \times \frac{3}{5}$ d) $3\frac{1}{5} \times 1\frac{4}{10}$

   e) $3\frac{3}{4} \times 1\frac{2}{5}$ f) $2\frac{4}{6} \times 6\frac{3}{7}$ g) $12\frac{3}{5} \times 8\frac{3}{4}$ h) $(2\frac{3}{5})^2$

3) Find the product of these numbers. Write as mixed numbers where possible and simplify.

   a) $4\frac{3}{7}$ and $\frac{1}{3}$ b) $10\frac{1}{5}$ and $\frac{3}{5}$ c) 2 and $8\frac{5}{6}$ d) 9 and $6\frac{3}{4}$

4) If it takes $\frac{1}{5}$ of a minute to fill a bucket with cold water, what fraction of a minute will it take to fill $15\frac{1}{2}$ buckets with cold water?

5)

A tin of sweetcorn weighs $\frac{1}{5}$kg.
137 of the tins are packed in a bag.

a) What is the total weight of the sweetcorn?

b) If the bag used in packaging weighs $1\frac{3}{5}$ kg, what is the total weight of the sweet corns and bag?

3.10 DIVIDING FRACTIONS

Our mantra would be "**Keep, Change, Flip**."

Keep the first fraction, **change** the division to multiplication (×) and **flip** the second fraction. Once we have successfully applied the mantra, we then multiply out the fractions (See section 5.9).

Example 1: Work out $\frac{5}{6} \div \frac{3}{4}$

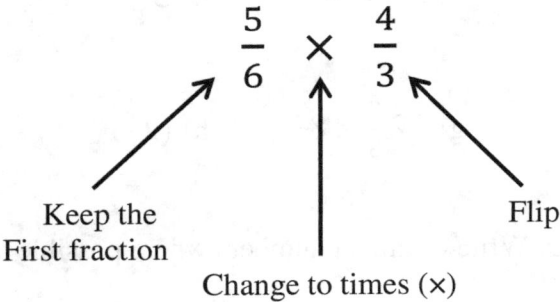

Keep the First fraction

Change to times (×)

Flip

Multiply the two fractions

$$\frac{5 \times 4}{6 \times 3} = \frac{20}{18}$$

Since $\frac{20}{18}$ is not in its simplest form, we cancel down by dividing by 2.

We then have $\frac{20 \div 2}{18 \div 2} = \frac{10}{9}$

Also, the answer is an improper fraction, we then change back to a mixed number (refer to section 5.4)

Therefore, $\frac{10}{9} = 1\frac{1}{9}$ ✓

The same system applies to dividing mixed numbers. However, we must change to improper fractions before applying the *keep, change flip*. See example 2.

Example 2: Work out $3\frac{2}{5} \div 1\frac{3}{4}$

Change the mixed numbers to improper fractions (refer to section 5.3).

$3\frac{2}{5}$ to an improper fraction is $\frac{(5 \times 3) + 2}{5} = \frac{17}{5}$

$1\frac{3}{4}$ to an improper fraction is $\frac{(4 \times 1) + 3}{5} = \frac{7}{4}$

Now, rewrite the fractions using the improper fractions as $\frac{17}{5} \div \frac{7}{4}$

Remember **Keep, change, flip**

$$= \frac{17}{5} \times \frac{4}{7}$$

$$= \frac{17 \times 4}{5 \times 7} = \frac{68}{35} = 1\frac{33}{35} \checkmark$$

Example 3: Work out $10 \div 3\frac{2}{7}$

Change $3\frac{2}{7}$ to improper fraction which will equal $\frac{(7 \times 3 + 2)}{7} = \frac{23}{7}$

Now, rewrite the fractions using the improper fraction as $10 \div \frac{23}{7}$

Remember **Keep, change, flip**

$$= 10 \times \frac{7}{23}$$

$$= \frac{10 \times 7}{23} = \frac{70}{23} = 3\frac{1}{23} \checkmark$$

Example 4: How many quarters are there in 6?

This means $6 \div \frac{1}{4}$

$$= 6 \times \frac{4}{1}$$

$$= \frac{6 \times 4}{1} = 24 \checkmark$$

EXERCISE 3G

1) Work out the fractions and give your answers as a mixed number when possible.

   a) $\frac{1}{2} \div \frac{1}{4}$

   b) $\frac{4}{5} \div \frac{2}{3}$

   c) $\frac{7}{10} \div \frac{3}{5}$

   d) $\frac{6}{7} \div \frac{1}{5}$

   e) $7 \div \frac{1}{2}$

   f) $\frac{1}{3} \div \frac{1}{5}$

   g) $10\frac{1}{2} \div \frac{1}{2}$

   h) $6\frac{2}{9} \div \frac{5}{6}$

   i) $20 \div 1\frac{4}{5}$

   j) $2\frac{7}{9} \div 1\frac{7}{10}$

   k) $5\frac{2}{3} \div 2\frac{7}{10}$

   l) $3\frac{3}{7} \div 1\frac{1}{10}$

2) How many thirds are there in

   a) 5   b) 7   c) 9   d) 15?

3) How many tenths are there in

   a) 6   b) 8   c) 11   d) 20?

4) The length of the top of a rectangular table tennis table is $2\frac{7}{10}$ m. The area of the top is $4\frac{1}{20}$ m². Work out the width of the top of the table top. Leave your answer as a mixed number.

5) A perimeter fence has 7 panels. The width of one panel is $1\frac{8}{10}$ m long. What is the total width of the panels?

6) Work out $15\frac{3}{4} \div 1\frac{3}{4}$, give your answer as a mixed number

7) Work out $12\frac{2}{5} \div 5\frac{7}{8}$, give your answer as a mixed number.

Chapter 3 Review Section
Assessment

1) Look at the picture. What fraction of these students are:

   a) males

   .... **1 mark**

   b) not males

   .... **1 mark**

   c) wearing white shirts?

   .... **1 mark**

2) What fraction of the months of the year starts with the letter J?

   .... **1 mark**

3) What fraction of an hour is 20 minutes?

   .... **1 mark**

4) Write the following as equivalent fractions with a denominator of 40.

   a) $\frac{1}{4}$ b) $\frac{3}{8}$ c) $\frac{4}{5}$ d) $\frac{9}{10}$

   .... **4 marks**

5) Change to improper fractions.

   a) $2\frac{1}{7}$ b) $1\frac{5}{8}$ c) $12\frac{1}{3}$ d) $20\frac{9}{11}$

   .... **4 marks**

6) Add $\frac{4}{7}$ to :

   a) $\frac{1}{7}$ b) $\frac{2}{3}$ c) $\frac{5}{6}$ d) $\frac{3}{11}$

   .... **8 marks**

7) Simplify each fraction.

   a) $\dfrac{4}{8}$ b) $\dfrac{9}{27}$ c) $\dfrac{18}{30}$ d) $\dfrac{144}{168}$

   .... **8 marks**

8) Copy and complete the pairs of equivalent fractions below.

   a) $\dfrac{4}{7} = \dfrac{?}{21}$ b) $\dfrac{12}{20} = \dfrac{?}{100}$ c) $\dfrac{6}{8} = \dfrac{54}{?}$ d) $\dfrac{?}{15} = \dfrac{24}{30}$

   .... **4 marks**

9) Change to mixed numbers.

   a) $\dfrac{7}{3}$ b) $\dfrac{23}{4}$ c) $\dfrac{59}{7}$ d) $\dfrac{204}{20}$

   .... **4 marks**

10) Work out and leave your answers as a mixed number where possible. Also, leave your answers in its lowest form where possible.

    a) $\dfrac{4}{17} + \dfrac{4}{17}$ b) $\dfrac{8}{9} - \dfrac{1}{3}$ c) $1\dfrac{4}{5} + 4\dfrac{4}{5}$ d) $\dfrac{2}{3} \div \dfrac{1}{5}$

    e) $3\dfrac{1}{3} - 1\dfrac{7}{9}$ f) $3\dfrac{1}{3} \times \dfrac{1}{4}$ g) $2\dfrac{1}{6} \times 2\dfrac{5}{9}$ h) $8\dfrac{1}{3} \div 5\dfrac{2}{5}$

    .... **16 marks**

11) Work out

    a) $\dfrac{2}{5}$ of 30 b) $\dfrac{3}{4}$ of 36 kg c) $\dfrac{5}{7}$ of ₦420 d) $160 \times \dfrac{7}{8}$

    .... **8 marks**

12) The diagram is divided into rectangles as shown.

    a) Work out the area of **each** rectangle.

    b) Work out the **total** area of Q, R, S and T

    c) What is the mathematical name of the shape formed by Q, R S and T? Give a reason for your answer.

    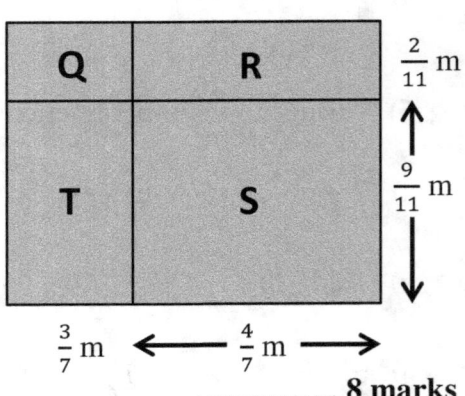

    .... **8 marks**

13) Simplify the following.

a) $\dfrac{16}{20} \div \dfrac{4}{10}$

b) $\dfrac{\frac{1}{2} \times \frac{6}{7}}{\frac{1}{7}}$

c) $\dfrac{5\frac{3}{5} + \frac{4}{7}}{\frac{2}{3}}$

14)

The weight of each pack of rice is 2kg.
9 packets of rice are packed into a travelling bag of weight $1\frac{5}{6}$ kg.

a) What is the total weight of the bag when loaded with the 9 packets of rice?

b) Kola bought 21 of such travelling bags. What is the total weight of the travelling bags?

c) Kola gave three travelling bags away to his friend, Tayo. What is the combined weight of Tayo's bags?

d) If a travelling bag costs ₦5 500, how much did Kola pay for the travelling bags?

4 Percentages 1

This section covers the following topics:

- Fractions, decimals and percentages
- Percentage of a quantity
- Chapter Review Section

LEARNING OBJECTIVES

By the end of this unit, you should be able to:

a) Understand the word 'percent.'
b) Change fractions to decimals
c) Change fractions and decimals into percentages
d) Change percentages into fractions and decimals
e) Write one quantity as a fraction of another

KEYWORDS

- Percentage
- Percent
- Quantity
- Fraction
- Decimal

4.1 UNDERSTANDING PERCENTAGES (%)

The word percent means **'out of a hundred.'** A percentage is a special type of fraction.

1% means 1 part per hundred. This can be written as $\frac{1}{100}$.

1% can also be written as a decimal **0.01**.

Likewise,

2% means $\frac{2}{100}$ or 0.02 as a decimal

10% means $\frac{10}{100}$ or 0.1 as a decimal

47% means $\frac{47}{100}$ or 0.47 as a decimal

100% means $\frac{100}{100}$ which is equal to **one whole** (1)

Percentages are equal to fractions with the denominator (bottom number) equal to 100.

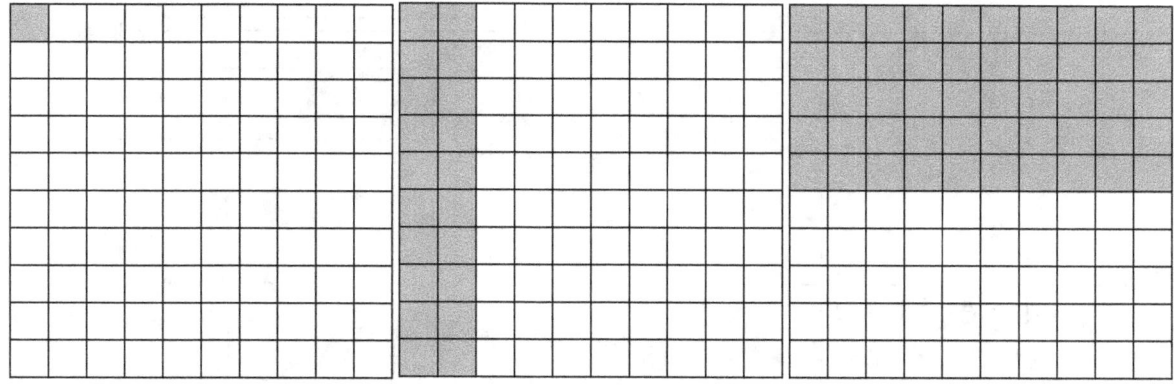

One percent is shaded   **Twenty** percent is shaded   **Fifty** percent is shaded
$\frac{1}{100}$ $\frac{20}{100}$ or $\frac{1}{5}$ or $\frac{2}{10}$ $\frac{1}{2}$ or $\frac{50}{100}$ or $\frac{5}{10}$ or $\frac{25}{50}$

Example 1: What percentage of each box has been shaded?

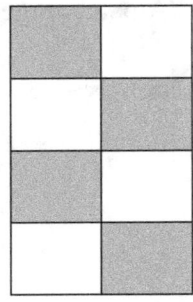

$\frac{1}{4} = 25\%$

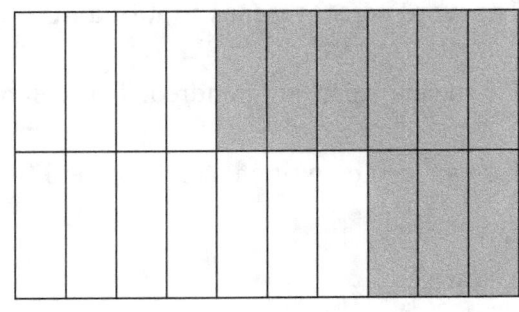

$\frac{4}{8} = \frac{1}{2} = 50\%$

$\frac{9}{20} = 45\%$

Some key percentages to memorise

Fractions	Percentages	
$\frac{1}{4}$	25%	
$\frac{1}{2}$	50%	
$\frac{3}{4}$	75%	
A whole (1)	100%	

Also note:

$\frac{1}{8} = 12\frac{1}{2}\%$

4.2 FRACTIONS TO DECIMALS

A fraction is simply a division. The knowledge in section 3.3 is vital. If you can divide a number, you have successfully converted a fraction to a decimal.

However, the knowledge of equivalent fractions is very important in changing fractions to decimals.

Example 1: Change $\frac{2}{5}$ to a decimal.

Write as an equivalent fraction with a denominator of 100.

$$\frac{2 \times 20}{5 \times 20} = \frac{40}{100} = \mathbf{0.4}$$

Not all numbers will have an equivalent fraction of 100. If that is the case, use other methods like dividing decimals or change the denominator to an equivalent fraction which you may multiply or divide by a number to make 100 or multiples of 10 (if it is easier). You may also cancel down (reducing to its simplest form) and the find an equivalent fraction with a denominator of 100.

Example 2: Convert $\frac{9}{12}$ to a decimal.

As you can see, 100 is not a multiple of 12, and it will not be possible to make an equivalent fraction with a denominator of 100.

Cancelling down gives $\frac{9 \div 3}{12 \div 3} = \frac{3}{4}$

It now possible to make an equivalent fraction with a denominator of 100

$$\frac{3 \times 25}{4 \times 25} = \frac{75}{100} = \mathbf{0.75}$$

Example 3: Change $\frac{13}{20}$ to a decimal.

$$\frac{13 \times 5}{20 \times 5} = \frac{65}{100} = \mathbf{0.65}$$

4.3 PERCENTAGES INTO FRACTIONS

To change a percentage to a fraction, write the percentage as a fraction of a hundred and reduce the fraction to its lowest terms when possible.

Remember, 3% means 3 out of a hundred or $\frac{3}{100}$

4% means 4 out of a hundred or $\frac{4}{100}$

27% means 27 out of a hundred or $\frac{27}{100}$

Example 1: Change 5% to a percentage.

Write as a fraction of 100.

$$5\% = \frac{5}{100}$$

However, $\frac{5}{100}$ is not in its lowest term, so we cancel down.

$$\frac{5 \div 5}{100 \div 5} = \frac{1}{20}$$

Example 2: Express 38% as a fraction in its lowest form.

Write as a fraction of hundred.

$$38\% = \frac{38}{100}$$

However, $\frac{38}{100}$ is not in its lowest term, so we cancel down.

$$\frac{38 \div 2}{100 \div 2} = \frac{19}{50}$$

EXERCISE 4A

1) Convert the following fractions to decimals.

   a) $\dfrac{3}{5}$ f) $\dfrac{1}{8}$ k) $\dfrac{30}{40}$ p) $\dfrac{50}{100}$

   b) $\dfrac{4}{5}$ g) $\dfrac{2}{8}$ l) $\dfrac{40}{50}$ q) $\dfrac{2}{20}$

   c) $\dfrac{3}{10}$ h) $\dfrac{5}{8}$ m) $\dfrac{7}{20}$ r) $\dfrac{7}{8}$

   d) $\dfrac{2}{20}$ i) $\dfrac{3}{50}$ n) $\dfrac{12}{25}$ s) $\dfrac{1}{50}$

   e) $\dfrac{7}{25}$ j) $\dfrac{19}{25}$ o) $\dfrac{6}{10}$ t) $\dfrac{16}{20}$

2) In their lowest terms, write the following percentages as fractions.

   a) 2%   f) 20%   k) 50%   p) 79%

   b) 3%   g) 25%   l) 55%   q) 80%

   c) 10%   h) 30%   m) 60%   r) 81%

   d) 15%   i) 35%   n) 70%   s) 90%

   e) 18%   j) 46%   o) 75%   t) 98%

4.4 FRACTIONS TO PERCENTAGES

Multiplying a fraction or a decimal by **100** changes it to a percentage

Example 1: Express $\frac{1}{4}$ as a percentage.

This means $\frac{1}{4} \times 100$ ……. (Refer to section 5.9 on how to multiply fractions and the fraction of an amount.)

$$\frac{1 \times 100}{4} = \frac{100}{4} = 25$$

Therefore, $\frac{1}{4}$ as a percentage is **25%**

Example 2: Write $\frac{2}{20}$ as a percentage.

$$= \frac{2}{20} \times 100$$

$$= \frac{2 \times 100}{20} = \frac{200}{20} = \mathbf{10\%}$$

Example 3: Change $\frac{5}{8}$ to a percentage.

$$= \frac{5}{8} \times 100$$

$$= \frac{5 \times 100}{8} = \frac{500}{8}$$

Refer to section 2.6 on dividing whole numbers

You may also reduce to its simplest form and then change to a mixed number.

$$\frac{500 \div 2}{8 \div 2} = \frac{250}{4} = \frac{250 \div 2}{4 \div 2} = \frac{125}{2} = \mathbf{62\frac{1}{2}} \%$$

EXERCISE 4B

1) Write the following fractions as percentages.

a) $\frac{1}{4}$ f) $\frac{3}{20}$ k) $\frac{16}{50}$ p) $\frac{4}{40}$

b) $\frac{1}{5}$ g) $\frac{6}{10}$ l) $\frac{30}{50}$ q) $\frac{7}{14}$

c) $\frac{2}{5}$ h) $\frac{3}{8}$ m) $\frac{6}{25}$ r) $\frac{16}{32}$

d) $\frac{3}{10}$ i) $\frac{7}{20}$ n) $\frac{16}{25}$ s) $\frac{13}{52}$

e) $\frac{4}{5}$ j) $\frac{9}{25}$ o) $\frac{13}{20}$ t) $\frac{16}{320}$

2) Draw each shape below and answer the question for each shape.

a)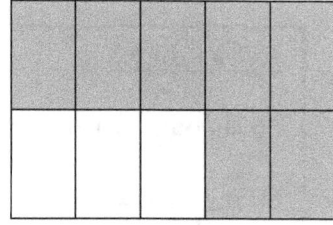

b)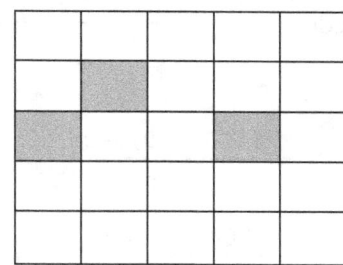

c)

 i) What fraction is shaded?
 ii) What percentage is shaded?
 iii) What fraction is unshaded?
 iv) What percentage is unshaded?

3) Copy and complete.

a) $40\% = \frac{?}{10}$ c) $\frac{2}{5} = \boxed{}\%$ d) $70\% = \frac{?}{10}$

b) $2\% = \frac{?}{100}$ d) $\frac{7}{20} = \boxed{}\%$ e) $60\% = \frac{?}{5}$

4.5 WRITING ONE NUMBER AS A FRACTION OF ANOTHER

Example 1: Express 4 as a fraction of 10.

This is written as $\dfrac{4}{10}$

Reduce the fraction to its lowest term.

$$\dfrac{4 \div 2}{10 \div 2} = \dfrac{2}{5} \checkmark$$

Example 2: Write 3 minutes 20 seconds as a fraction of an hour.

Remember: The quantities must be in the same units before we can successfully work out the calculations.

This is written as $\dfrac{3 \text{ min } 20 \text{ sec}}{1 \text{ hour}}$

Convert everything to seconds $= \dfrac{(3 \times 60) \text{ sec} + 20 \text{ sec}}{60 \times 60}$

$$= \dfrac{(180 + 20) \text{ sec}}{3600 \text{ sec}}$$

$$= \dfrac{200 \text{ sec}}{3600 \text{ sec}}$$

Remember:

60 seconds = 1 minute

60 minutes = 1 hour

1 hour = 60 × 60
= 3600 seconds

In its lowest term, first, divide both numbers by 100

$$= \dfrac{200 \div 100}{3600 \div 100}$$

$$= \dfrac{2}{36} \quad \text{(then divide both numbers by 2)}$$

$$= \dfrac{1}{18} \checkmark$$

EXERCISE 4C

1) Express 5 as a fraction of 7

2) Express 4 as a fraction of 20. Leave your answer in its lowest term.

3) What fraction of 5 cm is 2 m? Leave your answer in its lowest term.

4) What fraction of 30 cm is 3 m? Leave your answer in its lowest term.

5) Write the first number as a fraction of the second number. Leave your answers in their lowest term.

a) 12 week, 1 year

b) 30 minutes, 2 hours

c) 7 mm, 21 cm

d) $34, $100

e) 6 min 15 seconds, 1 hour

f) 16 cm, 10 m

g) 20m, 3 km

4.6 WRITING ONE QUANTITY AS A PERCENTAGE OF ANOTHER

Two steps to follow:

1) Write the first number as a fraction of the second number (See section 6.5).

2) Multiply the fraction or decimal formed by 100

Example 1: Write 6 as a percentage of 24

Write 6 as a fraction of 24: $\frac{6}{24}$

Multiply by 100 to convert to a percentage

$$= \frac{6}{24} \times 100$$

To make it easier, reduce $\frac{6}{24}$ to its lowest term which is $\frac{6 \div 6}{24 \div 6} = \frac{1}{4}$

Therefore, $\frac{1}{4} \times 100 = \frac{1 \times 100}{4} = \frac{100}{4} = \mathbf{25\%}$ ✓

Alternatively, $\frac{6}{24} \times 100 = \frac{6 \times 100}{24} = \frac{600}{24} = \mathbf{25\%}$

Example 2: Elizabeth scored 20 out of 25 in a maths test. Work out Elizabeth's percentage mark.

As a fraction, Elizabeth's mark $= \frac{20}{25}$

To convert to percentage, multiply by 100.

$$= \frac{20}{25} \times 100$$

A simpler way is to divide 100 by 25 and multiply by 20.
$$= 100 \div 25 = 4$$

$$4 \times 20 = \mathbf{80\%}$$

Example 3: A kettle is in a sale and reduced to ₦4000. What is the percentage reduction?

Find the reduction:
₦5 000 – ₦4 000 = ₦1 000

The reduction as a fraction is $\dfrac{1000}{5000}$

To its lowest term the fraction is $\dfrac{1}{5}$

Percentage reduction = $\dfrac{1}{5} \times 100$
= **20%**

EXERCISE 4D

1) Write

    a) 20 as a percentage of 50

    b) 5 as a percentage of 20

    c) £100 as a percentage of £500

    d) 12 as a percentage of 36

2) Express these times as a percentage of an hour.

    a) 6 minutes

    b) 30 minutes

    c) 15 minutes

    d) 60 minutes

3) Express 2 hours as a percentage of 5 days.

4) Express 300 ml as a percentage of 2 litres

5)  A polo shirt is reduced from ₦50 to ₦35 in a sale. What is the percentage reduction?

6) The Nigerian Football Association has 80 members as shown in the table below.

Female	Male
30	50

a) What percentages of the members are male?

b) What percentages of the members are female?

7) The results of a science test for three students are shown below.

a) What is Abdul's percentage score?

b) What is Amina's percentage score?

c) Okonkwo says "Uduak scored 71%."
Is Okonkwo correct?
Explain fully.

Names	Out of 40
Abdul	8
Uduak	28
Amina	20

Chapter 4 Review Section
Assessment

1) 73% of students walk to school. What percentage of students goes to school by other means?

.... **1 mark**

2) Copy and shade in

a) 25%

b) 50%

c) 75%

.... **3 marks**

3) Write 4cm as a fraction of 2m

.... **1 mark**

4) Write the following numbers as a percentage of the second.

a) 15 weeks, 1 year

b) 500 g, 2 kg

c) ₦4 000, ₦8 000

...............**6 marks**

5) Express these fractions as percentages.

a) $\frac{1}{2}$ b) $\frac{15}{25}$ c) $\frac{7}{10}$ d) $\frac{3}{20}$

.... **8 marks**

6) In summer examinations, Rebecca scored these marks:

   Maths: 17 out of 25
   Economics: 2 out of 40
   English: 30 out of 50.
   Physics: 7 out of 20

   a) Work out Rebecca's percentage for **each** subject.
   .... 8 marks

   b) Arrange the percentages in question 6a above in order of size, smallest first.
   .... 1 mark

7) Copy and complete the table below.

Fractions	Percentages	Decimals
$\frac{2}{5}$		
		0.35
	60%	
$\frac{6}{25}$		

   .... 8 marks

8) Arrange in order of size, highest first.

   $\frac{1}{2}$,   0.65, $\frac{3}{5}$,   51%

   .... 2 marks

9) Which is bigger, 30% or $\frac{8}{25}$?
   Explain your decision.
   .... 2 marks

10) There are 45 women and 15 men in a theatre.

    a) What percentage are women?   .... 2 marks

    b) What percentage are men?   .... 2 marks

5 Directed Numbers

This section covers the following topics:

- Directed numbers
- Understanding number lines
- Addition and subtraction of negative numbers

LEARNING OBJECTIVES

By the end of this unit, you should be able to:

a) Understand the meaning of directed numbers
b) Use a number line
c) Add and subtract negative numbers
d) Multiply and divide negative numbers
e) Apply negative numbers to real life situations
f) Understand the concept of integers

KEYWORDS

- Directed numbers
- Add
- Subtract
- Negative numbers
- Number line
- Positive numbers
- Integers
- Multiply

5.1 UNDERSTANDING NEGATIVE NUMBERS

Before we can successfully understand negative numbers, we must familiarise ourselves with directed numbers and integers.

Directed numbers are numbers which have a direction and a size. Positive and negative numbers fall into this category and as such, directed. The positive (+) or negative (−) sign indicates which direction to go from zero (0) until the position of the number is reached.

The numbers below are directed number:

$$-30, \quad -20.4, \quad -10, \quad -5.6, \quad -5 \quad 4, \quad 7, \quad 20$$

From the directed numbers above, only -30, -10, -5, 4, 7 and 20 are **integers**. Integers are positive or negative **whole** numbers **including zero**.

As a summary, if a directed number is a whole number, then it is called an integer.

All numbers above zero or to the right of zero are positive numbers.
All numbers below zero or to the left of zero are negative numbers
Zero (0) is neither positive nor negative. It is a **neutral** number.

-5 is read as a negative five. It is 5 less than zero. 0 - 5 = **-5**

We can show positive and negative integers on a **number line** below.

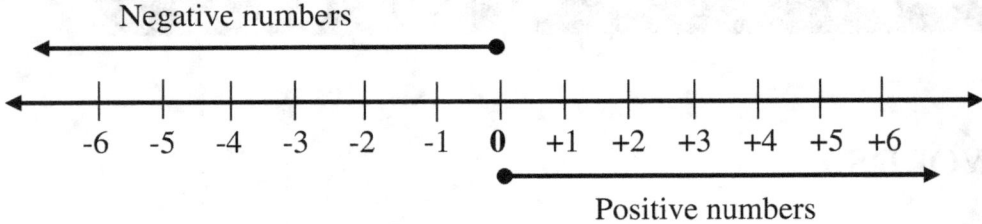

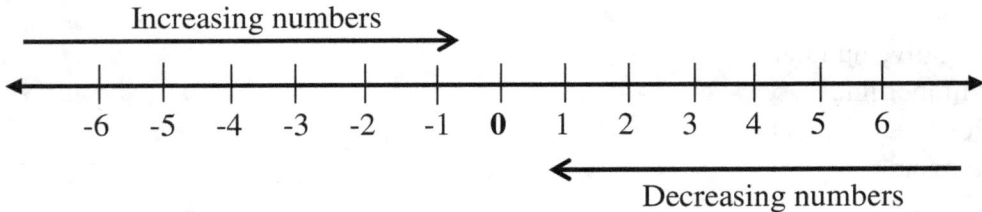

Do not forget:

-1 is greater than -2, -6 is greater than -7, +2 is greater than -2, and +6 is greater than -6

We often use negative numbers with temperatures. During winter in some countries, the weather is very cold, and the temperature can go as low as -10°C. This means that the temperature is 10 below zero.

Water freezes at 0°C approximately. Temperatures are given in degrees Celsius (°C).

Example 1: Suppose the temperature of a room at 3 am was -3°C and it rises **to** 5°C at 9 am. What is the difference in temperature between 3 am and 9 am?

Solution: we may use a number line to help us.

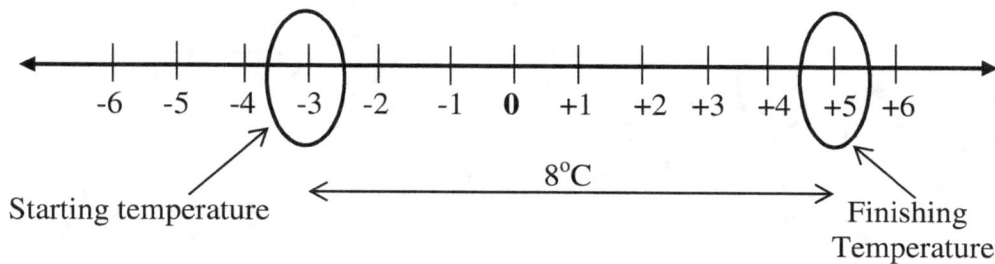

The difference in temperature is clearly **8°C** because +5 - (-3) = 5 + 3 = 8 ✓

Example 2: Suppose the temperature is -4°C and it rises **by** 7°C. What is the new temperature?

Solution: Use the number line to help. Start from -4 and count 7 to the right.

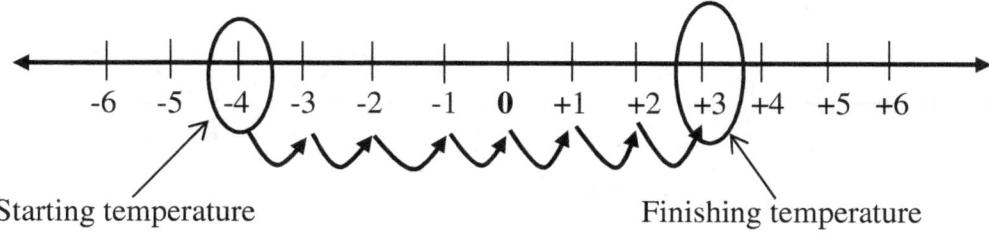

The new temperature is **3°C**. ✓

Example 3: Suppose the temperature is 4°C and it falls by 6°. What is the new temperature?

Solution: Start from 4°C and count six units to the left. The new temperature is **-2°C**.

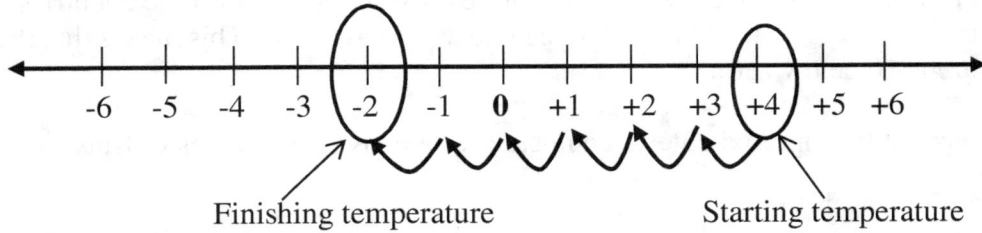

Alternatively, (+4) - (+6) = +4 - 6 = **-2**

Example 4: The diagram shows a thermometer with two arrows.

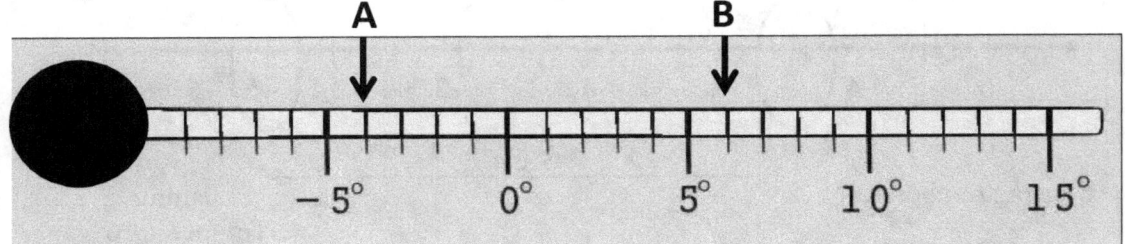

a) Which temperature is cooler?
b) What is the difference between the two temperatures?

Solution:
    a) **A (-4°C)** is cooler as it is lower than B (6°C)
    b) 6 - (-4) = 6 + 4 = **10°**……………..The difference is 10°.
    (Alternatively, you may count the units between A and B.)

Example 5: Use a number line to work out the value of -2 + 5.

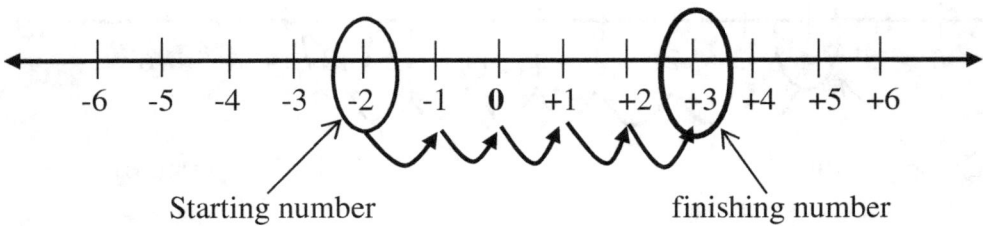

Start from -2 and go to the right five times. The answer is **3**.

EXERCISE 5A

1) Write these temperatures from hottest to coldest. All number in °C.

a) -10, 4, 5, -8, -3, 9
b) 10, -7, -15, 9, 40, -2
c) 42, -20, -10, -15, 17
d) -1, -7, 3, 5, -5,
e) -8, 8, -9, 9, -4, 4
f) -16, 14, -12, 10, -18, 30

2) What temperature in Celsius is shown by the alphabets below?

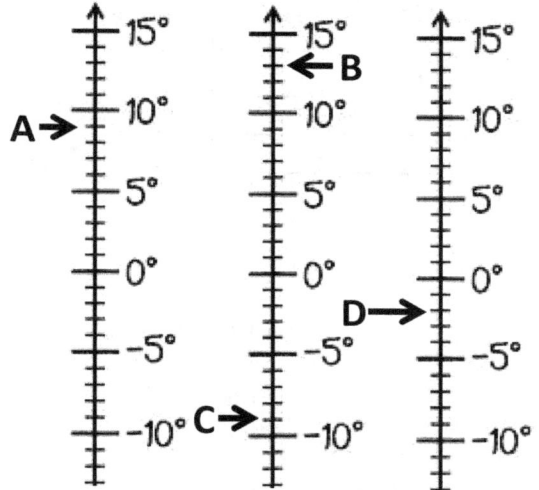

3) What would the temperatures in question 2 above be if it was
a) at C and fell 3°?
b) at A and rose 5°?
c) at B and fell 9°?
d) at D and rose 11°?

4) In the statement below, write <, > or = in the box.

a) 4.5 ☐ 2
b) -3 ☐ -4
c) -10 ☐ 40
d) -6 ☐ -6

5) Find the differences between each pair of temperatures in °C.

a) -3, 5
b) -13, 3
c) -7, -10
d) -3, -8
e) 0, -3
f) -2, -5
g) 10, -3
h) 2, -4

6)

Which of the temperatures above are
a) below freezing point
b) above freezing point

Which of the temperatures above is
c) the hottest
d) the coldest?
e) What is the difference between the coldest and hottest temperatures?

7) For each question below, state whether the temperature has fallen or risen and by how many degrees.

a) It was -4°C and now -10°C
b) It was -5°C and now 10°C
c) It was 7°C and now -3°C
d) It was 2°C and now 25°C

8) Use a number line to perform the following calculations.

a) 2 + 3
b) -2 + 3
c) -4 + 2
d) -1 - 3
e) - 4 - 5
f) 7 - 3

EXERCISE 5 B

1) Write the next two numbers in the sequences below.

a) -4, -6, -8, _, _

b) 7, 4, 1, _, _

c) 10, 5, 0, _, _

d) 20, 14, 8, 2, _, _

e) -7, -5, -3, _, _

f) -3, -5, -7, _, _

2) Put each set of numbers in order, lowest first.
a) -4, 5, -9, 0, -3
b) -50, 55, -30, 40, 20,
c) -7, -3, 1, -2, -1, 0, 4

3) Copy and complete the table below.

Temperature °C	Change °C	New Temp. °C
-3	+4	
	-3	-6
-7	-3	
	+6	12
-9	-9	

4) Find the range of these temperatures in °C.
a) -3 and -5
b) 2 and -7
c) 13 and -15

5) The table shows different temperatures in different towns on one day.

Town	Temperature °C
Lagos	+25
Onitsha	+27
Essex	-3
Manchester	-1
Kano	+32
Abuja	+30

a) Which town is the warmest?
b) Which town is the coldest?
c) Write down the names of towns in order, from coldest to warmest
d) At 3 pm on the same day, the temperatures in each town are 4°C higher. Write down the new temperatures for Essex and Onitsha.

5.2 ADDING AND SUBTRACTING DIRECTED NUMBERS

Number lines are useful in adding or subtracting a directed number.

Example 1: -3 + 4
This means start from -3 and go **right** four times.

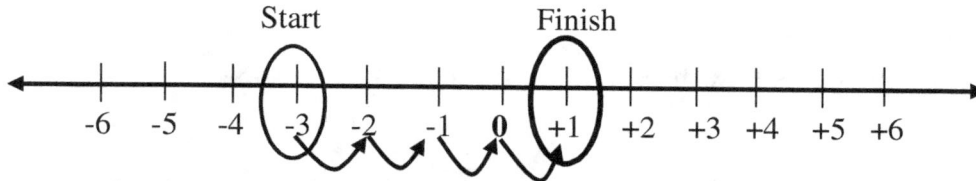

The answer is **+1** or **1**

Example 2: 5 - 7
This means start from +5 and go **left** seven times. The answer is -2

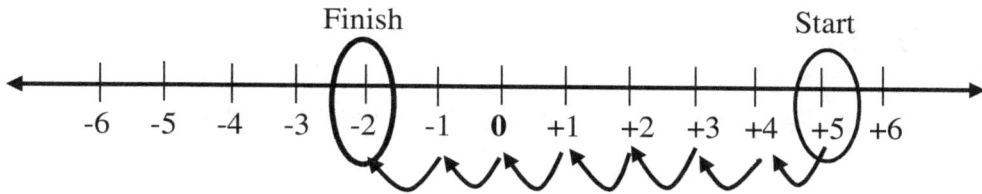

Example 3: -2 + (-3)
To add a negative number, go left from the starting number. The answer is -5.

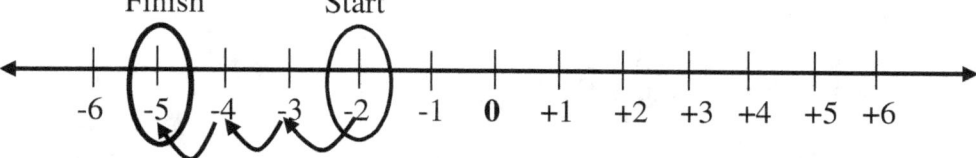

When you have two sighs together, use this rule:

$$+ \, + = +$$
$$- \, - = +$$
} Same signs

$$+ \, - = -$$
$$- \, + = -$$
} Different signs

Example 4: 4 - (-2)

        Replace **- -** with **(+)**   It becomes 4 + 2 = **6**

Example 5: -5 + (-2)

        Replace **+ -** with **(-)**   It becomes -5 - 2 = **-7**

* You may use a number line to help by starting from -5 and go left 2 times.

Example 6: -3 - (+2)

        Replace **- +** with **(-)**   It becomes -3 - 2 = **-5**

Example 7: 5 + (-6)

        Replace **+ -** with **(-)**   It becomes 5 - 6 = **-1**

Example 8: -6 - (-4)

        Replace **- -** with **(+)**   It becomes -6 + 4 = **-2**

EXERCISE 5C

Work out without using a calculator. You may use a number line to help.

1) a) 8 + 3   b) 8 – 8   c) 4 – 5   d) 6 – 10   e) 5 – 9   f) 8 + (-5)
   g) 7 + (-2)   h) 3 + (-7)   i) 9 + (-6)   j) 2 + (-4)   k) 3 + (-8)   l) 9 + (-2)
   m) 6 + (-3)   n) 7 + (-5)   o) 10 + (-8)   p) 1 + (-1)   q) 7 + (-7)   r) 3 + (-2)

2) a) -3 + 5   b) -8 + 10   c) -6 + 3   d) -5 + 4   e) -2 + 2   f) -8 + 8
   g) -10 + 3   h) -8 + 5   i) -6 + 5   j) -12 + 5   k) -9 + 1   l) -3 + 7
   m) -10 + 5   n) -13 + 20   o) -9 + 3   p) -7 + 1   q) -3 + 3   r) -6 + 9

3) a) -1 -5   b) -6 -10   c) -7 -3   d) -4 - 2   e) -2 - 1   f) -8 - 8
   g) -10 -3   h) -8 - 5   i) -6 - 5   j) -12 - 5   k) -7 - 4   l) -3 -7
   m) -10 -5   n) -1 - 8   o) -2 - 3   p) -5 - 7   q) - 4 - 6   r) -20 - 3

4) a) -3 + (-8)   b) -6 + (-9)   c) -2 + (-5)   d) -4 + (-8)   e) -10 + (-12)   f) -7 + (-3)
   g) -9 + (-6)   h) -10 + (-5)   i) -8 + (-2)   j) -3 + (-2)   k) -9 - (+5)   l) -6 - (+10)
   m) -3 - (+5)   n) -4 - (+7)   o) -6 - (+3)   p) -5 - (+1)   q) -1 - (+4)   r) 8 - (-7)

5) a) -7 - (-5)   b) -3 - (-2)   c) -6 - (-4)   d) -10 - (-8)   e) -5 - (-3)   f) -7 - (-1)
   g) -6 - (-13)   h) -7 - (-3)   i) -13 - (-20)   j) -6 - (-7)   k) -100 + (-30)

6) Work out the missing numbers.

a) 18 - ☐ = 5   b) 15 - ☐ = 10   c) 7 - ☐ = 3   d) 19 - ☐ = 7

e) 112 - ☐ = 93   f) 7 - ☐ = -3   g) 6 - ☐ = -4   h) ☐ + 8 = 3

7) Copy and complete the addition square below.

+	-4	-3	-2	-1	0	+1	+2
+4							
-5							
+3							
-2							
0							
+8							
-10							
-7							
-6							

8) Simplify

a) -8 + (-5) - 1

b) -5 - (- 4) - 1 (-2)

c) 9 + (-6) - 7

d) -10 - (- 4) - - (+5)

9) Tunde's bank account reads ₦54 283. He owes Babalola ₦76 542. How much will Tunde be overdrawn if he writes a cheque for the whole amount he owes Babalola?

5.3 MULTIPLYING AND DIVIDING WITH NEGATIVE NUMBERS

RULES:
1) Multiplying two numbers with the **same** sign always gives a **positive (+)** answer.

Example 1: a) $(+4) \times (+5)$ this is the same as $4 \times 5 = 20 =$ **20**
b) $(-3) \times (-5)$ the answer is $3 \times 5 =$ **15**
c) $(-6) \times (-2)$ the answer is $6 \times 2 =$ **12**
d) $(-4)^2 = -4 \times -4 =$ **16**

2) Multiplying two numbers with **different** signs always gives a **negative (-)** answer.

Example 2: a) $-3 \times 4 = -12$ b) $-6 \times 10 = -60$ c) $3 \times -4 = -12$ d) $6 \times -10 = -60$

3) The rules for multiplication are the same for the division.

Example 3: a) $(+4) \div (+2) = 2$ b) $(-6) \div (-2) = 3$ *Same signs always give positive answers*

c) $(-20) \div (+5) = -4$ d) $(+6) \div (-3) = -2$ *Different signs always give negative answers*

When multiplying negative/directed numbers,

(+) × (+) = Positive (+)
(-) × (-) = Positive (+)
(+) × (-) = Negative (-)
(-) × (+) = Negative (-)

Also, the same rules apply exactly when dividing with negative numbers.

(+) ÷ (+) = Positive (+)
(-) ÷ (-) = Positive (+)
(+) ÷ (-) = Negative (-)
(-) ÷ (+) = Negative (-)

EXERCISE 5D

Work out the following:

1) a) -2 × (-3) b) -10 × (-2) c) -7 × (-1) d) -6 × (-5)
   e) -3 × (-7) f) -5 × (-5)   g) -12 × (-10) h) -15 × (-3)
   i) -7 × (-2) j) -8 × (-9)   k) -2 × (-4)   l) -8 × (-7)

2) a) -2 × 3   b) -7 × 10   c) -3 × 5   d) -6 × 6
   e) -9 × 4   f) -4 × 2   g) -10 × 3 h) 6 × -3
   i) 9 × -7   j) 3 × -2   k) 4 × -4   l) 11 × -3

3) a) -16 ÷ (-8) b) -9 ÷ (-3) c) -30 ÷ (-10) d) -6 ÷ (-2)
   e) -50 ÷ (-5) f) -8 ÷ (-2) g) 9 ÷ (-3)   h) 40 ÷ (-4)
   i) 8 ÷ (-2)   j) -8 ÷ 2   k) -45 ÷ 9   l) -35 ÷ 7

4) Copy and complete the multiplication grid below.

×	8	-2	-5	7	-3
-3					
4			-20		
5					
-10					
-2					
-7					

5) Match the following calculations with the answers.

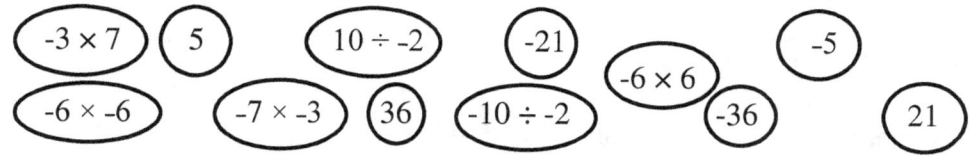

6) Work out without a calculator.

   a) 2 ÷ 2   b) -18 ÷ (-3)   c) -10 ÷ (-5)   d) 40 ÷ (-4)

   e) -4 × 8   f) 9 × -3   g) -7 × (-8)   h) -126 ÷ 14

Chapter 5 Review Section
Assessment

1) Which is greater -5 or 2? .... **1 mark**
2) What do you subtract from -4 to make -9? .... **1 mark**
3) Simplify the following
    a) (+ 3) + (- 6) .... **1 mark**
    b) (- 7) - (- 4) .... **1 mark**
    c) - 18 ÷ 6 .... **1 mark**
    d) - 9 - (+5) .... **1 mark**

4) The temperature in a room in winter was -5°C. The temperature fell by a further 4°C. What is the new temperature? .... **1 mark**

5) Look at the number line below.

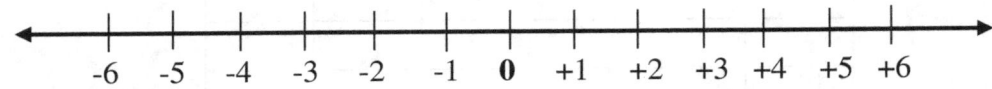

Find the difference between
    a) -5 and 0   b) -2 and -4   c) 3 and -6   d) -6 and 5
.... **4 marks**

6) Write these temperatures (°C) in ascending order.
    a) -2, -3, -5, 7, 10, -1
    b) 5, -5, 3, -4, -7, 8, 0
    c) -2, 1, -17, 4, -5, 5, 2

7) Copy and complete the multiplication table below.

×	-4	+3	-8	+2
-2				
-7				
+8				
+5				

6 Percentages 2

This section covers the following topics:

- Percentage of a quantity
- Proportions using percentages
- Find percentage change
- Simple Interest
- Reverse Percentages
- Compound interest

LEARNING OBJECTIVES

By the end of this unit, you should be able to:

a) Find percentage of a quantity
b) Compare proportions using percentages
c) Find a percentage increase
d) Find a percentage decrease
e) Find percentage change
f) Find Simple Interest
g) Find reverse percentages
h) Find compound interest

KEYWORDS

- Percentage Change
- Increase and Decrease
- Quantity
- Proportion
- Simple Interest
- Compound Interest

6.1 PERCENTAGE OF A QUANTITY

Finding the percentage of a quantity is the same as finding the fraction of a quantity. Remember, 10% means $\frac{10}{100} = \frac{1}{10}$, therefore, finding 10% of a number simply means **dividing the number by 10**.

Example 1: Find 10% of 20.
Since 10% means $\frac{1}{10}$, we have to find $\frac{1}{10}$ of 20 which means $\frac{20}{10} = $ **2**

Example 2: Find 10% of 70……………………..It means 70 ÷ 10 = **7**
Example 3: Find 10% of 635…………………..It means 635 ÷ 10 = **63.5**

From 10%, a lot of percentages can be obtained.
- To find 20%, find 10% and multiply by 2
- To get 30%, find 10% and multiply by 3
- To get 70%, find 10% and multiply by 7, and so on.
- To get 5%, find 10% and divide by 2.
- To get 50%, find 10% and multiply by 5 **or** simply divide the number by 2.

As you get familiar with percentages, there are other ways you could work out any percentage of an amount. For example: to get 70%, you could work out 50% and then add it on to 20%.

Also remember, 1% is the same as $\frac{1}{100}$. Therefore, finding 1% of a number simply means **dividing the number by 100.**

Example 3: Find 1% of 34…………………....It means $\frac{34}{100}$ = **0.34**
Example 4: Find 1% of 436………………….It means $\frac{436}{100}$ = **4.36**

Mixed examples

Example 5: Find 20% of 70
Find 10% and multiply by 2.   10% of 70 = 7   7 × 2 = **14**

Example 6: Find 40% of ₦120   10% of 120 = 12   12 × 4 = **₦48**

Example 7: Find 5% of 30   10% of 30 = 3   3 ÷ 2 = **1.5** or $1\frac{1}{2}$
Example 8: Find 15% of 80
10% of 80 = 8   5% of 80 = 10% ÷ 2 = 8 ÷ 2 = 4   **15%** of 80 = 8 + 4 = **12**

EXERCISE 6A

Work out without a calculator. Show all working out.

1) a) 10% of 30   b) 10% of 4   c) 10% of 60   d) 10% of 8
   e) 10% of 16   f) 10% of 40   g) 10% of ₦80   h) 10% of ₦300
   i) 20% of 30   j) 20% of 4   k) 20% of 60   l) 20% of 8
   m) 40% of 40   n) 50% of 36   o) 80% of 20   p) 100% of 50

2) a) 5% of 20   b) 5% of 140   c) 60% of 100   d) 1% of 200
   e) 5% of 38   f) 1% of 24   g) 50% of 2440   h) 35% of 80kg

Example 9: Find 13% of 30

Method 1
10% of 30 = 30 ÷ 10 = 3
+ 1% of 30 = 30 ÷ 100 = 0.3
1% of 30 = 30 ÷ 100 = 0.3
1% of 30 = 30 ÷ 100 = 0.3
13% of 30 = **3.9**

Method 2
13% means $\frac{13}{100}$

Therefore, $\frac{13}{100} \times 30$

= 0.13 × 30 = **3.9**

Example 10: Find 7% of $150
1% of 150 = $\frac{1}{100} \times 150 = \frac{150}{100} = 1.5$

1.5 × 7 = **$10.50**

Example 11: Find 11% of ₦200

10% of ₦200 = 200 ÷ 10 = 20
+ 1% of ₦200 = 200 ÷ 100 = 2
11% of ₦200 = **₦22**

Method 2
11% means $\frac{11}{100}$ = 0.11
0.11 × 200 = 22
Therefore, 11% of 200 = **₦22**

Example 12: Work out 2.3% of 19.35
2.3% means $\frac{2.3}{100}$ = 0.023   0.023 × 19.35 = **0.44505**

Example 13: Work out $4\frac{1}{2}$% of 90
$4\frac{1}{2}$% means $\frac{4.5}{100}$ = 0.045   0.045 × 90 = **4.05**

EXERCISE 6B

Work out the following without using a calculator.

1) a) 1% of 130 b) 3% of 35 c) 20% of 80 d) 80% of 30
   e) 21% of 33 f) 14% of 55 g) 4% of 130 h) 7% of ₦300
   i) 55% of £60 j) 90% of 750 k) 15% of 8 l) 35% of 80
   m) 61% of 40 kg n) 7% of 890 o) 23% of ₦2000 p) 4% of 9

2) a) 2.7% of £70 b) 3.2% of 56 c) 2.8% of 200 d) 17.5% of 460
   e) $2\frac{1}{2}$% of 30 f) $7\frac{1}{2}$% of 120 g) 3.5% of 68kg h) $6\frac{1}{2}$% of ₦5000

6.2 COMPARING PROPORTIONS USING PERCENTAGES

Okechukwu sat tests in Physics, Chemistry, Biology and Mathematics. His results were:

Physics: $\frac{20}{30}$,   Chemistry: $\frac{77}{90}$,   Biology: $\frac{16}{25}$,   Mathematics: $\frac{18}{20}$

Which subject did he do best?

Solution: Before making any decision(s), we can write each fraction as a percentage. (Refer to section 6.4).

Physics: $\frac{20}{30} \times 100 = 66.7\%$ Biology: $\frac{16}{25} \times 100 = 64\%$

Chemistry: $\frac{77}{90} \times 100 = 85.6\%$ Mathematics: $\frac{18}{20} \times 100 = 90\%$

From the above calculations, Okechukwu did best in Mathematics test.

6.3: PERCENTAGE INCREASE AND DECREASE

A quantity can be increased or decreased by a percentage.

Example 1: Increase 100 by 10%

First, find 10% of 100 which is $\frac{100}{10} = 10$.

Therefore, increasing 100 by 10 means

100 + 10 = **110**

Example 2: Decrease ₦510 by 20%
First, find 20% of ₦510
 Find 10% and multiply by 2
10% of ₦510 = $\frac{510}{10}$ = ₦51
20% of 510 = 51 × 2 = ₦102
Decreasing ₦510 by 20% means 510 - 102 = **₦408**

6.4 FINDING A PERCENTAGE CHANGE

Percentage change is calculated by using the formula below.

$$\text{Percentage change} = \frac{\text{Change (increase or decrease)}}{\text{Original amount}} \times 100$$

Example 1: The price of a TV set was increased from £80 to £120. Work out the percentage increase.

Solution: First work out the increase.
$120 - 80 = 40$
Percentage increase = $\frac{40}{80} \times 100 = \mathbf{50\%}$

Example 2: The original price of a carpet was ₦2 400. The price is now ₦2 000. Find the percentage decrease.

Solution: Decrease = $2400 - 2000 = ₦400$

Percentage decrease = $\frac{400}{2400} \times 100 = \mathbf{16.7\%}$

EXERCISE 6C

1) Increase the following amounts by 10%.

a) 20 c) ₦200 e) 840
b) 80 d) ₦340 f) 3000

2) Increase the amounts in question 1 by 15%.

3) Decrease the following quantities by 20%.

a) 40 kg c) £123 e) 4000 g
b) 92 litres d) $183.52 f) 245 cm

4) Nonso earns ₦25 000 for selling CD's a day. How much **extra** will he get if he has a 10% salary increase?

5) The price of a Renault car is £4 200. There is a discount of 25% on all Renault cars. How much will Tochukwu pay for a Renault car?

6) Which of the following statements represent a 40% change?

a) 35 to 40 b) 50 to 70 c) 70 to 80.

7) *Shoprite* is having a 35% sale on all men's shirts. The original prices for three shirts are £15, £32 and £45 respectively. Work out the sale price for each shirt.

8) A dog weighs 2 kg and while running away from a fox, it loses one of its legs. As a result, the weight was reduced by 3%. What is the weight of the dog now?

9) Emma buys a car for ₦540 000 and sold it two weeks later for ₦720 000. Work out the percentage profit.

10) Increase ₦5500 by 2%
11) Decrease ₦6000 by 13%
12) Increase 72 kg by 15%
13) Decrease £780 by 1%

6.5 SIMPLE INTEREST

If you deposit money in a building society or bank, they will pay you **money (interest)** on this money you deposited.

Also, if you borrow money from a bank, you also have to pay interest on the borrowed money. Simple interest is normally calculated on a yearly basis (annual) and depends on the **interest rate**.

Example 1: Andy saves £500 in a bank for a year. If the interest rate is 5% per annum, a) Calculate the interest on the amount. b) The total amount Andy will have after a year.
Solution
a) Interest = 5% of £500
   = $\frac{5}{100} \times 500$ = **£25**
b) After one year, Andy will
   £25 + £500 = **£525**.

Interest paid out this way is known as **simple interest**.

2) Henry's bank account pays interest at the rate of 4.5% per annum. If he deposits ₦20 000 into his account, how much simple interest will he receive after a) 1 year b) 6 years
Solutions
a) $\frac{4.5}{100} \times$ ₦20 000 = **₦900**
b) ₦900 × 6 = **₦5 400**

Generally, the formula for simple interest is $I = \frac{PRT}{100}$, where I is the simple interest, P = amount invested, R = rate and T = time invested.

EXERCISE 6D

1) Work out the simple interest on the following:

a) £700 at 3% per annum for 1 year
b) £8 000 at 4% per annum for 1 year
c) ₦30 000 at 2% per annum for 4 years
d) ₦25000 at 5% per annum for 2 years
e) ₦2000 at 7.5% per annum for 6 years
f) $5500 at $2\frac{1}{2}$% per annum for 3 years
g) $7000 at 3.5% per annum for 2 years
h) ₦45 000 at 5% per annum for 4 years

2) Chibogu saves ₦5000 in a bank for a year. If the interest rate is 2.5% per annum, calculate
a) the interest on the amount.
b) the total amount Chibogu will have after a year.

3) Sarah applied for a £35 000 car loan and was successful. If the interest is 5% per annum, how much simple interest will she pay after a) 1 year b) 3 years?

4)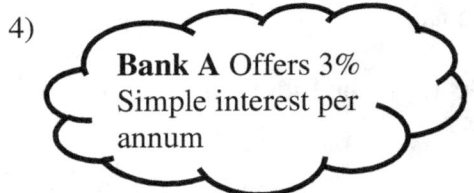

Bank A Offers 3% Simple interest per annum

Bank B offers 5% interest for the first year and then 2% simple interest per annum for the following years.

Nicole wants to invest £10 000 for 5 years. Which bank will she deposit her money to earn more interest?

5) Ajayi borrowed ₦3 500 000 from Bank X for a second-hand car. The agreement with Bank X was for three years at a simple interest of 8% per annum.

a) What is the simple interest on the borrowed money at 8% per annum for three years?
b) What is the total amount Ajayi would pay back at the end of 3 years?

c) If Ajayi was to pay the total amount due in monthly instalments over three years, how much would that be each month?

VALUE ADDED TAX CALCULATIONS (VAT)

VAT is a form of tax that everybody has to pay when buying goods and services. The rates vary from country to country, and in the UK, it is currently 20%.

Example 1: How much will a customer pay for the microwave below?

Solution: VAT = 20% of £30
10% of £30 = £3.
Therefore, 20% = 2 × £3 = £6
Amount to pay = £30 + £6 = **£36**

EXERCISE 6E

1) A bicycle is on sale at £80 + 20% VAT.
a) Work out the amount of VAT to be paid.
b) Work out the total cost of the bike.

2) A bill for garden clearance is £75 + 20% VAT. How much is the total bill?

3) Calculate the total price of the following items:
a) Dinning table: £350 + 20% VAT   b) Lexus Car: £12 000 + 20% VAT
c) 40-inch TV: £900 + 20% VAT

4)

a) How much would the government receive if VAT rate is 20%?

b) How much does the customer pay for the laptop?

PROBLEM SOLVING

1) A dining table originally costs £200 but a $3\frac{1}{2}$% discount is given for cash payments.
a) What is the cash price for the dining table?
b) Mark wants to buy two dining tables. He has up to £1000 in his credit card and £195 cash.
i) How much will Mark pay for the two dining sets if he must use the cash and credit card?
ii) How much change will he receive in cash?

2) Bradley wants to buy the Lexus car but decided to buy on credit. He deposited 40% of the cash value and agreed on £500 monthly plan for 12 months.

a) How much did Bradley deposit?
b) How much is the total of the monthly payments?
c) How much will it cost Bradley to buy on credit?

£8 500
Cash Price

3) Iconic Concepts Limited paid £240 electricity bill **inclusive** of VAT for 6 months usage at the rate of 20%. How much will the government receive as VAT?

4)

OPTION 1
Cash PRICE: £1500

OPTION 2
25% deposit and 12 monthly payments of £175

a) How much is the total monthly instalments?
b) If not paying by cash, how much is the deposit?
c) How much more is buying on credit than paying with cash?
d) How much is the total amount for option 2?

6.6 COMPOUND INTEREST

When interest is given on the interest, we talk about **compound interest**.

Example 1: Jude invests £300 in a building society which pays 3% interest each year. What is the value of the investment after 3 years?
Solution:
1st year: 3% of £300 = **£9**

2nd year: £300 + £9 = £309
3% of £309 = **£9.27**

3rd year: £9.27 + £309 = £318.27
3% of £318.27 = £9.55
9.55 + 318.27 = **£327.82** ✓

However, the value after 3 years could have been worked out in one calculation.

3% + 100% = 103% = 1.03
1.03 is now the multiplier.

£300 × 1.03³ ← Number of years
= **£327.82** ✓

You may use the formula for compound interest which is

$$A_n = \left[1 + \frac{r}{100}\right]^n A_0$$

Where A_n = Total amount after n years
r = Interest rate (%)
A_0 = Initial amount invested

Example 2: Sandra's father invested ₦100 000 in a bank for her at 20% compound interest. **a)** Find the total amount due after 15 years. **b)** Find the total interest on the investment for the 15 years.

Solution:
a) 100% + 20% = 120% = 1.2
₦100 000 × $(1.2)^{15}$ = **₦1 540 702.16**

b) Interest = ₦1 540 702.16 − 100 000
    = **₦1 440 702.16** ✓

Example 3: How much do you need to invest now to receive £20 000 in 8 years at 7% interest rate?

Solution: This is working backwards.

100% + 7% = 107% = 1.07 (multiplier)
Working backwards:
Present value = $\dfrac{20\ 000}{(1.07)^8}$ = $\dfrac{20\ 000}{1.71818618}$

= **£11 640.18** ✓

DEPRECIATION

Depreciation is a reduction in the value of an asset due to wear and tear or other factors.

Example 4: Tom bought a car for £8500 and its value depreciates (decreases) by 7% each year. Work out the value of the car after 5 years.

Depreciation: 100% − 7% = 93% = 0.93
After 5 years: £8500 × 0.93^5
= **£5 913.35** ✓

EXERCISE 6F

1) Callum invests £2800 in a bank which earns 3% compound interest per annum. Find the value of Callum's investment after:
a) 2 years   b) 3 years   c) 7 years.

2) Find the compound interest on £3000 invested for 3 years at 8%.

3) £13500 is invested at $7\frac{1}{2}$% per annum compound interest for 6 years. Find
a) the interest at the end of 6 years.
b) the total amount payable at the end of 6 years.

4) The price of a car depreciates at the rate of 15% per annum. Find the value of the car after 5 years if the price of the car is £12 000.

5) The population of Nigeria is 200 million and growing at 1.5% per annum.

a) In how many years' time will Nigerian population exceed 210 million?
b) What size will the population be in 5 years' time?

6) Leanne, a head of mathematics department in a UK school, **borrows** £8500 for 5 years at 4% simple interest. Niamh, Leanne's second in charge borrows the same amount at 4% per annum compound interest.

a) Who pays more interest?
b) How much more will the person pay back?

6.7 REVERSE PERCENTAGES

Reverse percentages simply means working backwards to find the **original** amount after an increase or a decrease.

Example 1: A laptop sells for £350 after a 40% increase in the cost price. What is the cost of the laptop before the increase?

Solution: 100% + 40% = 140% = 1.4
Cost price = $\frac{350}{1.4}$ = **£250**

Example 2: John was offered 15% discount when buying a new sofa. The discounted price is £595. What is the full price of the sofa?

Solution: Full price means the price before the discount or the original/cost price.

Because it is discount, we subtract the percentages: 100% - 15% = 85% = 0.85.

Full price = $\frac{595}{0.85}$ = **£700**

Example 3: Dylan invests some funds in a bank at 5% interest per annum. After 5 years, the value of Dylan's investment is £25000. Calculate the amount Dylan invested.

100% + 5% = 105% = 1.05

Amount invested = $\frac{25\,000}{(1.05)^5}$
= **£19 588.15**

EXERCISE 6G

All questions to 2 decimal places where possible.

1) After an increase of 2%, the price of air ticket is £725. What was the price before the increase?

2) The volume of a metal increased by 5% to 258.9 cm^3 after heating. What is the volume of the metal before heating?

3) The price of a dining table is £600 including VAT at 20%. What is the price of the dining table without VAT?

4) The sale prices of two shirts are shown below in a sale of 25% off.

Find the original prices of the shirts.

5) Kola visited London and bought some items inclusive of VAT at 20%.

Laptop	£115.20
Electric kettle	£45
Samsung mobile phone	£420

a) Find the total cost of the items Kola bought without VAT.
b) How much can Kola reclaim in VAT?

Chapter 6 Review Section
Assessment

1) Work out without a calculator.
   a) 10% of 130   b) 1% of 34   c) 30% of 600   d) 21% of ₦6000
   .... **6 marks**

2) Decrease ₦4500 by 20%   .... **2 marks**

3) Increase 45kg by 50%   .... **2 marks**

4) Work out the percentage change for each of the following.

   a) Increasing ₦200 by ₦50
   b) Decreasing ₦5 000 by ₦200
   c) Increasing 72 kg by 3 kg
   .... **6 marks**

5) A man bought a Ferrari car. The value depreciates from £200 000 when new to £180 000 two years later. Calculate the percentage depreciation.   .... **2 marks**

6) A price of ₦960 is increased by 5% and then three weeks later, it increased by a further 6%. Find the final price.   .... **3 marks**

7) The price of a bicycle is reduced by 30% in a sale. The original price was ₦35 000.

   a) What is the sale price?

   b) What would 10 bicycles cost before the sale?

   c) Nneamaka bought 3 bicycles while on sale, what would be her loss if she had bought them before the sale?

.... **6 marks**

8) A bicycle is on sale at £130 + 20% VAT.
   a) Work out the amount of VAT to be paid.
   b) Work out the total cost of the bike.
   ....4 marks

9) A bill for garden clearance is £95 + 20% VAT. How much is the total bill?
   .... 2 marks

10) Calculate the total price of the following items:

a) Dining table: £300 + 20% VAT

b) Volvo Car: £15 000 + 20% VAT

c) 50-inch TV: £1100 + 20% VAT

.........................6 marks

11) £13500 is invested at $7\frac{1}{2}\%$ per annum compound interest for 6 years. Find:
a) The interest at the end of 6 years. b) The total amount payable at the end of 6 years.
.........................4 marks

12) William invests some funds in a bank at 6% interest per annum. After 3 years, the value of William's investment is £35000. Calculate the amount William invested.
.........................3 marks

7 Algebra 1

This section covers the following topics:

- Order of operations
- Substitution
- Identifying equations
- Solving equations
- Forming and solving equations
- Additive inverse
- Inequalities

LEARNING OBJECTIVES

By the end of this unit, you should be able to:

a) Remove brackets from basic algebraic expressions
b) Expand/multiply algebraic terms
c) Divide algebraic terms
d) Work out calculations using BIDMAS
e) Identify and solve equations with brackets
f) Understand additive inverse
g) Solving inequalities

KEYWORDS

- Brackets
- Algebraic terms
- Multiply
- BIDMAS
- Algebraic expressions
- Equations

7.1 ORDER OF OPERATIONS

When there are several operations in one calculation, the correct order must be followed to get the calculation right.

$4 + 2 \times 3$ will have two answers; 18 and 10 if the correct order is not followed. To make sure that everybody is doing the same thing, we use BIDMAS or BODMAS to help us.

Brackets
Other things **or** **I**ndices like (3^2), (2^3), √, ...
Division and **M**ultiplication
Addition and **S**ubtraction

In a calculation, do the brackets first, followed by other things like powers (3^2) and roots ($\sqrt{}$), then division and multiplication comes next – *do them in the order they appear*, and finally, addition and subtraction - **do them in the order they appear.**

Also, you may use **bracket(s)** to indicate the calculation you want to perform first.

Example 1: Work out the values of

a) $4 + 2 \times 3$
   Multiplication (×) comes first before addition (+). $4 + (2 \times 3) = 4 + 6 =$ **10**
   Notice that 4 retained the position on the left even though we had to multiply first.

b) $4 - 2 \times 3 = 4 - (2 \times 3) = 4 - 6 =$ **-2**

c) $6 + 3 \times 5 - 4 = 6 + (3 \times 5) - 4 = 6 + 15 - 4 =$ **17**

d) $6 - 4 + 3$ ...Since the operations are addition and subtraction, perform the calculation in the order they appear in the question. $6 - 4 = 2$ and $2 + 3 =$ **5**

e) $20 - 6 + 3 \times 4 = 20 - 6 + (3 \times 4) = 20 - 6 + 12 =$ **26**

f) $7 \times 10 \div 2$ ...Since the operations are multiplication and division, perform the calculation in the order they appear in the question. $7 \times 10 = 70$ and $70 \div 2 =$ **35**

g) $3x + 8x \div 4 = 3x + (8x \div 4) = 3x + 2x =$ **5x**

h) $24 \div 2 + 5 - 3 \times 4 = (24 \div 2) + 5 - 3 \times 4$
   $= 12 + 5 - 3 \times 4 = 12 + 5 - (3 \times 4) = 12 + 5 - 12 =$ **5**

EXERCISE 7A

1) Calculate
a) $4 \times 3 + 10$
b) $3 + 4 \times 2$
c) $5 \times 10 - 3$
d) $5 + 18 \div 6$
e) $3 + 2 - 1$
f) $3 - 2 + 4$
g) $10 \div 5 \times 9$
h) $3 \times 4 + 5 \times 2$
i) $5 \times 8 - 3 + 7$
j) $4 + 4^2 \div 2 - 3$
k) $9 + 3 \times (17 - 7)$
l) $54 \div (20 - 5 + 3)$
m) $27 - (3^2 \div 1)$
n) $7 - 4 \div 2 + 17$
o) $35 - (7 \times 3)$
p) $4 \times 5 - 5 - 10 \div 2$
q) $5^2 + (8 - 2 + 4) - 20$
r) $2 + 3 \times (8 - 5)$
s) $6 - 4 \times 2$
t) $100 + (34 - 14) \times 10$

2) Copy the calculations below and put in **brackets** where necessary to make the answer correct.
a) $5 + 6 \div 2 = 8$
b) $21 \div 3 + 4 = 3$
c) $2 \times 7 - 4 = 6$
d) $8 - 4 \div 4 = 7$
e) $9 + 3 + 3 \times 2 = 21$
f) $40 - 8 \times 7 = 224$

3) Simplify
a) $4w \times 3 + 10$
b) $3 + 4 \times 2y$
c) $5 \times 10c - 3c$
d) $5 + 18x \div 6$
e) $3n + 2n - 1$
f) $3w - 2 + 4w$
g) $10s \div 5s \times 9$
h) $3 \times 4m + 5 \times 2$
i) $5p \times 8 - 3 + 7p$
j) $4w + 4^2 \div 2 - 3$
k) $9 + 3 \times 17x - 7$
l) $200x \div 20 - 5 + 3$
m) $27p - 3^2 \div 1$
n) $7 - 4n \div 2 + 17$
o) $35k - 7 \times 3$
p) $4 \times 5 - 5 - 10k \div 2k$
q) $8b \times 8 - 5 \times 4b$
r) $2 + 3g \times 8g - 5$
s) $6x - 4x \times 2$
t) $4v \times 3 + 4 \times 6v - 4v \times 2$

7.2 SUBSTITUTION AND FORMULAE

Substitution simply means putting numbers in place of letters in expressions and formulae and performing the calculation(s).

For example, $6x + 3y$ is an algebraic expression. $6x$ and $3y$ cannot be added together because they are not like terms. However, if the values of the letters x and y are known, we can then substitute (replace) them for the letters in the expression and find its value.

Remember: $p \times p \times p = p^3$, $p + p + p = 3p$, $2y = 2 \times y$, $abc = a \times b \times c$, $2x^2 = 2 \times x \times x$

Example 1: If $x = 10$ and $y = 2$, the value of the expression $6x + 3y$ will be
$6 \times 10 + 3 \times 2 = 60 + 6 = \mathbf{66}$

Example 2: If $n = 3$, $y = ½$ and $t = 9$, find the values of
a) $t - n$ $\longrightarrow$ $9 - 3 = \mathbf{6}$
b) n^2 $\longrightarrow$ $3 \times 3 = \mathbf{9}$
c) $4t$ $\longrightarrow$ $4 \times 9 = \mathbf{36}$
d) $4(n + 9)$ $\longrightarrow$ $4 \times (3 + 9) = 4 \times 12 = \mathbf{48}$
e) $8y$ $\longrightarrow$ $8 \times ½ = \mathbf{4}$

EXERCISE 7B

1) If m = 3, n = 10 and y = ½, work out the value of
a) 2n
b) 10m
c) m + n
d) n - m
e) 20y
f) 4y + m
g) n^2
h) 6n + ½y
i) y^2
j) $2m^2$
k) 13n
l) 15 - n

2) If q = 7 and c = 0, work out the value of
a) 3q
b) 40 – q
c) 145c + 37
d) q^2
e) q - q
f) ⅓c + 10q
g) 5q + c + 7
h) 8 – 2q
i) 3(q + c)
j) 16 + 10c - q
k) 91 – 6q
l) 20 - ¾c

3) Work out the perimeter of the shapes below if
   i) x = 3 cm   ii) x = 1.5 m

a)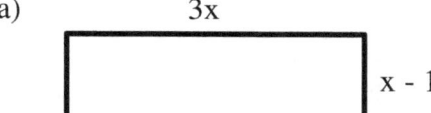

b)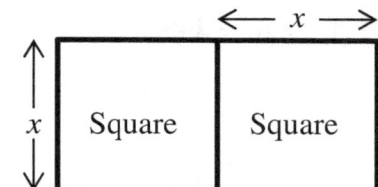

c)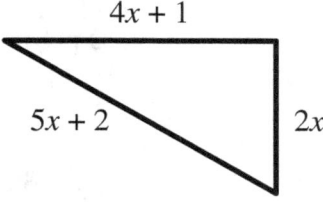

d)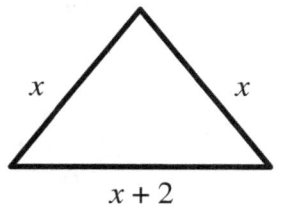

4) If c = 4 and d = 10, find the value of
a) 3(d - c)
b) 4c + 10d
c) 40 ÷ c
d) 10(d + c)
e) $\dfrac{42}{c+d}$
f) c(c - d)
g) $\dfrac{cd}{20}$
h) $\dfrac{100}{d}$

5) If x = 2 and y = 3, calculate
a) $(y + x)^2$
b) $\sqrt{2x + 5}$
c) $x^2 + y^3$
d) $(3x)^2$
e) $y^0 + x^0$
f) 10(x – y)

6) If s = 10, t = 3 and u = 0, calculate the value of:
a) $\dfrac{st}{3}$ b) $t^3 s$ c) $3st - t^3$ d) $(s + t + u)^2$

7) If c = $\dfrac{1}{2}$, d = 0.25 and e = -6, work out
The value of:
a) 20c   b) 8c + d   c) c^3 d) d + e

7.3 SUBSTITUTION INVOLVING NEGATIVE NUMBERS

In algebra, it is important to understand the operations with negative numbers. A negative number multiplied by a positive number always gives a **negative result**. A negative number multiplied by another negative number always gives a **positive result**.

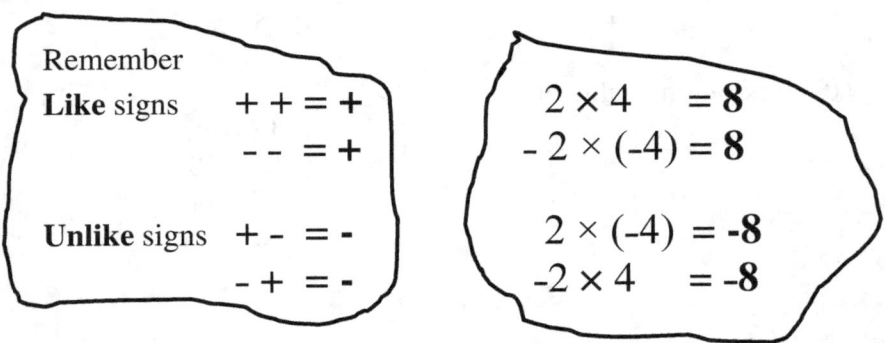

Example 1: If $x = 10$ and $y = -5$, find the value of

a) $x + y$ $10 + (-5) = 10 - 5 = \mathbf{5}$
b) $2x - y$ $2 \times 10 - (-5) = 20 + 5 = \mathbf{25}$
c) y^2 $(-5) \times (-5) = \mathbf{25}$
d) x^2 $10 \times 10 = \mathbf{100}$
e) $4x^2$ $4 \times x^2 = 4 \times x \times x = 4 \times 10 \times 10 = \mathbf{400}$

EXERCISE 7C

1) If $p = -3$, find the value of
    a) $3p$ d) $10 - p$ g) $p - 100$ j) $p - p$
    b) $4p + 6$ e) $2p - 4$ h) $p + p + p$ k) $p \div p$
    c) $p + 1$ f) $18 - 2p$ i) $p^2 + 12p$ l) $16p - 10$

2) If $x = 3$ and $y = -5$ find the value of
    a) $x + y$ d) $2x + y$ g) $70 - 5y$ j) $y + 3x$
    b) $3x + 4y$ e) $x + 1 + y$ h) $3x + 10y + 20$ k) $x^2 + y^2$
    c) $6x - 2y$ f) $x - y$ i) $9y - 3$ l) $19 + y - y$

3) If $a = -12$ and $b = -6$, work out the value of
    a) $a + b$ d) $b - a$ g) $3b + 5$ j) $a \div b$
    b) $b + a$ e) $a - b$ h) $3a - 2b$ k) $a \div 2b$
    c) $a + a + b$ f) $2a$ i) $3 - b$ l) $40 + b$

7.4 SUBSTITUTION INTO A FORMULA

A **formula** is a rule to work out a value.

Example 1: In the formula V= IR, you may work out the value of V if the values of I and R are known. Likewise, you may work out the values of I or R if the remaining values are known.
When I = 3 and R = 5, the value of V in the above formula will be 3 × 5 = **15**.

Example 2: The formula for area of a circle is given as $A = \pi r^2$, where r is the radius.
If r = 5 cm and
$\pi = 3.142$,
A = 3.412 × 5^2
= 3.142 × 25
= **78.55 cm²**

Example 3: Find the value of P is r = 3 from the formula P = 40r - 6
P = (40 × 3) – 6 = **114**

EXERCISE 7D

1) $S = \frac{P}{6} - 8$. Find the value of S when
a) P = 18   b) P = 48

2) $T = \frac{3u-2}{5}$. Find T when u = 4

3) S = 2t + 30. Find the value S when
a) t = 10   b) t = -30

4) V = u + at. Find the value of V when u = 7, a = 3 and t = 4.

5) $C = \frac{T}{4} - 3$. Find C when T = 28.

6) S = 200 + 3y. Find the value of S when a) $y = \frac{1}{3}$ b) y = -10

7) Volume of an object is given by the formula $V = r^2h$. Find
a) V when r = 5.7 and h = 2.5
b) h when V = 294 and r = 7

8) Copy the diagram below and find the value of each expression using y = 3 in the middle.

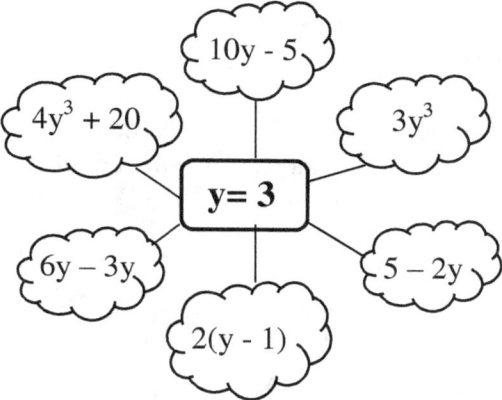

9) $f = \frac{1}{u} + \frac{1}{v}$. Find the value of F when
a) u = 2 and v = 3
b) u = 0.6 and v = 0.3

10) The equation of a straight line is in the form **y = mx + c**, where m is the gradient and c the intercept on the y – axis. Find

a) y when m = 3, x = 2 and c = -5
b) y when m = -2, x = 4 and c = 0.4

7.5 SOLVING EQUATIONS

In expressions, there are **no equal** signs. Examples $5x$, $2x + 5$, $n + 6y$....
In equations, there are equal signs and letters stand for a particular number - the solution of the equation.

Linear equations have only the first power of the unknown quantity.
Examples of linear equations are: $2x = 10$, $x + 5 = 20$...

RULES FOR SOLVING EQUATIONS

Do the same thing to both sides. You may
- Add the same number to both sides
- Subtract the same number from both sides
- Multiply both sides by the same number
- Divide both sides by the same number

Example 1: Solve $x + 5 = 9$

To solve means to find the value of the unknown. The left side of the equation MUST equal the right side. $x + 5$ must equal 9. To make the left side equal to 9, find the opposite of **+ 5**, which is -5. Subtract 5 from both sides as explained below.

So in $x + 5 = 9$
          (-5)   (-5) ………………….. Subtract 5 from both sides
$x + 5 - 5 = 9 - 5$
         $x = 4$ ✓

Check that $x = 4$ is the right answer by replacing x with 4 in the original equation. If it equals 9, then 4 is the solution. From the original equation
$x + 5 = 9$
$4 + 5 = 9$
   $9 = 9$…. this is the confirmation needed. The left side is now equal to the right side.

Example 2: Solve $2x = 10$
This means 2 multiply by a number gives 10. We must find that number.
Find the opposite of times 2 (×2). This is (÷ 2). We divide both sides by 2.

$$\frac{2x}{2} = \frac{10}{2}$$
$$x = 5 \checkmark$$

Or $2x = 10$
   (÷2)   (÷2)
      $x = 5$

Check your answer by multiplying 2 by 5 in the original equation. $2 \times 5 = 10$

Example 3: Solve $\dfrac{x}{5} = 7$

Multiply both sides by 5 to make a whole number.

$\dfrac{x}{\cancel{5}} \times \cancel{5} = 7 \times 5$ Therefore, $x = 35$ ✓

Check: From the original equation, $35 \div 5 = 7$

Example 4: Solve $x - 3 = -7$
Add 3 to both sides. $x - 3 + 3 = -7 + 3$ $x = -4$ ✓
Check: $-4 - 3 = -7$

Example 5: solve $40 = 8y$
Divide both sides by 8 because 8y means $8 \times y$. $5 = y$ which implies that $y = 5$ ✓
Check: $8 \times 5 = 40$. Left side equals right side.

Example 6: Solve $7 = \dfrac{n}{3}$

A number divides 3 and the answer is 7. To solve this, find the opposite of divide by 3. It is multiply (×) by 3. So we multiply both sides by 3.

$7 \times 3 = \dfrac{n}{\cancel{3}} \times \cancel{3}$ $21 = n$ Therefore, $n = 21$ ✓

Check: $21 \div 3 = 7$

Example 7: Solve $2x = -8$

Divide both sides by 2 $\dfrac{\cancel{2}x}{\cancel{2}} = \dfrac{\overset{-4}{\cancel{-8}}}{\cancel{2}}$ $x = -4$ ✓
Check: $2 \times -4 = -8$

Equations could be thought of as a set of balanced scales. The middle section represents the equal sign.

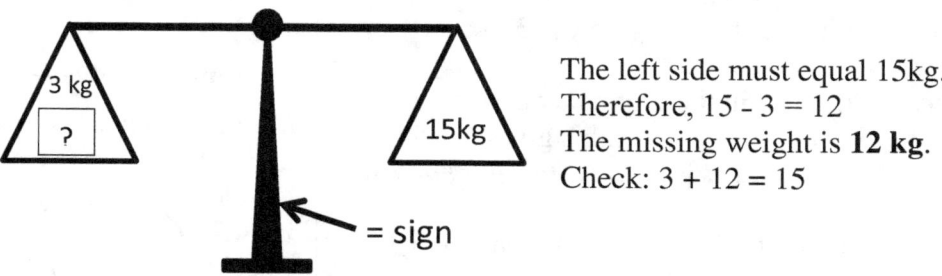

The left side must equal 15kg.
Therefore, $15 - 3 = 12$
The missing weight is **12 kg**.
Check: $3 + 12 = 15$

EXERCISE 7 E

1) Solve the equation below.
a) $c + 5 = 7$
b) $c - 4 = 9$
c) $c + 5 = 20$
d) $9 - x = 8$
e) $x + 11 = 22$
f) $x - 90 = 1$
g) $x + 19 = 37$
h) $w + 6 = 16$
i) $w - 6 = 16$
j) $w + 7 = -2$
k) $u + 5 = -8$
l) $6 + x = -10$
m) $u + 0 = -7$
n) $r - 49 = 4$
o) $g - 3 = -8$
p) $w - 7 = 16$
q) $13 + y = -3$
r) $56 = r - 8$
s) $4 + y = 34$
t) $n - 7 = 15$
u) $9 + e = 2$
v) $20 = x - 4$
w) $c - 20 = -10$
x) $x + 6.8 = 13.6$
y) $x - 5.5 = 2.3$
z) $u + 256 = 600$

EXERCISE 7F

1) Solve
a) $8x = 16$
b) $7x = 21$
c) $42 = 7x$
d) $63 = 9x$
e) $5x = 55$
f) $13w = 39$
g) $3w = 27$
h) $-2c = 4$
i) $-7c = 14$
j) $20 = -5w$
k) $8x = 64$
l) $-7x = 70$
m) $-x = -7$
n) $-3n = -9$
o) $\frac{n}{7} = 6$
p) $\frac{x}{3} = 10$
q) $\frac{w}{9} = -2$
r) $\frac{3}{n} = 1$
s) $-9x = -63$
t) $-n = 9$
u) $18 = -9n$
v) $-20m = -60$
w) $\frac{w}{1.5} = 3$
x) $9z = 3$
y) $67n = 670$
z) $\frac{y}{2} = -47$

7.6 SOLVING EQUATIONS IN TWO-STEPS

Example 1: Solve $2x + 5 = 11$
Solution: Get rid of (+5) by doing the exact opposite. Therefore, subtract 5 from both sides.

$2x + 5 - 5 = 11 - 5$
$2x + 0 = 6$ Therefore, $2x = 6$
Divide both sides by 2 to give $x = 3$ ✓

Check: Put $x = 3$ in the original equation
$2 \times 3 + 5 = 11$ ……………..this gives $6 + 5 = 11$
$11 = 11$ (*Left hand side is equal to right hand side*)

So, the solution to the above equation is $x = 3$

Example 2: Solve $7x - 1 = -8$
Add 1 to both sides
$7x - 1 + 1 = -8 + 1$
$7x \quad = -7$
Divide both sides by 7
$x = -1$ ✓

Example 3: Solve $\frac{2x}{5} = 10$
Multiply both sides by 5 to get
$2x = 50$
Divide both sides by 2
$x = 25$ ✓
Check: $2 \times 25 = 50$, $50 \div 5 = 10$

EXERCISE 7G

1) Solve the equations

a) $2c + 5 = 13$
b) $5c + 10 = 20$
c) $5x - 10 = 10$
d) $4w + 1 = 9$
e) $5 + 3x = 8$
f) $8 + 2x = 6$
g) $7x + 5 = 26$
h) $20w - 7 = 13$
i) $8y + 3 = 51$
j) $6x - 16 = -4$
k) $3u - 7 = -10$
l) $10x - 10 = -20$
m) $4y + 6 = 14$
n) $16y - 5 = 155$
o) $\frac{4n}{3} = 8$
p) $\frac{6v}{9} = 2$
q) $7 = 6x - 5$
r) $19 = 5x + 4$
s) $8c + 9 = 49$
t) $40w - 4 = 76$
u) $3x + 20 = 44$
v) $2n - 5 = 9$
w) $8w + 2 = 66$
x) $8 + 4x = 20$
y) $13c - 6 = 46$
z) $4 + 17x = -13$

7.7 FRACTIONAL EQUATIONS

Example 1: Solve $\frac{x}{3} + 7 = 5$
Subtract 7 from both sides

$\frac{x}{3} + 7 - 7 = 5 - 7$

$\frac{x}{3} = -2$

Multiply both sides by 3
$x = -6$ ✓

It could also be viewed as a function machine

$x \rightarrow \boxed{\div 3} \rightarrow \frac{x}{3} \rightarrow \boxed{+7} \rightarrow \frac{x}{3} + 7$

Like all function machines, the inverse of the above operation is subtract 7 then multiply by 3.

$\frac{x}{3} + 7 = 5$
(-7)   (-7)............Inverse of +7 is -7

$\frac{x}{3} = -2$

$\frac{x}{3} \times 3 = -2 \times 3$ ....Inverse of ÷3 is ×3

$x = -6$

Example 2: Solve $\frac{2x}{5} = 10$
Multiply both sides by 5 to get
$2x = 50$
Divide both sides by 2
$x = 25$ ✓

Check: $2 \times 25 = 50$, $50 \div 5 = 10$

Example 3: Solve $\frac{c}{2} = \frac{1}{4}$

Multiply both sides by 2 and 4 to make the numbers horizontal.

$\frac{c}{2} \times 2 \times 4 = \frac{1}{4} \times 2 \times 4$

$\frac{8c}{2} = \frac{8}{4}$ ....cancelling down gives

$4c = 2$....divide both sides by 4

$c = \frac{2}{4}$

$c = \frac{1}{2}$ ✓

EXERCISE 7H

1) Solve the equations below

a) $\frac{n}{5} + 1 = 4$

b) $\frac{n}{4} - 3 = 7$

c) $\frac{n}{2} + 5 = 10$

d) $8 = \frac{40}{n}$

e) $\frac{n}{-3} = 3$

f) $\frac{n}{10} - 8 = 22$

g) $\frac{n}{-7} = 9$

h) $\frac{3n}{5} = 3$

i) $6 + \frac{n}{7} = 8$

j) g) $\frac{n}{8} + 8 = 23$

k) $\frac{4n}{3} = 4$

l) $\frac{n}{6} - 2 = 11$

m) $8 = \frac{1}{4}n$

n) $\frac{n}{6} = \frac{2}{3}$

o) $\frac{n}{5} = \frac{8}{20}$

p) $\frac{3n}{6} = \frac{2}{5}$

q) $\frac{n}{-10} = \frac{2}{5}$

r) $\frac{n}{6} + 5 = 2$

s) $15 + \frac{n}{5} = 16$

t) $\frac{20}{n} = 5$

u) $\frac{35}{n} = 7$

v) $3 - \frac{3}{n} = 0$

w) $\frac{1}{2} = \frac{4n}{8}$

x) $\frac{n}{6} = \frac{2}{3}$

y) $20 = 6 + \frac{n}{2}$

z) $\frac{n}{4} - 1 = -3$

EXTENSION QUESTIONS

1) The rectangle below has a perimeter of 26 cm. Work out the length of the longest side of the rectangle.

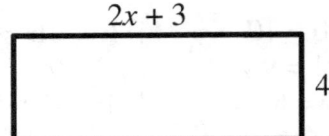

2) Work out the area of the rectangle in question 1 above.

3) Solve the equations below.

a) $\frac{d}{5} - 3 = -6$

b) $\frac{4}{5}w = -8$

c) $\frac{7-n}{3} = 2$

7.8 EQUATIONS WITH BRACKETS

Example 1: Solve $3(2x + 2) = 30$

Expanding Method
$3(2x + 2) = 30$
$(3 \times 2x) + (3 \times 2) = 30$
$6x + 6 = 30$
(Now subtract 6 from both sides)
$6x = 24$
Divide both sides by 6
$x = \frac{24}{6} = \mathbf{4}$

Dividing by multiplier method

$\frac{\cancel{3}(2x+2)}{\cancel{3}} = \frac{30}{3}$

$2x + 2 = 10$
  $(-2)\quad (-2)$
$2x = 8$
$x = \frac{8}{2} = \mathbf{4}$

Check: Put $x = 4$ in the original equation. Left hand side (LHS) = 30 and right hand side (RHS) = 30.

Example 2: Solve $8 = 4(5x - 1)$

Using the expanding method,
$\quad 8 = 20x - 4$
$\quad (+4) \quad\quad (+4)$
$\quad 12 = 20x$
$\quad (\div 20) \quad (\div 20)$
$\quad \frac{12}{20} = x$
$\quad \frac{3}{5} = x$ or $x = \frac{3}{6}$ or $\mathbf{0.6}$

Example 3: Solve $3(x+2) + 4(x-2) = 12$

$3x + 6 + 4x - 8 = 12$
$7x - 2 = 12$
$7x = 12 + 2$
$7x = 14$
$x = \frac{14}{7} = \mathbf{2}$

Check: Put $x = 2$ in the original equation.

$3(\mathbf{2}+2) + 4(\mathbf{2}-2) = 12$
$3 \times 4 \quad + 4 \times 0 \ = 12$
$12 \quad\quad\quad\quad\quad\quad = 12$
LHS = RHS = 12

EXERCISE 7 I

Solve the equations

1) $2(3b - 7) = 4$
2) $3(2t + 5) = 21$
3) $3(x - 4) = 3$
4) $2(12 - x) = 0$
5) $3(5 - c) = 6$
6) $21 = 3(2p + 5)$
7) $4(2x + 5) = 19$
8) $6(x - 3) = 32$
9) $6 = 2(4 - x)$
10) $4(x - 1) = 16$
11) $11(4y - 2) = 43$
12) $1 + 3(d - 1) = 4$
13) $3(4 - x) = 6$
14) $3(x - 2) = -3$
15) $5(x + 2) = 60$
16) $7(x + 3) = -7$

Solve the equations

17) $3(2y + 1) + 2(y - 1) = 25$
18) $4(2 - 4x) - 2(3 + 5x) = 28$
19) $6x - (3 - x) = 0$
20) $0.2(3x - 4) + 0.3(5x + 2) = 30$

7.9 EQUATIONS WITH UNKNOWN ON BOTH SIDES

Sometimes, the unknown terms are on both sides of the equality sign. It is best solved by collecting the unknown terms either to the left or right of the equality sign.

Example 1: Solve $3n - 5 = 2n + 8$

A faster approach would be to take 5 over to the right side and take 2n to the left of the equality sign. However, we must follow the rules of moving terms and numbers. *If it is a positive term or number, subtract on both sides. If it is a negative term or number, add on both sides.*
$$3n - 5 = 2n + 8$$
To take 5 to the right side, add 5 on both sides.
$$3n - 5 + 5 = 2n + 8 + 5$$
$$3n = 2n + 13$$
Now subtract 2n on both sides.
$$3n - 2n = 13$$
$$n = 13 \checkmark$$

Point to note: You have a choice of getting rid of any of the terms or numbers first, but it must be systematic (One after another). You will still have the same answer. For example, you may decide to get rid of 2n first.
$$3n - 5 = 2n + 8$$
$$3n - 5 - 2n = 2n + 8 - 2n$$
$$n - 5 = 8$$
Now add 5 on both sides
$$n - 5 + 5 = 8 + 5$$
$$n = 13$$

Example 2: Solve $3(x + 5) = 2(4 - x)$

Expand first to give $3x + 15 = 8 - 2x$
Now, add 2x to both sides
$$3x + 15 + 2x = 8 - 2x + 2x$$
$$5x + 15 = 8$$
Take 15 away from both sides
$$5x + 15 - 15 = 8 - 15$$
$$5x = -7$$
Dive both sides by 5
$$5x \div 5 = -7 \div 5$$
$$x = \frac{-7}{5} = -1\frac{2}{5} \checkmark$$

Exercise 3: Solve $\frac{3}{4}y - 5 = 9 - y$

Solution: When fractions are involved in an equation, remove the denominator. To remove the denominator, multiply both sides by the denominator, 4.

$$4 \times (\frac{3}{4}y - 5) = 4 \times (9 - y)$$
$$3y - 20 = 36 - 4y$$
Now, add 20 to both sides
$$3y - 20 + 20 = 36 - 4y + 20$$
$$3y = 56 - 4y$$
Add 4y to both sides
$$3y + 4y = 56 - 4y + 4y$$
$$7y = 56$$
Divide both sides by 7
$$\frac{7y}{7} = \frac{56}{7}$$
$$y = 8 \checkmark$$

As usual, you may check your answer by replacing y in the original equation with 8. LHS = RHS = 1.
Therefore, y = 8 is the correct solution.

Example 4: Solve $\dfrac{2x-1}{3} = \dfrac{3x+1}{4}$

If we have a fraction, the first step is to remove the denominator(s). Since there are two denominators, multiply both sides by the LCM of the two numbers. The LCM of 3 and 4 is 12.

$$\dfrac{\cancel{12}^{4}(2x-1)}{\cancel{3}} = \dfrac{\cancel{12}^{3}(3x+1)}{4}$$

$$4(2x-1) = 3(3x+1)$$

Expand
$$8x - 4 = 9x + 3$$
Add 4 to both sides
$$8x - 4 + \mathbf{4} = 9x + 3 + \mathbf{4}$$
$$8x = 9x + 7$$
Subtract 9x from both sides
$$8x - \mathbf{9x} = 9x + 7 - \mathbf{9x}$$
$$-x = 7$$
Divide both sides by -1

Therefore, **x = -7** ✓

EXERCISE 7 J

Solve the equations below.

1) $5y - 8 = 3y$
2) $2a - 7 = 8 - 3a$
3) $2x + 15 = 8 - 4x$
4) $18 - 5x = 3x + 2$
5) $11x - 5 = x + 25$
6) $y + 5 = 3y + 9$
7) $4 - x = 8x + 20$
8) $3w - 15 = 8 - 4w$
9) $y + 9 = 3y + 16$
10) $7c - 11 = 2c + 4$
11) $d + 3 = 14 - 3d$
12) $3t = 2t + 7$
13) $1 + 7x = 4 - x$
14) $15x = 10 - 5x$
15) $y - 3 = 3y + 7$
16) $4x = 3x + 6$
17) $7 - 3u = 5 - 2u$
18) $6x - 2 = 1 - 3x$
19) $f - 16 = 16 - 2f$
20) $14 - 3d = d + 3$
21) $4x + 5 = x + 5$
22) $k - 3 = 3k - 2$
23) $8c - 1 = c - 5$
24) $6w + 8 = w + 1$

EXERCISE 7K

Find the solution to
1) $5(x + 2) = 2(x + 6)$
2) $8(y - 3) = 2y$
3) $7(3a - 5) = 2(5a - 1)$
4) $4(3x + 1) = 32 - 2x$
5) $x(3 + 5) = 6 - 4x$
6) $3(2x - 1) = 8x + 1$
7) $2(c - 3) - (c - 2) = 5$
8) $5(g + 1) = 2g + 3 + g$
9) $4(1 - 2x) = 3(2 - x)$
10) $5y - 3(y - 1) = 39$

11) $\dfrac{2f - 1}{3} = \dfrac{f}{2}$

12) $\dfrac{t-1}{5} - \dfrac{t-1}{3} = -2$

13) $\dfrac{d}{2} - 3 = \dfrac{d}{5}$

14) $\dfrac{3}{x+3} = \dfrac{9}{x+5}$

15) $\dfrac{3}{2x-1} = \dfrac{4}{3x+1}$

FORMING AND SOLVING EQUATIONS

Example 1: I subtract 3 from a number and then multiply the result by 10, the answer is 100. a) Write an **equation** and b) **solve** it to find the original number.

Solution: Let the number be **y**.
If I subtract 3, it will give y – 3.
If I multiply (y - 3) by 10, it will give 10(y – 3). **a)** Since the answer is 100, the equation is **10(y – 3) = 100.** ✓

b) Expand the bracket to give
10y – 30 = 100
   (+30)   (+30)
10y = 130
 (÷10) (÷10)
y = 13. The original number is **13.** ✓

Example 2: Conrad doubles a number, added 12 and then divides the result by 5. He got the same answer when he multiplied the number by 1.
a) Write an equation for Conrad's number. **b)** Solve the equation to find Conrad's number.
Solution
Let Conrad's number be **n**. He doubles the number to give 2n. He added 12 to give 2n + 12. He then divides 2n + 12 by 5 to give $\frac{2n + 12}{5}$. The result is the same as 1 × n = n.
a) So, the equation is $\frac{2n + 12}{5} = n$ ✓
b) Multiply both sides by 5 to give
2n + 12 = 5n
(Subtract 2n from both sides)
12 = 3n. Therefore, **n = 4**
Conrad's number is **4** ✓

EXERCISE 7 L

In questions 1 – 5, Helen is thinking of a number. Write an equation and hence, solve it to find Helen's number.

1) She doubles the number and adds 10. The answer is 44.

2) She adds 3 to the number and multiplies the result by 7. The answer is 70.

3) She multiplies the number by 3 and adds 6. She then divides the result by 3. She got the same number by multiplying the number by 2.

4) She is thinking of a fraction. She subtracts 6 from the number and divides the result by 4. She got the same answer when she subtracts 4 from the number.

5) Helen trebles the number, adds 2 and multiplies the result by 2. She got the same answer when she multiplied the number by 2 and add 12.

6) The sum of three consecutive numbers is 66. Find the numbers.

7) Philip is t years old. His mother is 25 years older.

a) How old will Philip be in 10 years' time?
b) How old will Philip's mother be in 10 years' time?
c) How old was Philip 5 years ago?
d) If the mother is 50 years old, how old is Philip?

8) The perimeter of the rectangle below is 22 cm.

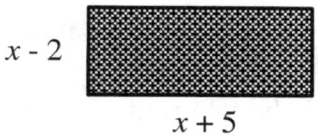

a) Write an equation for the perimeter.
b) Hence, solve the equation to find the value of x.
c) Work out the length and width of the rectangle.
d) Calculate the area of the rectangle.

9) Dr Brown has £x. Jude, his eldest child, has $\frac{2}{3}$ of his money. Leanne, his only daughter, has $\frac{1}{5}$ of his money. If the total amount of money Jude and Leanne have is £13 000, work out how much their father have in Pounds.

10) I add 38 to a number. I then divide the result by 2. The overall result is the same as ten times the number. Work out the original number by first forming an equation.

11) I am thinking of a number. I added 6 to the number and multiplied the result by 7. I then subtract 40, and the result is 16.
a) Write an equation to find my number.
b) Hence, solve the equation to find the number.

12) Joe is thinking of a number. He subtracts 9 from the number and then doubles it. He got the same answer as dividing the number by 5. Write an equation and solve it to find Joe's number.

13) For each diagram, write down an equation involving w and solve it.

a)
b)
c)

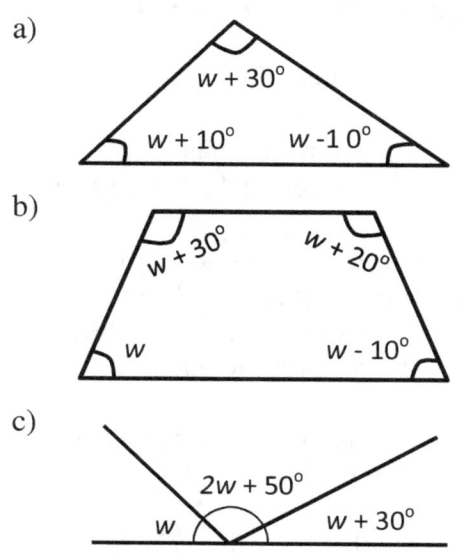

14) Tony added four consecutive numbers and the total was 450.
a) Write down an equation. $4x + 6 = 450$
b) Solve your equation to find the consecutive numbers. 111,112,113,114

15) The triangle below has a perimeter of 49 cm.

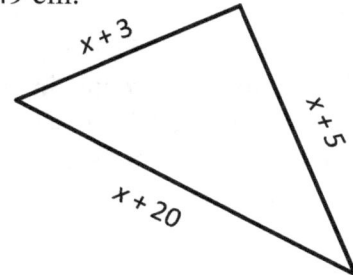

a) Write an expression for the perimeter of the triangle. $3x + 28$
b) Form an equation. $3x + 28 = 49$
c) Solve your equation. $x = 7$
d) Work out the lengths of all the sides. 10 cm, 12cm, 27cm

7.10 INEQUALITIES

The symbols for inequalities are:

<	Less than
≤	Less than or equal to
>	Greater than
≥	Greater than or equal to

$x < 3$ means that the value of x is less than 3, but **not equal** to 3. The values that satisfy this inequality could be 2, 1, 0, -2,…..but 3 is **not** included.

$x \leq 3$ means that the value of x **must** be 3 or less. For example, 3, 2, 1, 0,……

$x > 3$ means that the value of x **must** be greater than 3 and **not equal** to 3. Examples are 4, 5, 6, ….

$x \geq 3$ means that the value of x is greater than or equal to 3. The values that satisfy this inequality must be 3 itself and above. For example, 3, 4, 5….

We encounter inequalities in everyday life.

- "The number of students in class A is more than 10." We can write as **s > 10** where s is the number of students.

- "The penalty for stealing must be at least 6 months in prison." We can write **p ≥ 6** months.

- "The speed limit in Harlow is 30 m.p.h." We can write **s ≤ 30**, where s is the speed limit.

INEQUALITIES ON A NUMBER LINE

To successfully represent inequalities on a number line, the following concepts should be followed.

1) **Open circle** (O) means that the number is **not included**. For example,

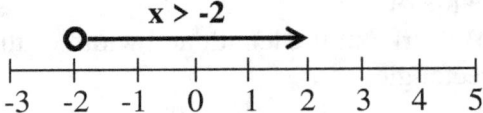

The direction of the arrow represents the values covered by the inequality. $x > -2$, means that -2 is not included because of the open circle. The integer values that satisfy the inequality $x > -2$ are -1, 0, 1, 2, 3…..

2) Closed circle (●) means that the number **is included**. For example,

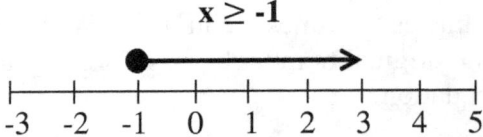

The integer values that satisfy the inequality $x \geq -1$ are -1, 0, 1, 2, 3……

3) $-3 < x \leq 2$ represents all the numbers between -3 and 2 but -3 is not included. The integer numbers that satisfy this inequality are -2, -1, 0, 1 and 2. Notice that -3 is not included. On a number line, $-3 < x \leq 2$ is

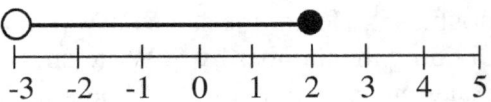

EXERCISE 7 M

Using x for the variables, write down the inequalities shown.

1)
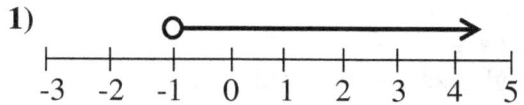

2)
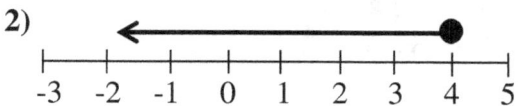

3)
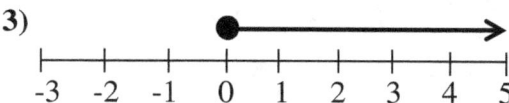

4)
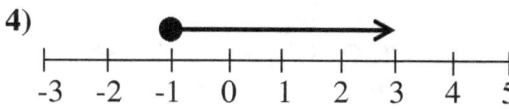

5)
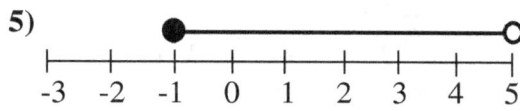

6)
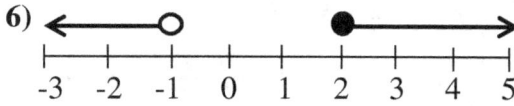

7)
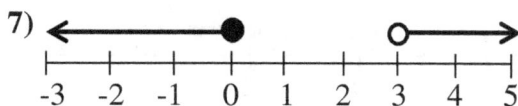

Unlike equality sign where $x = 5$ is the same as $5 = x$, inequalities are different. $x < 5$ is **not** the same as $5 < x$. $x < 5$ is **the same** as $5 > x$. The inequality sign is reversed if we want to start with the number.

The reason for this is simple; for $x < 5$, let's assume that x is 4. $4 < 5$ is correct. If we want to start from 5 and write $5 < 4$. This is not correct as 5 is not less than 4. Therefore, we reverse the sign to give **5 > 4,** which is now correct.

EXERCISE 7N

1) Using inequality symbols, rewrite the following statements.

a) x is less than 5
b) x is less than 3
c) x is greater than 1
d) x is less than or equal to -2
e) y is greater than or equal to 1

2) By drawing a number line, display the inequalities below.

a) $x < -3$ d) $x \geq 5$
b) $x < 3$ e) $w > 6$
c) $x \leq -4$ f) $w < 5$

3) Write **true** or **false** for each statement.

a) $5 < 6$ e) $0 < 7$
b) $4 < 3$ f) $4 \geq 4$
c) $5 \leq 7$ g) $4^3 < -20$
d) $10 \geq 10$ h) $-40 > -10$

4) Write down the **smallest** integer that satisfies the following inequalities.
a) $x > 1$ b) $x > -2$ c) $x \geq 5$ d) $x \geq -9$

5) Write down the **largest** integer that satisfies each of these inequalities.
a) $x < 2$ b) $x \leq 1\frac{1}{2}$ c) $x < -0.5$
d) $x \leq -3\frac{3}{4}$

SOLVING INEQUALITIES

If we solve normal equations like $2x = 6$, the solution is **one** value which is 3 in this case. However, when solving inequalities, we find **a range of values** while following most of the rules of solving linear equations.

RULES FOR SOLVING INEQUALITIES

1) Add or subtract the same thing to both sides.

2) Multiply or divide both sides by the same positive number while retaining the inequality sign.

3) If we must multiply or divide by a **negative** number, then the inequality sign must be **reversed**.

Observe the inequality $3 < 4$. If we multiply both sides by a negative number say (-2), we have -6 and -8.

$$3 < 4$$
$$\times (-2) \quad -6 > -8$$
Sign is reversed

If we retain the inequality sign, <, the statement would be incorrect. -6 is not less than -8. It is greater than -8. Therefore we reverse the sign to greater than (>). **-6 > -8.**

Same principle for dividing by a negative number! ***Reverse the sign.***

Example 1: Solve the inequalities
a) $x + 2 > 9$ d) $-2x \geq 9$
b) $x - 1 \leq 3$ e) $4 < x + 2 \leq 7$
c) $3x > -9$

Solutions:
a) $x + 2 > 9$
Subtract 2 to both sides, **x > 7**

b) $x - 1 \leq 3$
Add 1 to both sides, **x ≤ 4**

c) $3x > -9$
Divide both sides by 3, **x > -3**

d) $-2x \geq 9$
Divide both sides by -2, **x ≤ - 4.5**
Remember, the inequality sign is reversed because we divided by a negative number.

e) $4 < x + 2 \leq 8$
Subtract 2 from all sides (to have only x in the middle)
$4 - 2 < x + 2 - 2 \leq 8 - 2$
This gives **2 < x ≤ 6**

We can represent the inequality on a number line like

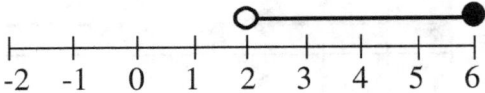

Example 2: Write down the integer values that satisfy the inequality
$-4 \leq x - 3 < 2$

Add 3 to all sides to give $-1 \leq x < 5$. The integer values are **-1, 0, 1, 2, 3, 4**

Notice that 5 is not included.

EXERCISE 7 O

Solve the inequalities below.

1) $x + 2 < 7$
2) $x - 3 < 2$
3) $x + 2 \leq -4$
4) $x - 4 \geq 8$
5) $6x + 3 > 9$
6) $3x - 3 > -2$
7) $4 + x < 2$
8) $6 - x \leq 8$
9) $-4 - x > 9$
10) $9 - 2x < 4$
11) $-7n \geq -28$
12) $-2w + 3 < 2$

Solve the inequalities and represent on a number line.

13) $-5n \leq 10$
14) $\frac{n}{4} \geq -2$
15) $\frac{n}{3} < 1$
16) $\frac{n}{10} < \frac{3}{5}$
17) $2 < x + 1 < 3$
18) $3x + 3 > 4$
19) $\frac{x+5}{4} > -2$
20) $-15 < x - 2 \leq 3$
21) $-6 \leq 3x < 3$
22) $-2 < \frac{1}{2}x < 3$

Where **possible**, find the range of values of **n** for which both inequalities are true.

23) $n > -2$ and $n < 1$
24) $n > 5$ and $n \leq 10$
25) $n < 2$ and $n > 4$

Two pairs of inequalities are given below. Solve them and write down integer values that satisfy both inequalities.

26) $x + 5 < 10$ and $2x + 3 \geq 1$

27) $3x - 1 > 5$ and $4x < 20$

28) $-4x < 12$ and $x + 3 < 9$

INEQUALITIES AND REGIONS

When two or more variables are involved, the inequality can be represented by a **region** on a graph. The region is an area where **all the points** obey a given rule.

RULES FOR THE REGIONS
1) We use a **broken** line for inequalities of $<$ or $>$. The points on the broken line are **not included** in the region.

2) We use a **solid** line for inequalities of $\leq$ or $\geq$. The points on the solid line are **included** in the region.

Example 1: Represent the regions marked by the inequalities below.
a) $x < 1$ b) $y \geq 1$ c) $x + y \geq 1$

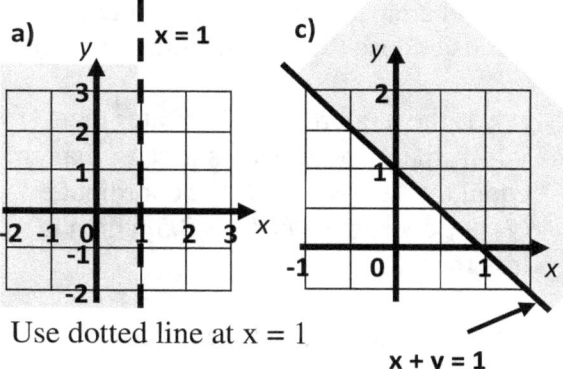

Use dotted line at $x = 1$

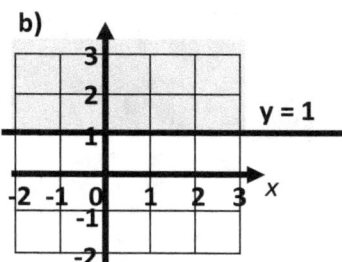

Use solid line at $y = 1$

Example 2: Shade the region which satisfies the inequalities x > 1, y ≥ 1 and x + y < 4.

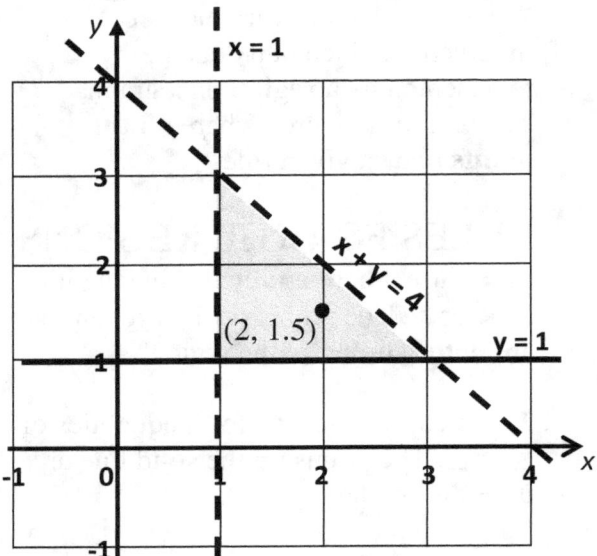

To **check** if the shaded region is correct, choose a coordinate point **inside** the shaded area and **not** on any of the lines. Let us choose (2, 1.5), see diagram. For

x > 1, the x coordinate is greater than 1.
y ≥ 1, the y coordinate is greater than or equal to 1 and x + y < 4, *x* coordinate (2) + *y* coordinate (1.5) is 3.5 which is less than 4.

Therefore, the shaded area is correct.

EXERCISE 7P

Draw diagrams to show the regions which satisfy the inequalities below.

1) x > 3
2) x ≤ 5
3) x > -1
4) y > 5
5) y > x + 2
6) x + y ≤ 3
7) x + y ≥ -2
8) y > 3 – 2x
9) x > 1, y ≤ 3, y ≥ x - 3

Find the inequalities which describe the **unshaded** regions.

10) 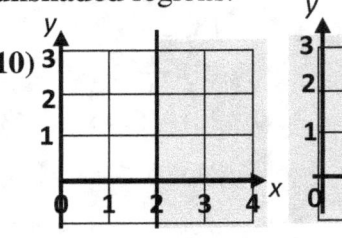 11)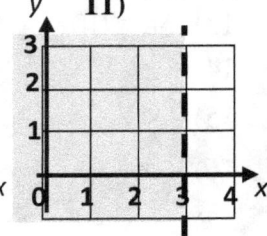

12) 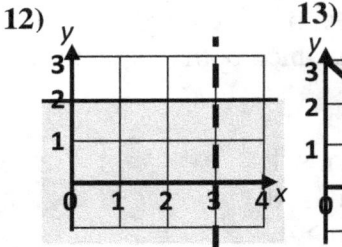 13)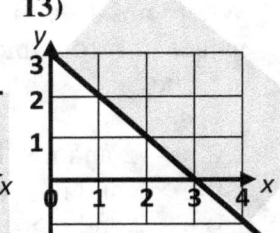

14) Arinze's supermarket employs people as security men or cleaners. Every evening, a maximum of 8 people are employed, with at least 3 people as security and one person as a cleaner.

a) If *s* = the number of security men and c = the number of cleaners, form inequalities to represent the information.
b) Draw a diagram to represent the region which satisfies these inequalities.

Chapter 7 Review Section
Assessment

1) Simplify by removing brackets
a) 8 + (13 – 5)   b) 10 – (6 + 7)   c) (4 + 2) + (13 – 4) - 8
..........................3 marks

2) Write without brackets and simplify where possible.
a) (c- d) + (e – f)   b) 9w + 5n + (n + 4w)   .... 2 marks

3) Simplify
a) (5f – b) – (4f – 5b)   b) 7a – (3d – 2e)
c) 8 × w^2 d) 30a^{3}b^2 ÷ 10ab   ....4 marks

4) Work out the values of
a) 6 + 2 × 3   b) 5 × 8 – 3 + 7   c) 4^2 + (5 – 2 + 4)   .... 3 marks

5) If *x* = 5 and y = -2, work out the value of
a) x + y   b) x – y   c) y^2 d) 7 – y   .... 4 marks

6) Solve the equations
a) n + 8 = 10
b) 2 + n = 13
c) n – 7 = 20
d) 6x = 24
e) x + 4 = - 5
f) 3x = -3
g) 17 – y = 3
h) x – 14 = - 4
.... 8 marks

7) Solve the equations
a) $\frac{n}{5} = 3$

b) $\frac{x}{7} = 10$

c) $\frac{2x}{7} = 2$

d) $\frac{x}{9} = -5$

e) $\frac{x}{20} = -3$

f) $13 + \frac{x}{3} = 19$

g) $11 - \frac{y}{6} = 3$

h) 5x – 4 = 11   .... 16 marks

8) Solve the equations

a) $3(2y + 1) + 2(y - 1) = 25$

b) $y + 9 = 3y + 17$

c) $7c - 11 = 2c + 4$

d) $4(2 - 4x) - 2(3 + 5x) = 80$

e) $7 - 5d = d + 3$

f) $8x + 5 = x + 4$

g) $\dfrac{4}{x + 4} = \dfrac{8}{x + 6}$

h) $\dfrac{1}{4x - 1} = \dfrac{3}{3x + 1}$

................24 marks

9) Dorothy is thinking of a fraction. She subtracts 6 from the number and divides the result by 4. She got the same answer when she subtracts 4 from the number. Find Dorothy's fraction.

.................3 marks

10) The perimeter of the rectangle below is 30 cm.

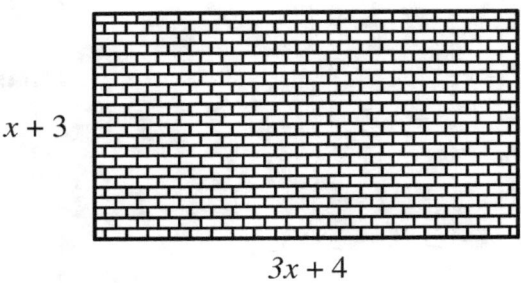

$x + 3$

$3x + 4$

a) Write an equation for the perimeter. ....2 marks

b) Hence, solve the equation to find the value of *x*. ....3 marks

c) Work out the length and width of the rectangle. ....2 marks

d) Calculate the area of the rectangle. ....2 marks

11) Complete the following inequalities by inserting > or < in each box.

a) 4 ☐ 5   b) 8 ☐ 5   c) 9 ☐ -3   d) -5 ☐ -7   e) -10 ☐ -9

…………5 marks

12) Show the following inequalities on a number line.

a) x < -5   b) x ≥ 3   c) x < -7   d) -2 < x < -1   e) 0 < x ≤ 5

…………5 marks

13) Solve the inequalities.
a) 6x – 2 > 5 …………2 marks
b) x + 7 ≤ -3 …………2 marks
c) -7w ≤ 49 …………2 marks
d) $\frac{a-5}{3}$ < -10 …………2 marks
e) -3 < x + 5 < 7 …………2 marks

14) List all the integer (whole number) values that satisfy the following inequalities.

a) 3 < x < 9 …………2 marks
b) -4 ≤ n ≤ 2 …………2 marks
c) -2 < 2(n – 3) < 7 …………2 marks
d) -8 ≤ 4x < 4 …………2 marks

15) Shade the region that satisfies the inequality y ≥ 2x – 3 …………3 marks

16) Write down the inequalities that describe the **shaded** regions.

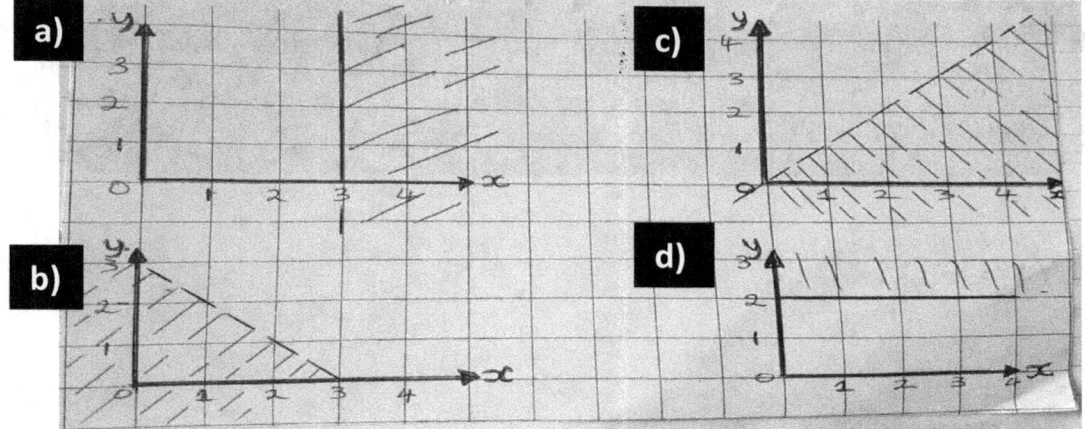

…………8 marks

8 Polygons and Circles

This section covers the following topics:

- Geometrical properties of 2D shapes
- Angles in 2D shapes
- Polygons
- Circles

LEARNING OBJECTIVES

By the end of this unit, you should be able to:

a) Understand and use properties of lines and angles
b) Calculate angles on a straight line
c) Calculate angles at a point
d) Work out vertically opposite angles
e) Understand geometrical properties of triangles, quadrilaterals and polygons
f) Calculate angles in parallel lines
g) Calculate angles in triangles and quadrilaterals
h) Understand corresponding and alternate angles
i) Understand the properties of circles

KEYWORDS

- Angles
- Parallel line
- Corresponding and Alternate angles
- Polygons
- Circles
- Quadrilaterals

8.1 LINES AND ANGLES

Angles on one side of a straight line always add to 180 degrees.

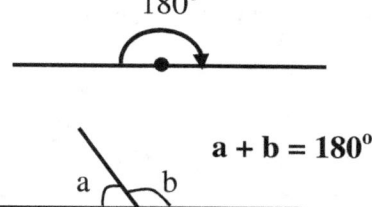

$a + b = 180°$

Example 1: Calculate the value of the missing angle.

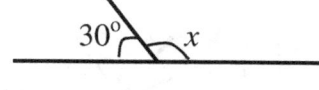

$30 + x = 180 \ldots\ldots\ldots x = 180 - 30 = \mathbf{150°}$

Check: $30° + 150° = 180$ …angles on a straight line

Example 2: work out the missing angle.

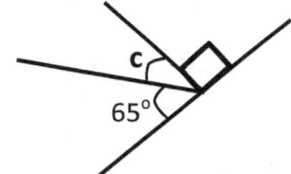

Since angles on a straight line add to 180°,
$65° + c + 90° = 180°$
$155° + c = 180°$
$c = 180 - 155 = \mathbf{25°}$

Check: $65° + 25° + 90° = 180°$

Example 3: Work out the value of x.

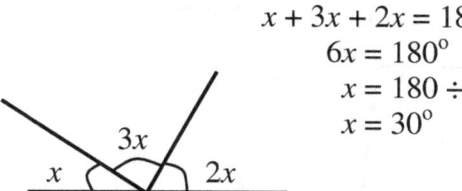

$x + 3x + 2x = 180°$
$6x = 180°$
$x = 180 \div 6$
$x = 30°$

8.2 ANGLES AT A POINT

The sum of angles at a point is 360°.
Note: There are 360° in a ***full*** turn.

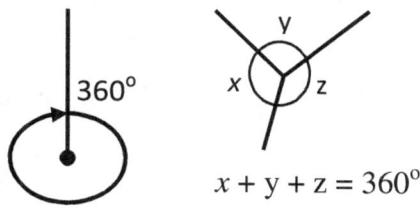

$x + y + z = 360°$

Example 1: Work out the missing angle

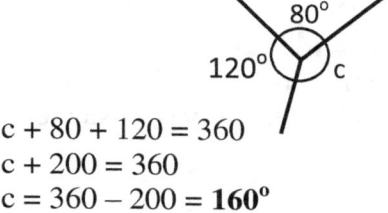

$c + 80 + 120 = 360$
$c + 200 = 360$
$c = 360 - 200 = \mathbf{160°}$

Example 2:

$w + w + w = 360°$
$3w = 360$
$w = 360 \div 3$
$w = 120°$

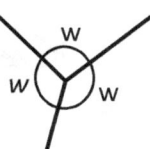

8.3 VERTICALLY OPPOSITE ANGLES

Vertically opposite angles are equal.

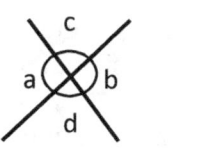 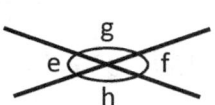

$a = b, c = d$ $\qquad$ $e = f$ and $g = h$

Example 1: Work out the missing angle

$x = 40°$....vertically opposite angles

Example 2 Work out the missing angles

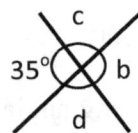

$b = 35°$ (vertically opposite angles)
$c = 180 - 35 = 145°$
  (Angle on a straight line)
d must be **145°** (Vertically opposite to angle c)

Example 3 Work out the missing angles

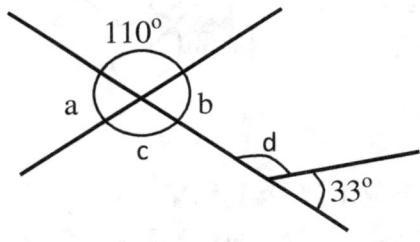

$c = 110°$......Vertically opposite angles
$b = 180 - 110$
  $= 70°$........angle on a straight line
$a = b = 70°$....Vertically opposite angles
$d = 180 - 33 = 147°$
  ....angle on a straight line

EXERCISE 8A

1) Calculate the value of the angles marked with letters in each diagram.

a)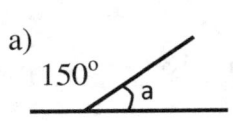

b)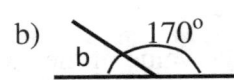

c)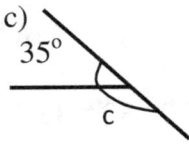

d)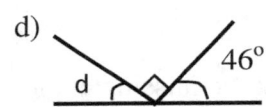

e)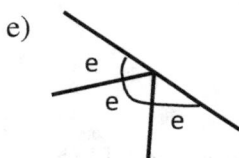

f)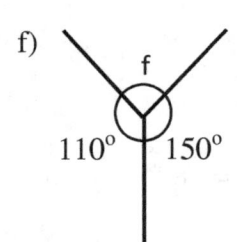

g)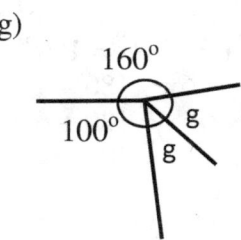

h)

i)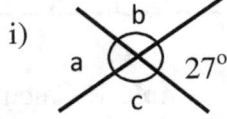

j)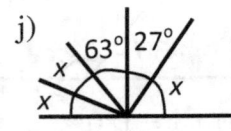

k)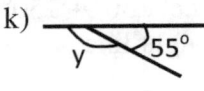

l)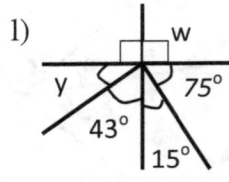

m)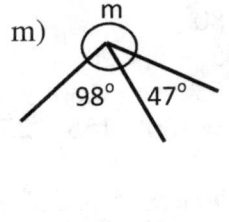

n)

8.4 TRIANGLES AND ITS PROPERTIES

A triangle is a two-dimensional shape made up of three sides and three angles. All the interior (inside) angles in a triangle add to **180°**.

There are basically four types of triangles.
1) Right-angled triangle
2) Isosceles triangle
3) Equilateral triangle
4) Scalene triangle

RIGHT-ANGLED TRIANGLE

A right-angled triangle has one angle that is 90°. It is usually marked as a square.

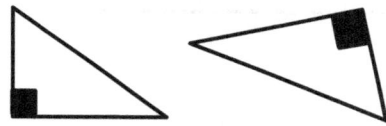

A right-angled triangle has **no line of symmetry** but has rotational symmetry of **order 1**.

However, if the right-angled triangle is also isosceles, there is only **one** line of symmetry.

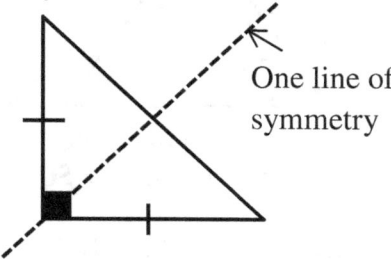

One line of symmetry

Right-angled isosceles triangle

ISOSCELES TRIANGLE

Isosceles triangle has two equal sides and two equal *base* angles. It has **one line of symmetry** and a rotational symmetry of **order 1**.

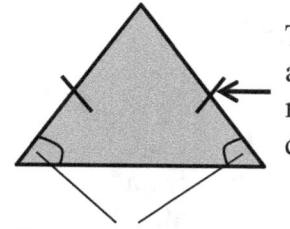

Two equal sides are usually marked with dashes

Base angles are equal

EQUILATERAL TRIANGLE

Equilateral triangle has three equal sides and three equal angles. All the three angles are 60° each.

It has **three lines of symmetry** and rotational symmetry of **order 3**.

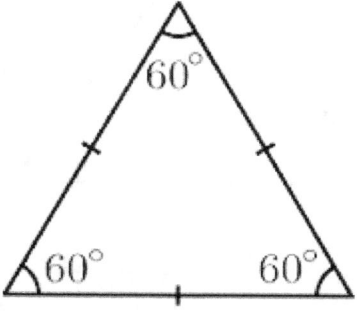

Most times, three dashes are used on the three sides indicating three equal sides.

SCALENE TRIANGLE

A scalene triangle has three unequal sides and angles

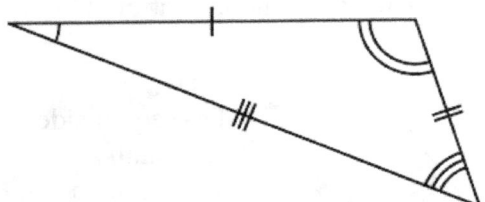

It has **no line of symmetry** and rotational symmetry of **order 1**.

Remember: Rotational symmetry is the number of times a shape fits exactly onto itself in a complete turn (360°).

CALCULATING ANGLES IN TRIANGLES

The interior (inside) angles in any triangle add up to 180°.

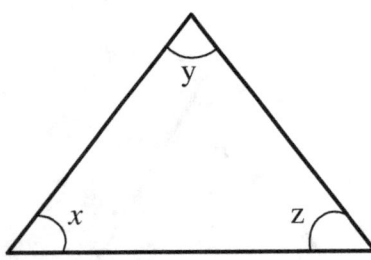

$x + y + z = 180°$

Once two angles are known, the third angle can be calculated.

Example 1: Calculate the missing angle

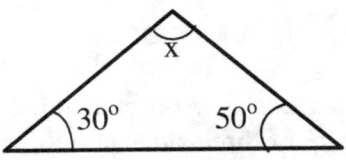

$x + 30 + 50 = 180°$
$x + 80 = 180°$
$x = 180 - 80 =$ **100°**

Check: $100 + 30 + 50 = 180$

Example 2: Calculate the missing angles $A\hat{C}B$ and $C\hat{A}B$

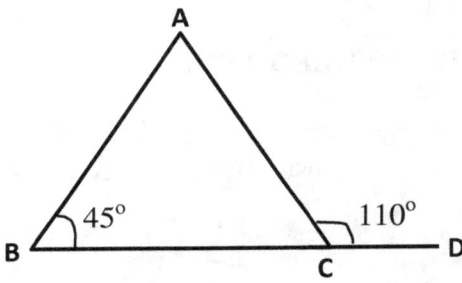

$A\hat{C}B = 180 - 110 =$ **70°**
(Angles on a straight line)

$C\hat{A}B = 180 - (45 + 70) =$ **65°**
(Angles in triangles add to 180°)

Example 3: Calculate the missing angles.

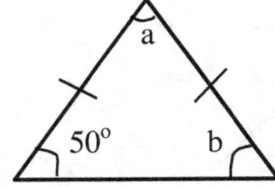

Triangle is isosceles; therefore the base angles must be the same. b = **50°**

Angles in a triangle add to 180°, so angle a = $180 - (50 + 50) =$ **80°**

EXERCISE 8B

All diagrams are not accurately drawn.
Calculate the size of angles marked by letters in the diagrams below.

1)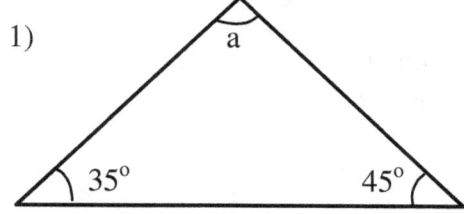

2)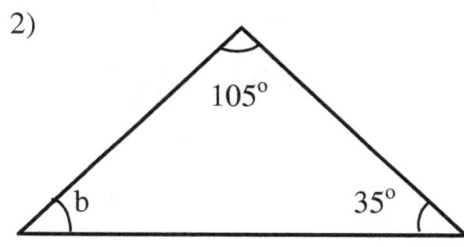

3)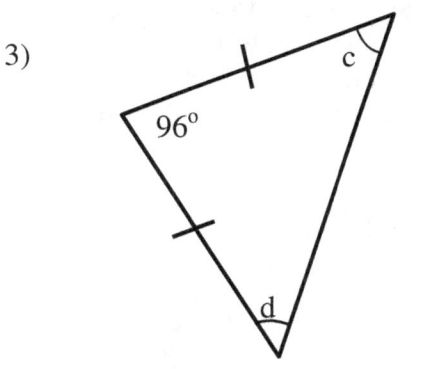

4)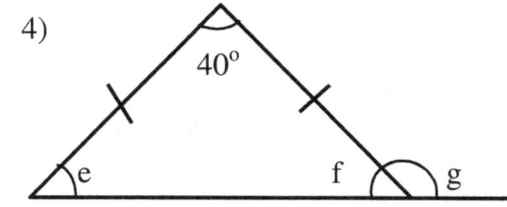

5)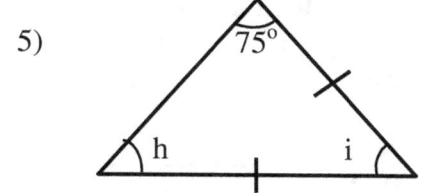

6)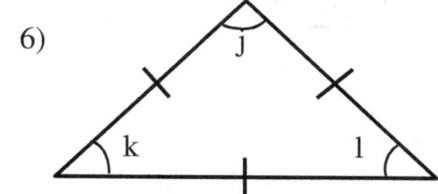

7)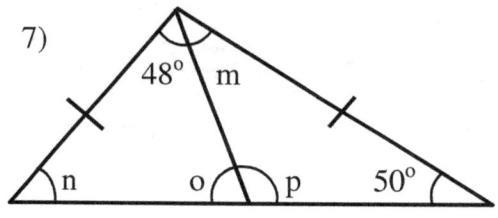

8)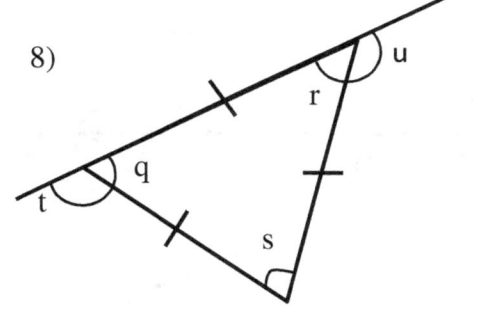

9)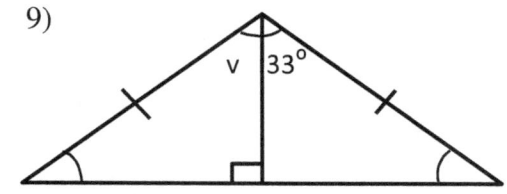

10) Indicate whether the triangles below are isosceles, right-angled, equilateral or scalene. *Give a reason for your answer.*

a) 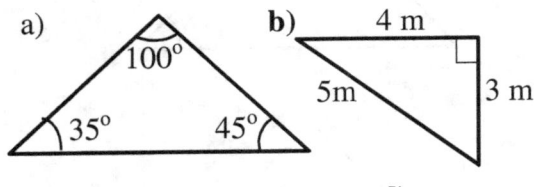 b)

c) 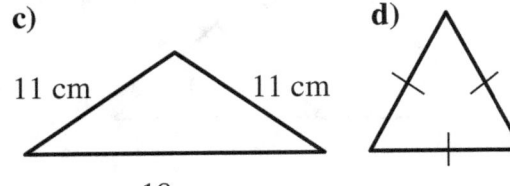 d)

e) 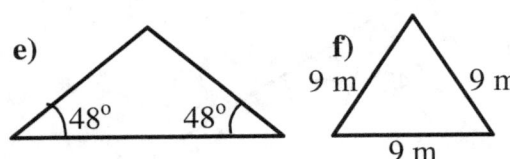 f)

11)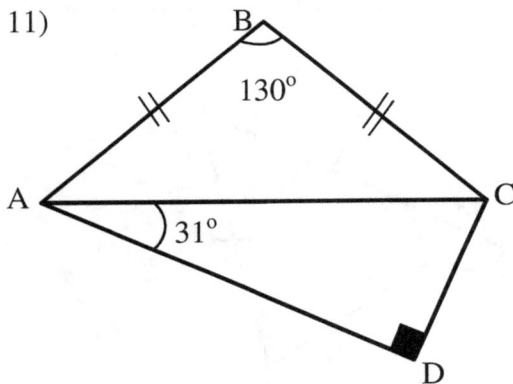

Work out the following angles.

a) $B\hat{A}C$

b) $A\hat{C}B$

c) $A\hat{C}D$

d) $B\hat{A}D$

e) $B\hat{C}D$

12) Obiora says "Angles 54°, 85° and 42° will form a triangle."
Is Obiora correct? Explain fully.

13) If two angles of a triangle are given below, calculate the third angle.

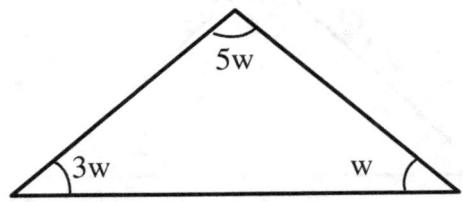

a) 30° and 56°   f) 60° and 60°
b) 130° and 20°   g) 45° and 45°
c) 23° and 76°   h) 20° and 79°
d) 114° and 56°   i) 56° and 85°
e) 100° and 40°   j) 55° and 70°

14) Look at the triangle below.

a) Form an equation in w
b) Sole the equation to find w
c) Work out the biggest angle.
d) Work out the range of the angles

15)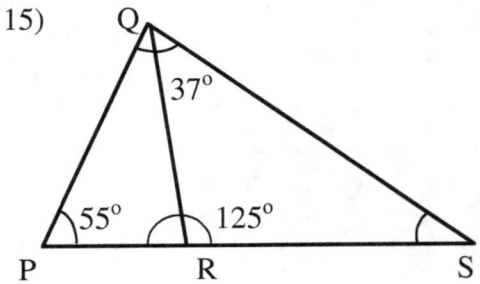

a) Calculate angles

i) $P\hat{R}Q$ ii) $P\hat{Q}R$ iii) $Q\hat{S}R$

b) What type of triangle is

i) $P\hat{Q}R$ ii) $Q\hat{R}S$?
Give a reason for each answer.

188

8.5 QUADRILATERALS

A quadrilateral is a two-dimensional shape with four straight sides and four angles.

A diagonal is a line joining two opposite corners. By using a diagonal, a quadrilateral may be divided into two triangles. The interior angles of a quadrilateral will always **add to 360°**.

Examples of quadrilaterals are square, rectangle, parallelogram, rhombus, trapezium and kite.

SQUARE

A square has four right angles. All the lengths are equal and the angles add up to 360 degrees.

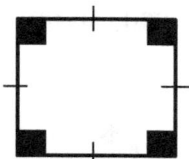

Also, opposite sides are parallel as shown with the arrows.

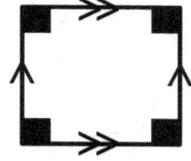

The diagonals are equal in length and bisect each other at 90 degrees.

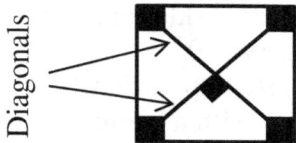

A square has four lines of symmetry and rotational symmetry of order 4.

RHOMBUS

A rhombus is a flat shape with four equal straight sides. None of the angles is 90°. They are like diamonds.

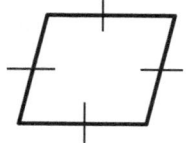

PROPERTIES

Four equal sides
Opposite angles are equal but not 90°
Opposite sides are parallel
Diagonals bisect each other at 90°
All angles add up to 360°
Two lines of symmetry
Rotational symmetry of order 2

RECTANGLE

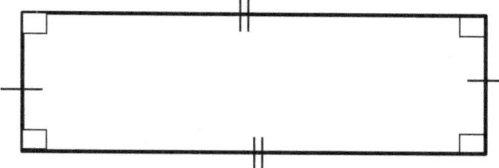

PROPERTIES

Opposite sides are parallel
Opposite sides are equal in length
All angles are 90° each
Diagonals bisect each other and are equal in length
All angles add up to 360°
2 lines of symmetry
Rotational symmetry of order 2

Remember: A square is a special type of rectangle with all the sides equal in length.

PARALLELOGRAM

A parallelogram is like a rectangle pushed out of shape. The angles **are not** 90° each.

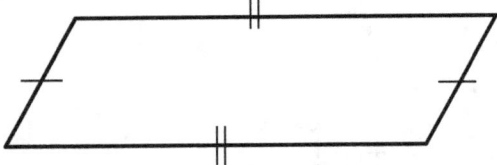

PROPERTIES

Opposite angles are equal
Opposite sides are equal in length
Opposite sides are parallel
Diagonals bisect each other
All interior angles add to 360°
Rotational symmetry of order 2
A general parallelogram like the shape above has **no line of symmetry**.

However, special parallelograms like rhombus, square and rectangle have lines of symmetry.

TRAPEZIUM

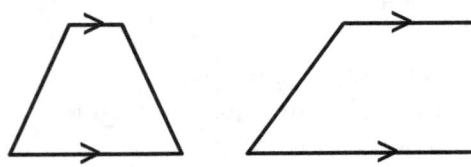

A trapezium has **one pair** of parallel sides.

PROPERTIES

One pair of parallel sides
No line of symmetry (unless it is an isosceles trapezium)
Rotational symmetry of order 1

A quadrilateral may also be called **isosceles trapezium** if the non-parallel sides are equal in length.

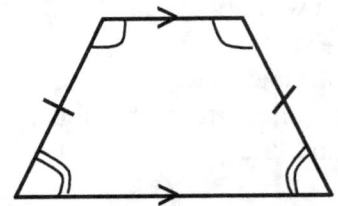

PROPERTIES

Two sets of equal angles
A set of equal sides
One line of symmetry
Rotational symmetry of order 1

KITE

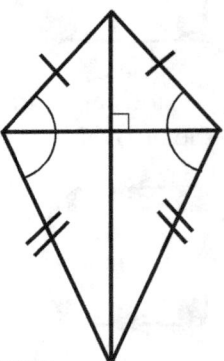

PROPERTIES

A pair of opposite angles is equal
Two pairs of adjacent sides are equal
Diagonal intersect at 90°
Note: A kite is made up of two isosceles triangles with a common base.

EXERCISE 8C

1) Write down the mathematical name for each shape below.

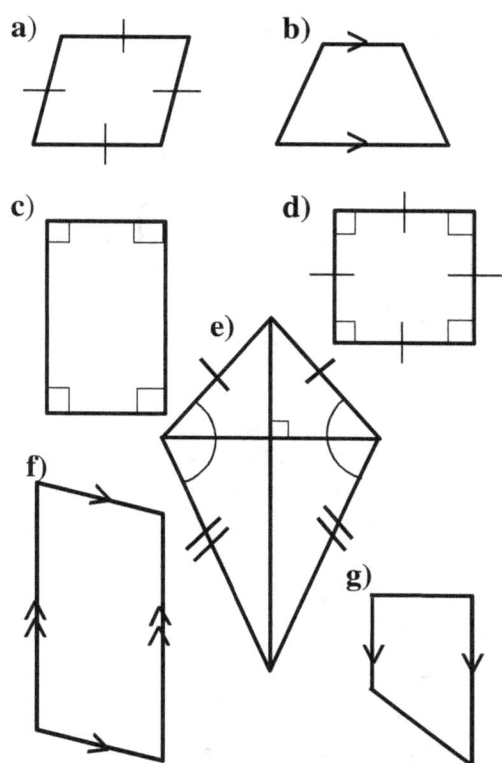

2) Show (by drawing) how two isosceles triangles can be joined together to form a kite.

3) Show by drawing how two of these shapes can be joined to make a rectangle.

4) Mention one difference between a square and a rhombus.

5) Is shape B a quadrilateral? Explain.

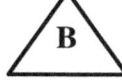

6) From the list of quadrilaterals below, copy and complete each statement.

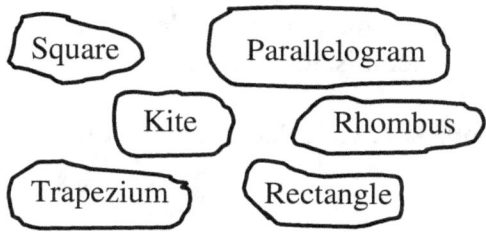

a) I have one pair of parallel sides. Therefore, I am a ……………………..

b) My diagonals bisect each other at right angles. All my sides are equal. All my angles are **not** 90°.
I am a …………………………………..

c) I have one pair of opposite angles equal. I am made up of two isosceles triangles. My name is a ………………

d) I am a ……………………........... because all my sides are equal and all my angles are 90°.

e) My opposite sides are parallel and equal in length. All my angles are 90° each. I am a ………………….. because I also have 2 lines of symmetry?

f) My opposite angles are equal. Also, my opposite sides are parallel and equal in length. However, my angles are not 90° and I have **no line** of symmetry. My name is a ……………………

7) Kunle says "The shape below is a trapezium because it has two sets of parallel sides." Is Kunle correct? Explain fully.

8.6 ANGLES IN QUADRILATERALS

The sum of the interior angles of any quadrilateral is 360 degrees.

Example 1: Calculate the size of the missing angle.

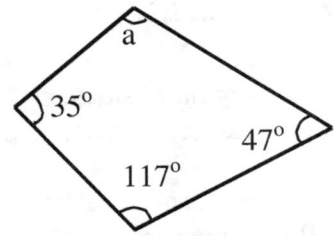

$a + 35° + 117° + 47° = 360°$
$a + 199° = 360°$
$a = 360° - 199° = \mathbf{161°}$ ✓

Check: $161 + 35 + 117 + 47 = 360$

Example 2: Calculate the missing angles.

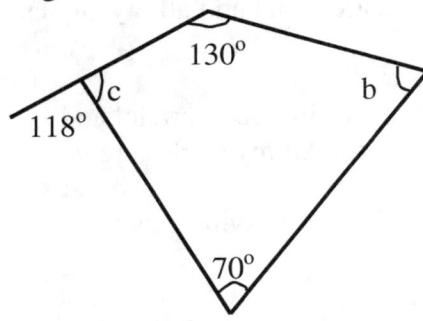

$c = 180 - 118 = \mathbf{62°}$ ✓
…..angle on a straight line add to 180°

$62 + 130 + 70 + b = 360°$
…angles in a quadrilateral add to 360°

$262 + b = 360$
$b = 360 - 262 = \mathbf{98°}$ ✓

Example 3
a) Form an equation in x.
b) Calculate the value of x.
c) Work out the value of $B\hat{C}D$

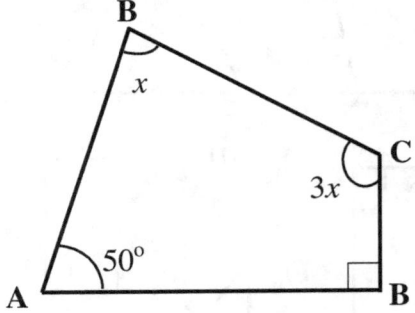

a) Form an equation in x.
$x + 3x + 90 + 50 = 360$
$4x + 140 = 360$
$4x = 360 - 140$
$\mathbf{4x = 220}$ ✓

b) $4x = 220$
$x = \frac{220}{4} = \mathbf{55°}$ ✓

c) $B\hat{C}D = 3x = 3 \times 55 = \mathbf{165°}$ ✓

Example 4 Calculate angles w, c and y.

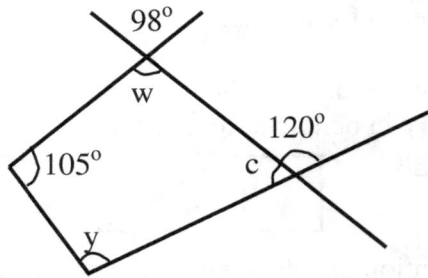

$w = \mathbf{98°}$…..vertically opposite angles
$c = 180 - 120 = \mathbf{60°}$…….straight line
$y = 360 - (105 + 98 + 60)$
   $= 360 - 263 = \mathbf{97°}$

EXERCISE 8D

1) Work out the size of the unknown angles in each diagram.

a)
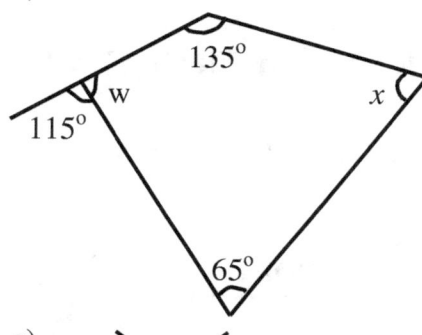

b)
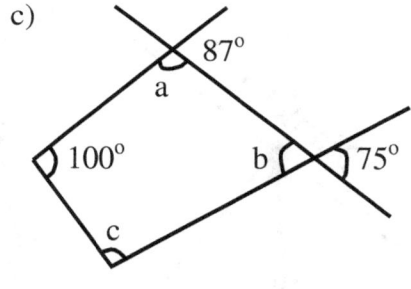

c)
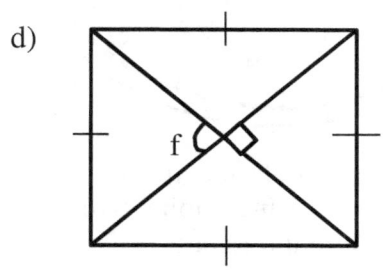

d)
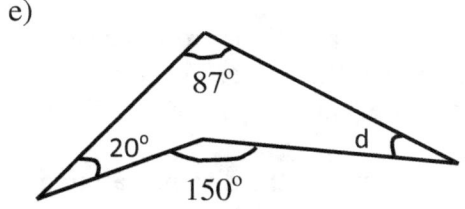

e)

2) The smallest angle is 40°. The opposite angle to the smallest angle is 60° more. The third angle is twice the smallest angle. Calculate the value of the remaining angle.

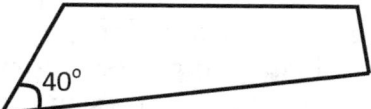

3) For the quadrilaterals drawn below, work out the missing angles by forming an equation first.

a)
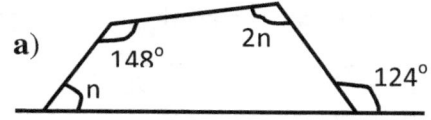

b)
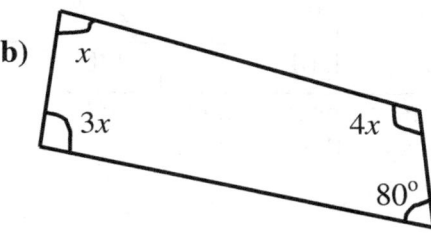

c)
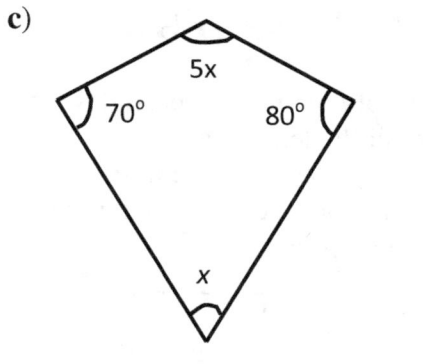

d)
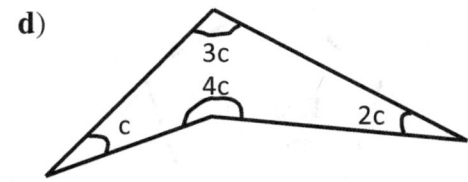

8.7 POLYGONS

Any close two-dimensional shapes with three or more straight sides are called **polygons**.

POLYGON	NUMBER OF SIDES	SUM OF INTERIOR ANGLES
Triangle	3	180°
Quadrilateral	4	360°
Pentagon	5	540°
Hexagon	6	720°
Heptagon	7	900°
Octagon	8	1080°
Nonagon	9	1260°
Decagon	10	1440°

Note: The number of interior angles goes up by 180° each time.

REGULAR POLYGON

A regular polygon has all its angles the same and all the lengths equal. Therefore, squares, equilateral triangles, regular pentagons etc. are all examples of regular polygons.

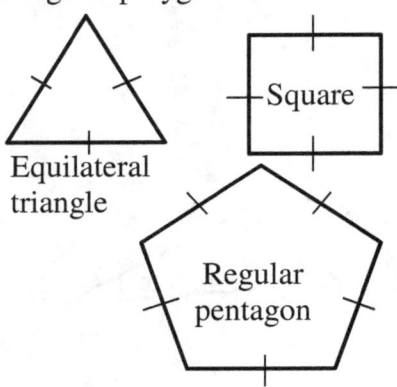

8.8 CIRCLES

We need a pair of compasses to draw a circle accurately.

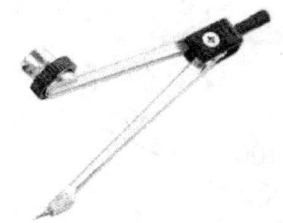

PARTS OF A CIRCLE

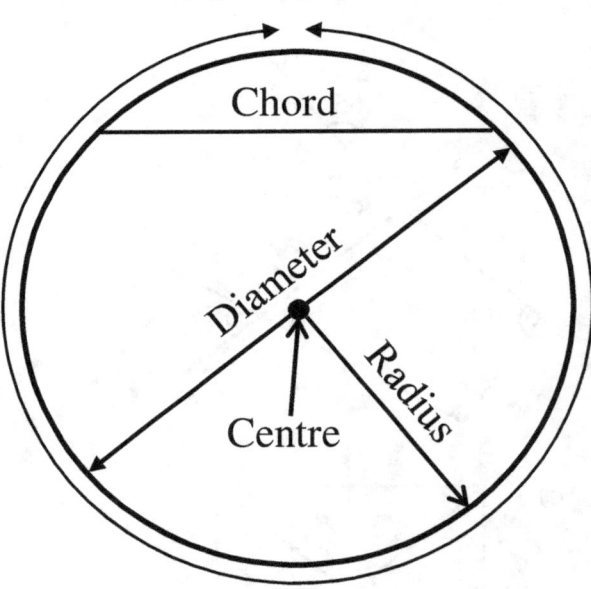

Diameter is a straight line passing through the **centre** and joining two points on the circumference

Radius is a straight line from the centre to the edge of the circle. Twice the radius will equal a diameter.

Circumference is the distance around the circle. It is similar to the perimeter.

Chord is a straight line joining two points on the circumference. If the chord passes through the centre, then it is a diameter.

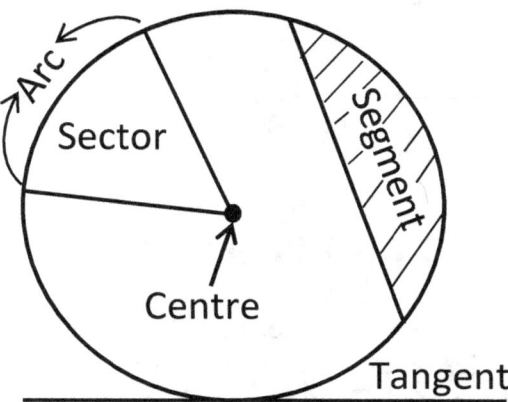

Arc is a part of the circumference of the circle.

The **sector** is the part of the circle lying between two radii and an arc.

The **segment** is the part of the circle between a chord and the circumference.

Tangent is a straight line that touches the outside of a circle at one point only.

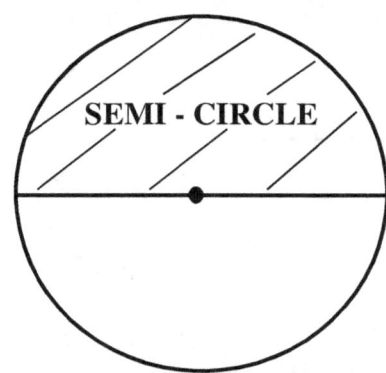

Two semi-circles = one full circle.

EXERCISE 8E

1) Is a circle a polygon? Explain fully

2) Name any four polygons together with the sum of their interior angles.

3) What would a shape with 12 sides be called? What would the sum of their interior angles add up to?

4) By using a pair of compasses, draw a circle with
a) radius of 4 cm
b) radius of 2.5 cm
c) diameter of 5 cm.

5) a) Draw a circle of diameter 12 cm.
b) Draw a chord and label it P.
c) Draw a tangent and label it Q.
d) Mark a point R on the circumference.

6) a) Draw a circle of radius 5.5 cm.
b) Draw a diameter of the circle.
c) Shade the area between the diameter and the circumference.
d) What is the name of the shaded part?

7) a) Draw a circle of radius 6 cm.
b) Identify and name any 6 parts of the circle.

8) Okoro says "A circle with a diameter of 20 cm must have a radius of 40 cm." Is Okoro correct? Explain fully

8.9 ANGLES AND PARALLEL LINES

Parallel lines are lines that will **never** meet, no matter how far they are extended.

Parallel lines exist everywhere in real life including rail tracks and the sides of a piece of paper. Also, in a square or a rectangle, the two opposite sides are parallel.

Small arrows are used to identify two or more parallel lines.

A line that cuts a pair of parallel lines is called a **transversal**.

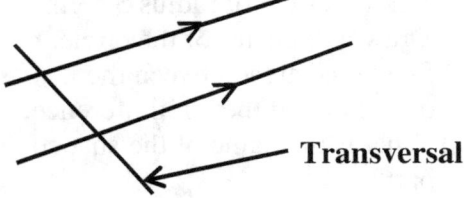

Transversal

Eight angles are formed when a line cuts through a pair of parallel lines.

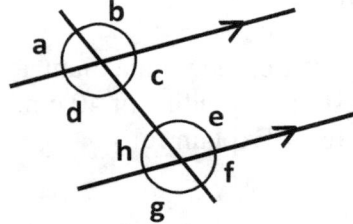

a = c = h = f   and   b = d = e = g

If the transversal is not perpendicular (at right angles) to the parallel lines, 4 acute and 4 obtuse angles are formed.

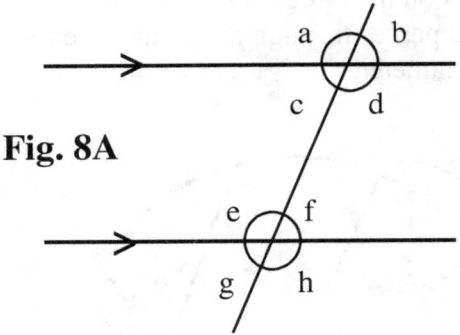

Fig. 8A

ALTERNATE ANGLES

Alternate angles are equal. They are angles on the opposite sides of the transversal. They form a **Z - SHAPE**.

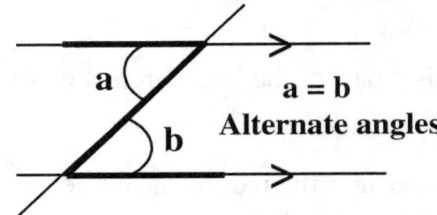

a = b
Alternate angles

In Fig 8A above, angles **c** and **f** are alternate angles and are equal.
Also, angles **d** and **e** are alternate angles and are equal too.

Example 1: Write down the value of angles *x* and *y*. Give a reason for your answer.

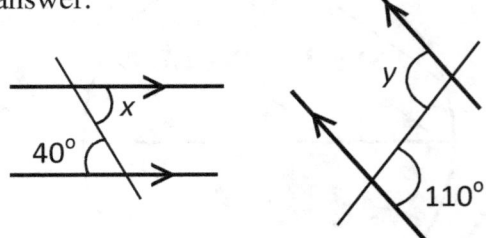

$x = 40°$ because they are alternate angles. $y = 110°$ because they are alternate angles.

CORRESPONDING ANGLES

The angles in **matching corners** when a line crosses two parallel lines are called **corresponding angles**.

Also, corresponding angles are **equal**. They also form an F- SHAPE.

In figure 8A above, angles **b = f, a = e, d = h** and **c = g** and all are corresponding angles.

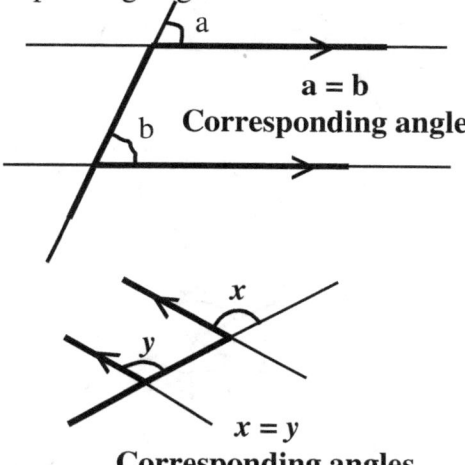

a = b
Corresponding angles

x = y
Corresponding angles

In the above examples, if **a** = 30°, **b** will also be 30° because they are corresponding angles.

Also, if **x** = 120°, **y** will be 120° because they are corresponding angles.

Note: Alternate and corresponding angles are always equal when the lines are parallel. If the lines are not parallel, **do not** assume that the angles formed are corresponding or alternate. They may not be equal.

EXERCISE 8F

1) Calculate the value of the angles marked by letters. Give a reason for your answer.

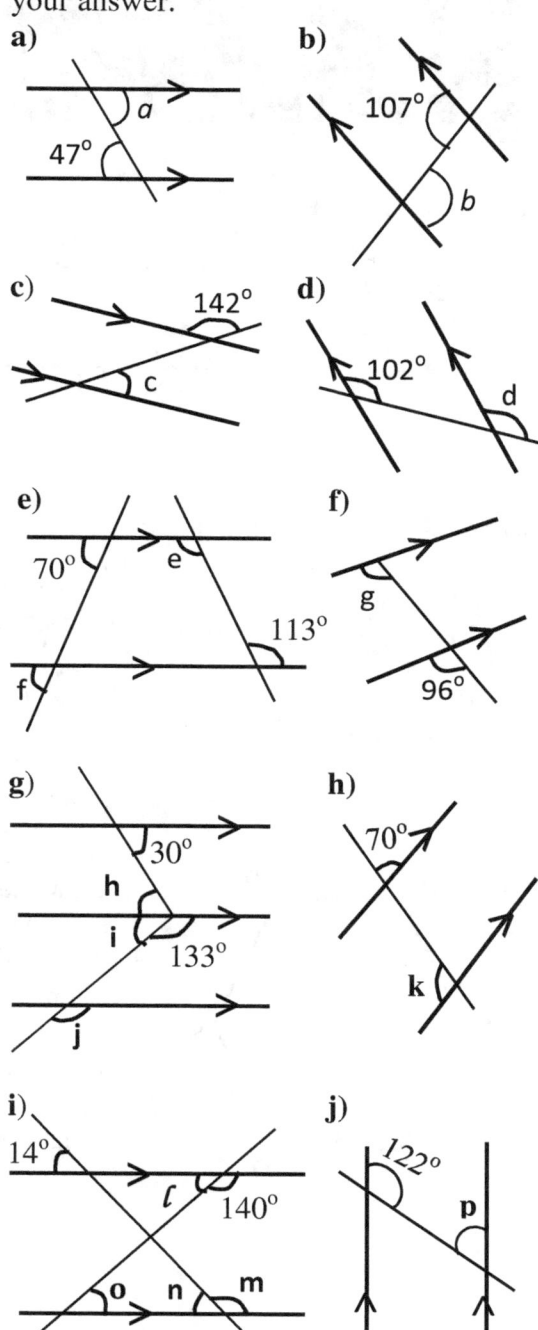

Chapter 8 Review Section
Assessment

1) Work out the missing angles
   a)
   b)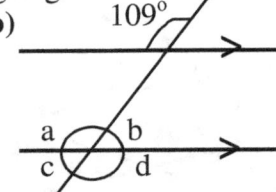
   c)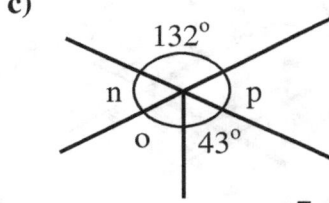

   ....10 marks

2) Work out the missing angles.
   a)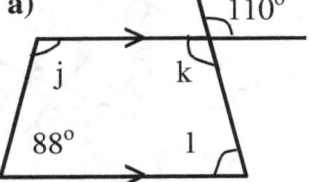
   b)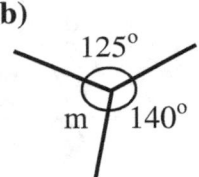
   c)

   ....7 marks

3)
   a) How many sets of parallel lines are there in the shape shown?
   b) Name the line which is parallel to AF.
   c) Name the line which is parallel to ED.
   d) What is the mathematical name for shape ABCDEF?
   e) How many lines of symmetry does shape ABCDEF have?
   f) What type of angle is EDC?

   .... 6 marks

9 Circles, Areas & Perimeters

This section covers the following topics:

- Area of circles
- Perimeter/circumference of circles and 2-d shapes
- Area and perimeter of compound shapes

LEARNING OBJECTIVES

By the end of this unit, you should be able to:

a) Calculate the area and perimeter of 2-d shapes
b) Calculate the area and circumference circular objects
c) Calculate the area and perimeter of compound shapes
d) Calculate area of shaded parts

KEYWORDS

- Circle
- Circumference
- Pi
- Area, perimeter
- Compound shape

9.1 CIRCLES AND PERIMETER

Circumference

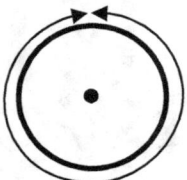

The total distance around the circle is called the circumference. Therefore, the perimeter of a circle is known as its circumference.

Since the outer surface of a circle is circular, it is difficult to use a ruler for measuring the circumference.

A more simplistic way of finding the circumference of a circle is by using a thread.

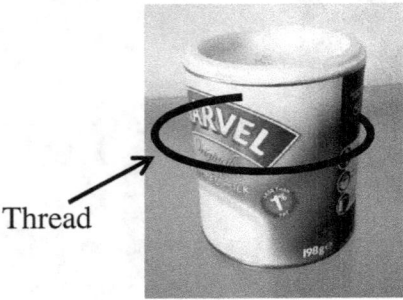

Thread

To find the circumference of the circular base of the cylindrical container, wind a piece of thread around the container once. Mark the thread where they cross each other. Pull the thread straight and measure against a ruler.

The length of the thread is the circumference of the circle.

Known fact: If you measure the distance around a circle (circumference) and then divide it by the distance across the circle through the centre (diameter), your answer will always come close to a particular value. This depends on the accuracy of the measuring instrument(s).

That particular value is approximately 3.141592653…….

In mathematics, we use the Greek letter π (pronounced as *pi*) to represent this number. In some textbooks, you might see π as 3.14, 3.141, $\frac{22}{7}$ or $3\frac{1}{7}$. All these numbers are not even accurate as it is impossible to express the number as an exact fraction or a decimal number.

Therefore,

Circumference (c) = π × diameter (d)

$C = \pi d$

We also know that twice the radius is the diameter of a circle. Therefore, the above formula could also be written as

$C = 2\pi r$

where r is the radius of the circle.

Example 1
Calculate the circumference of a circle with diameter of 14 cm. Use π as $\pi\frac{22}{7}$.

Solution: Using the formula, $C = \pi d$
$C = \frac{22}{7} \times 14$ cm
$C = 22 \times 2 =$ **44 cm**

Example 2: Calculate the circumference of the circle below. Use π as 3.14.

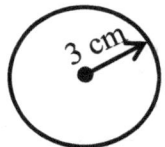

Solution: $C = 2\pi r$
$C = 2 \times 3.14 \times 3$
$C = \mathbf{18.84 \text{ cm}}$

Example 3:

The front wheel of the bicycle has a radius of 35 cm.

a) What is the circumference of the front wheel?
b) How many complete revolutions does the wheel make when the bicycle travels 500 **metres**?

a) $C = \pi d$ or $C = 2\pi r = 2 \times \frac{22}{7} \times 35$
   $C = \mathbf{220 \text{ cm}}$

b) 100 cm = 1 m and
   220 cm = 2.2 m

Complete revolution will be
$500 \div 2.2 = \mathbf{227}$

Example 4: Calculate the perimeter of the semi-circle below. Use π as 3.14.

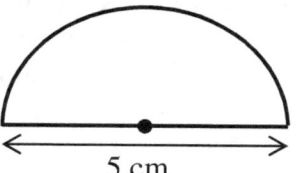

The perimeter of the semi-circle is the length of the circular face (circumference) plus the diameter.

Length of the circular face is the circumference ÷ 2.

$\frac{\pi \times d}{2} = \frac{3.14 \times 5}{2} = \frac{15.7}{2} = 7.85 \text{ cm}$

Perimeter of the semi-circle
$= 7.85 + 5 = \mathbf{12.85 \text{ cm}}$

Example 5: Calculate the perimeter of the object below. Use π as 3.14.

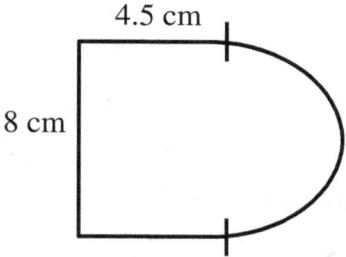

The total perimeter will be 8 cm + 4.5 cm + 4.5 cm + length of the circular end (semi-circle).

Length of the semi-circle
= circumference ÷ 2
$= (\pi \times 8) \div 2 = (3.14 \times 8) \div 2$
= 12.56 cm

Perimeter = 8 + 4.5 + 4.5 + 12.56
         = **29.56 cm**

EXERCISE 9A

1) **Group work:** Look around and find three circular objects. Measure the circular ends and record your answers. Remember to include the unit(s) you may have used.

In this exercise, you require a measuring instrument like a thread or a tape.

2) By measurements, work out the perimeter of the shapes below.

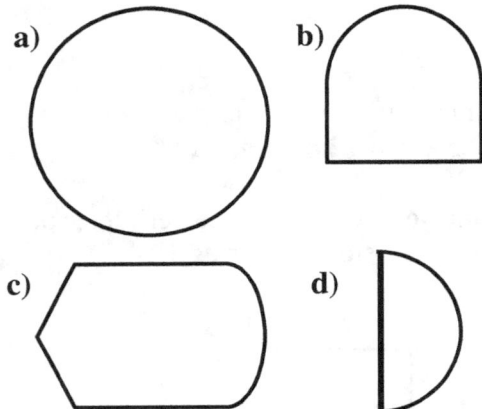

3) Calculate the circumference of the circles below. Use π as $\frac{22}{7}$.

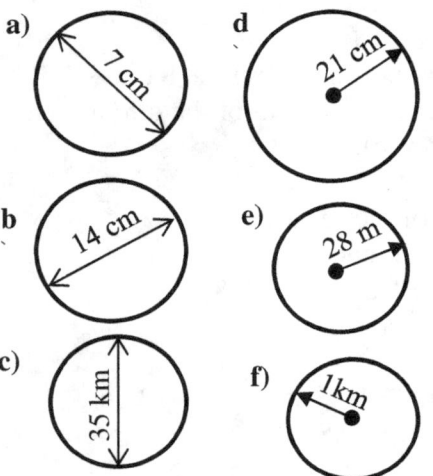

4) Calculate the perimeter of these shapes to one decimal place where possible and use π as $3\frac{1}{7}$.

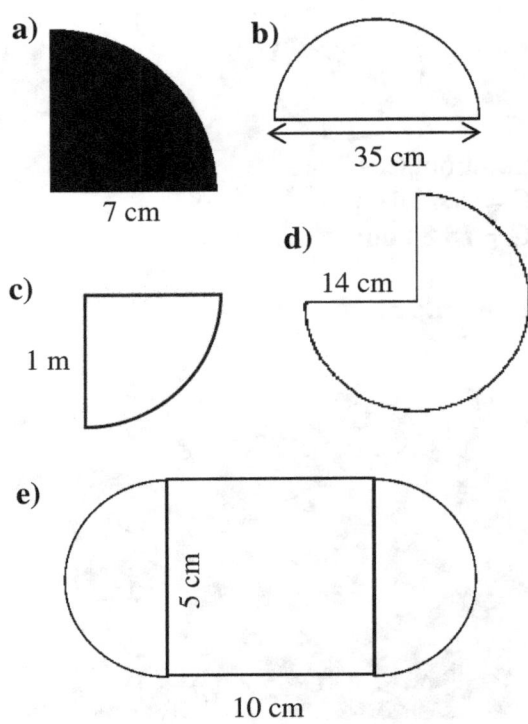

5) A circular piece of string has a radius of 21 cm. What is the circumference of the string? Use π as 3.14.

6) A duct tape is wound 20 times round a cylinder of diameter 8 cm. How long is the duct tape? Use π as 3.14

7) Calculate the diameter of a circle with a circumference of 7 cm. Take π as $\frac{22}{7}$.

8) A bicycle wheel has a diameter of 40 cm. How many complete revolutions does the wheel make when the bicycle travels 200 metres? Take π as 3.14.

9.2 AREA OF PLANE SHAPES

In simple terms, the **area** of a shape is the amount of **space** inside it.

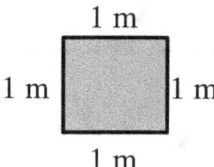

The square above has a length of 1 metre and covers an area of $1 \times 1 = \mathbf{1\ m^2}$.

Conventionally, the square is used as the shape for the unit of area. Similarly, squares will have units depending on the unit of its length. If a length is in centimetres, the unit of the area will be $\mathbf{cm^2}$.

AREA BY COUNTING SQUARES

When shapes are drawn on a centimetre square grid, the area can be worked out by simply **counting** the squares.

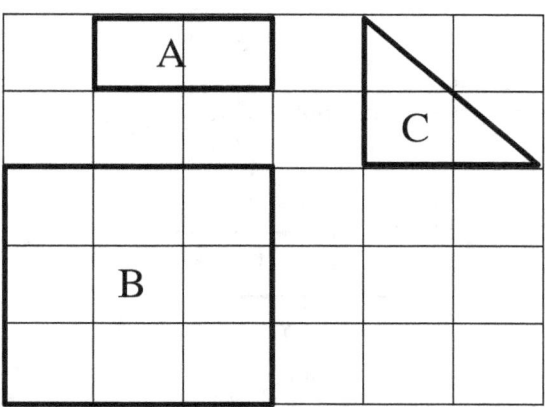

The area by counting the squares will be A = 2 cm², B = 9 cm² and C = 2 cm².

AREA BY ESTIMATE

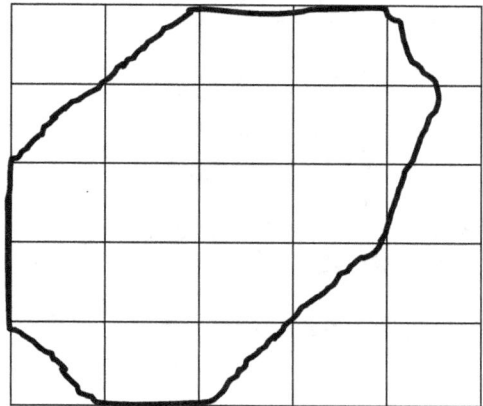

It will be almost impossible to find the exact area of the above shape. We estimate by counting the full squares and add up smaller parts to make up.

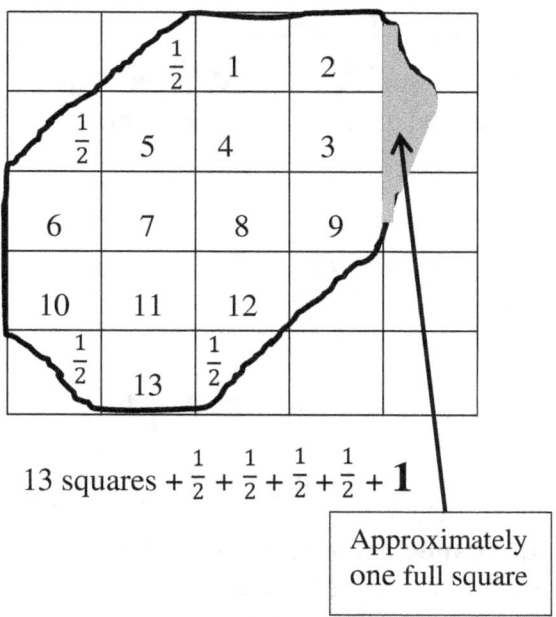

13 squares + $\frac{1}{2}$ + $\frac{1}{2}$ + $\frac{1}{2}$ + $\frac{1}{2}$ + **1**

Approximately one full square

The approximate area of the irregular shape above is **16 square units**.

AREA BY CALCULATIONS

AREA OF A RECTANGLE

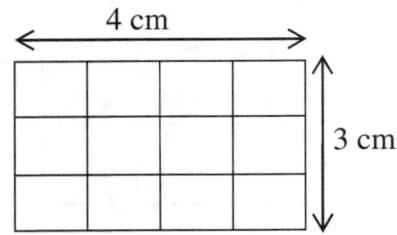

Area = length × breadth (width)
= 4 × 3 = **12 cm²**

AREA OF A SQUARE

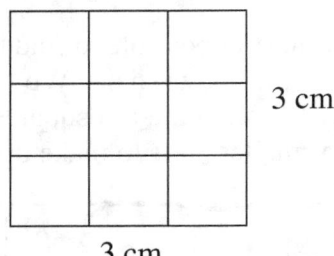

Area of a square = Length × length
= (Length)²
= 3 × 3 = 9 cm²

Example 1: The area of a rectangle is 42 cm². What is the width of the rectangle is the length is 7 cm?

Area = Length × breadth (width)
42 = 7 × width
Width = 42 ÷ 7 = **6 cm**

Example 2: Work out the area of a square with length 2 cm.
Area of a square = length × length
= 2 × 2
= **4 cm²**

Example 3: Work out the area of the shaded part.

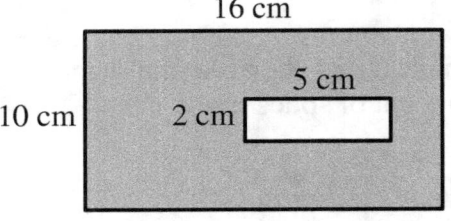

Area of big rectangle = 16 × 10
= 160 cm²
Area of small rectangle = 2 × 5
= 10 cm²
Area of shaded part = 160 − 10
= **150 cm²**

Example 4: Calculate the length of a side of a square with an area of 49 km².

The length = $\sqrt{49}$ = **7 km**

Check: 7 × 7 = 49

Example 5: Calculate the area of the shape below.

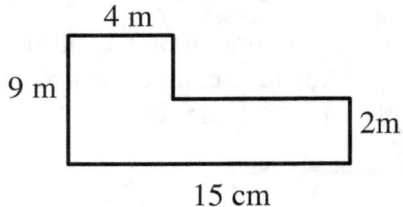

Solution
Split the shape into two rectangles.

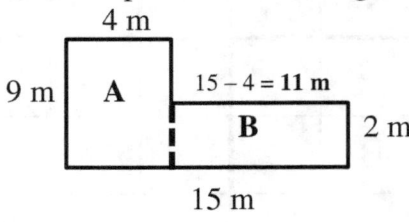

Area of A = 4 × 9 = 36 m²
Area of B = 11 × 2 = 22 m²
Total area = 36 m² + 22 m² = **58 m²**

EXERCISE 9B

1) Calculate the area of the rectangles and square below. All lengths in cm.

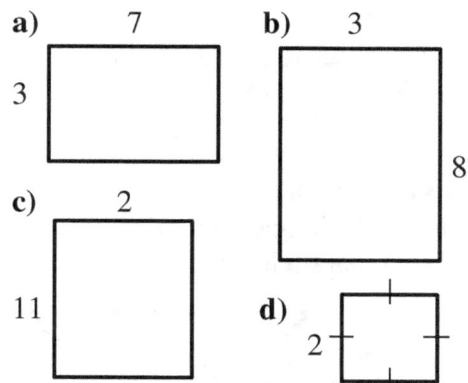

2) Below is a centimetre square grid. Work out the area of the shapes.

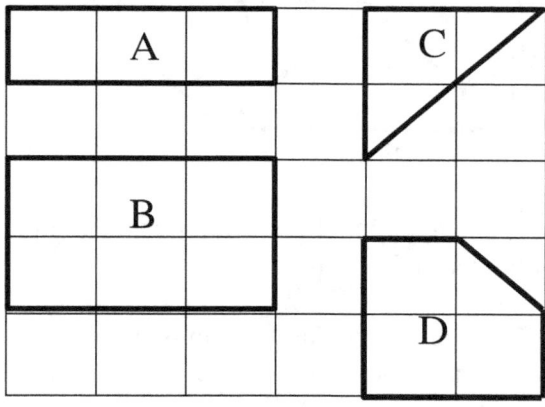

3) Estimate the area of the shapes drawn on a centimetre square grid below.

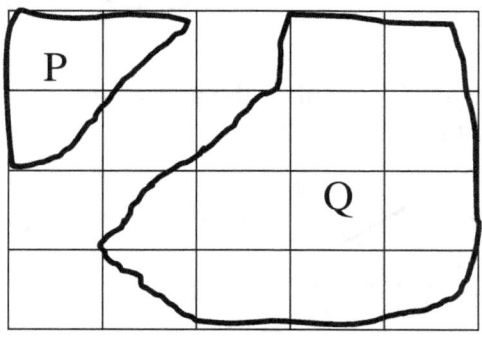

4) The area of a rectangle is 8 cm^2. What could be the length and breadth (width) of the rectangle?

5) A square has an area of 36 m^2. A length is 6 m, calculate the size of the other length.

6) Work out the area of the shaded region of the two rectangles.

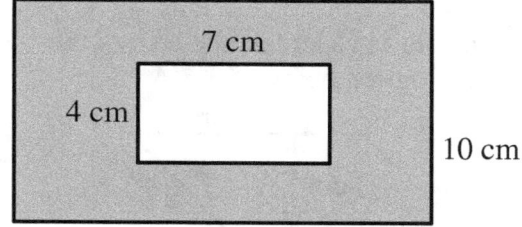

7) The floor below is to be carpeted.

a) Work out the area of the floor.
b) A vinyl carpet costs ₦1 500 per square metre. The cost of labour to lay the carpet is ₦4 300.
What is the total cost to successfully lay the carpet on the floor?

8) Complete the table of squares below.

Area	Length of side
1 cm^2	
9 m^2	
	8 cm
	4.3 cm
81 cm^2	

AREA OF A PARALLELOGRAM

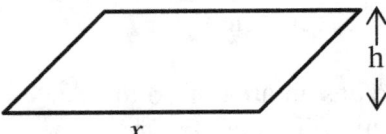

The area of a parallelogram is **base × perpendicular height**

Perpendicular means at 90°.

Example 1: Calculate the area of the parallelogram below.

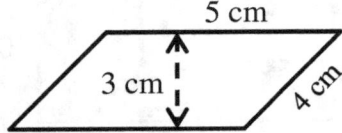

Area = 5 × 3 = **15 cm²**

Notice that 4 cm was not used. It is not perpendicular (at 90°) with the base.

Generally, do **not use** the slant height for calculating the area.

Example 2: Calculate the area of a parallelogram with a base of 10 cm and perpendicular height of 5 cm.

Area = 10 × 5 = **50 cm²**

Example 3: Calculate the height of the parallelogram below.

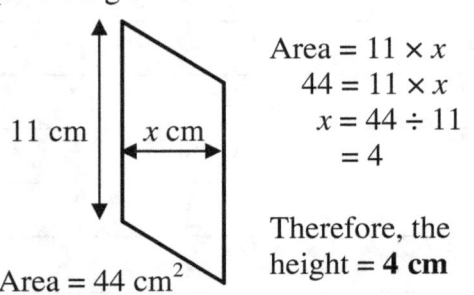

Area = 11 × x
44 = 11 × x
x = 44 ÷ 11
 = 4

Therefore, the height = **4 cm**

Useful formula:

Height of a Parallelogram = area ÷ its base

Base of a Parallelogram = area ÷ perpendicular height

AREA OF A TRIANGLE

Two identical triangles will always join to produce a parallelogram.

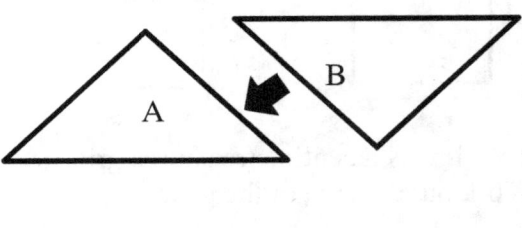

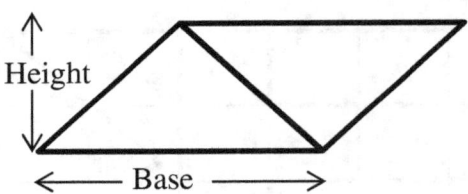

It goes to say that two triangles make up one parallelogram.

The area of a parallelogram is base × perpendicular height; therefore, the area of a triangle must be area of a parallelogram ÷ 2.

Area of a triangle = $\frac{1}{2}$ **base × perpendicular**

Example 4: Calculate the area of the area of the triangle below.

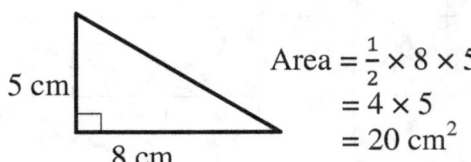

Area = $\frac{1}{2}$ × 8 × 5
 = 4 × 5
 = 20 cm²

Example 5: Calculate the area of triangle PQR.

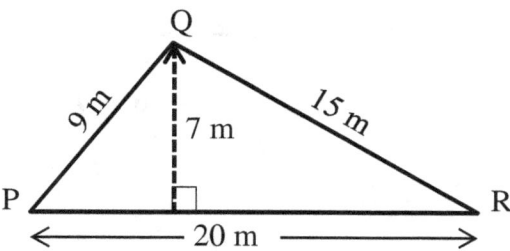

Area of PQR = $\frac{1}{2} \times 20 \times 7$
= **70 m²**

Notice that 9 m and 15 m were not used. Only the lengths that are perpendicular to each other are used for calculating the area of a triangle.

In this case, 7 m height is perpendicular to the base of 20 m.

Example 6: Work out the area of triangle STU

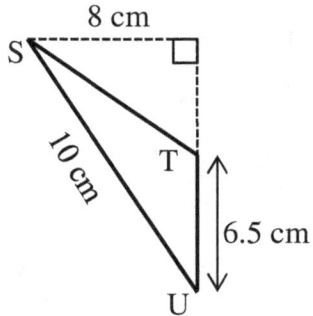

Area = $\frac{1}{2} \times 6.5 \times 8$
= **26 cm²**

EXERCISE 9C

1) All lengths are in cm. Work out the area of the triangles below.

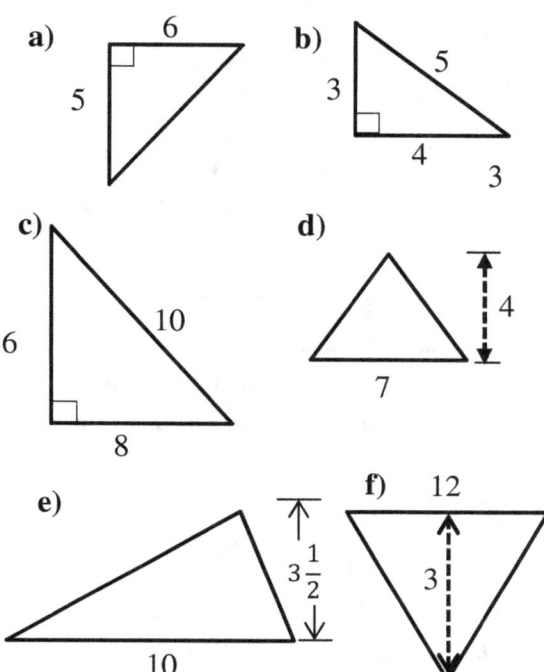

2) Calculate the total area of the quadrilateral ABCD. All lengths are in metres.

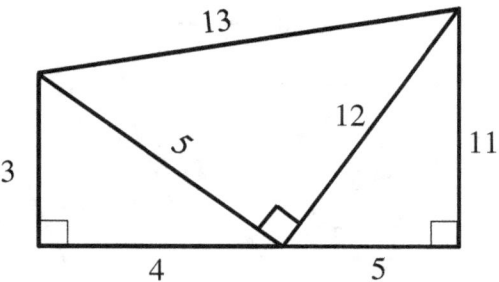

3) A triangle has an area of 30 cm² with a base length of 10 cm. Calculate the perpendicular height of the triangle.

4) List a possible base length and height of a triangle with area 28 cm².

5) All shapes below are parallelograms. Calculate the area of each shape. All lengths are in cm.

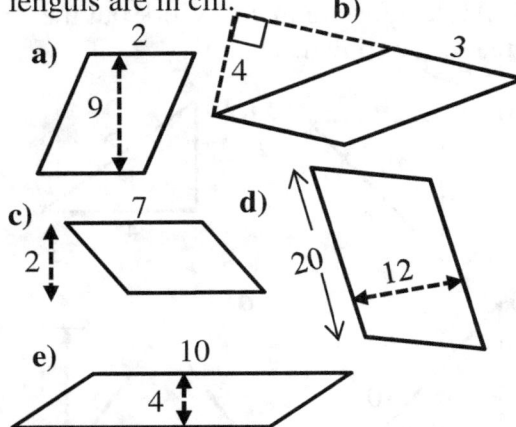

6) All lengths are in cm. Work out the area of the parallelograms and triangles below.

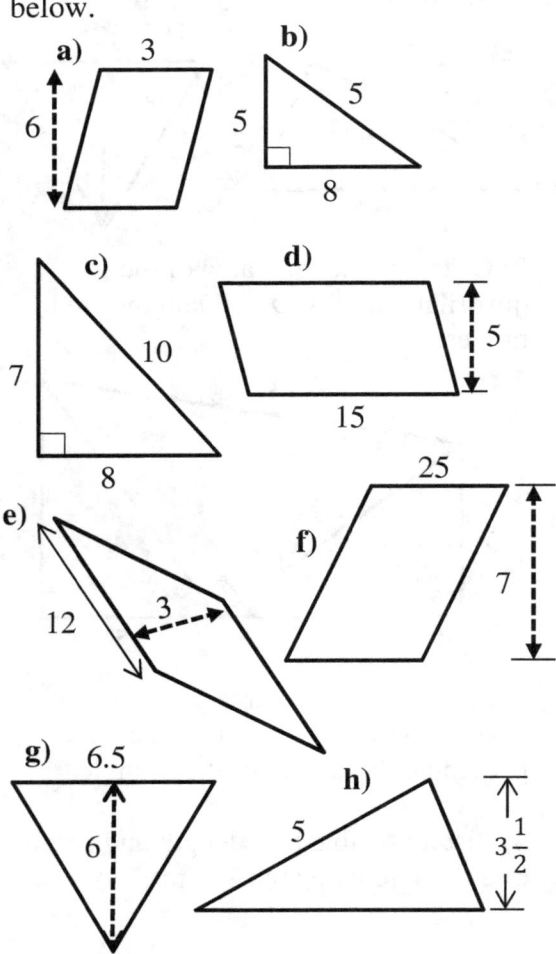

7) In questions 7a - d, calculate the base, y.

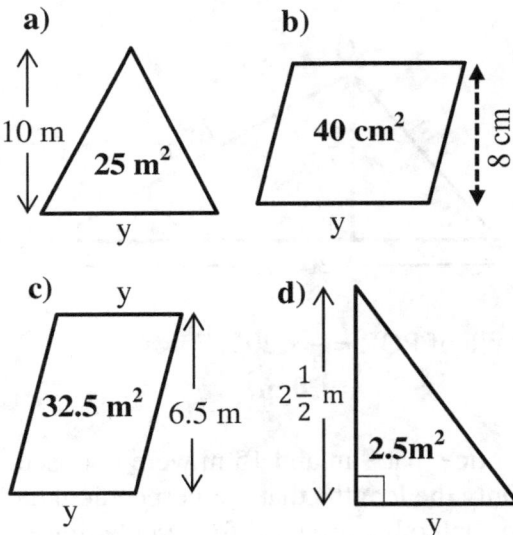

8) Calculate the area of the plane shapes below. All lengths are in cm.

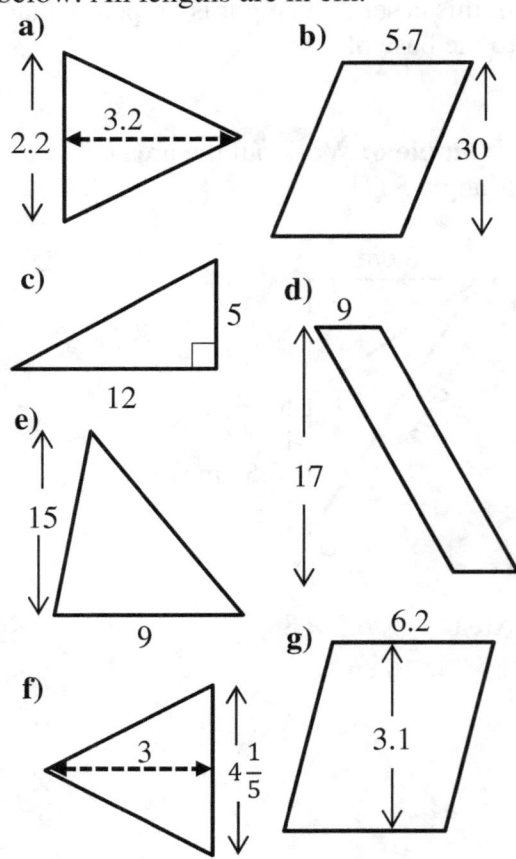

AREA OF A TRAPEZIUM

A trapezium is a four-sided straight shape with **one pair** of parallel sides.

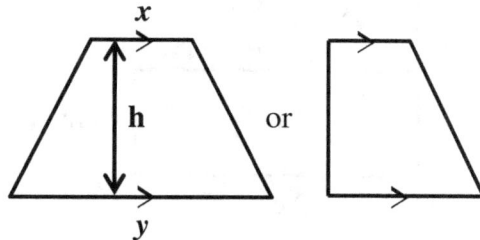

If two identical trapezia are put together, they form a parallelogram.

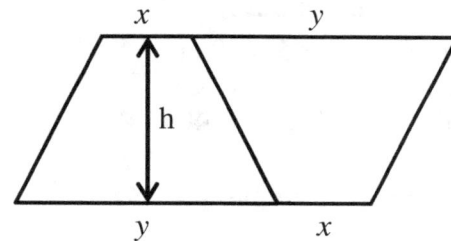

Remember:
Area of a parallelogram = base × perpendicular height $(x + y) \times h$

Since two trapezia make up one parallelogram, the area of a trapezium
$= \frac{1}{2} \times$ sum of parallel sides × height
$= \frac{1}{2} \times (x + y) \times h$

Example 1: Calculate the area of the trapezium below.

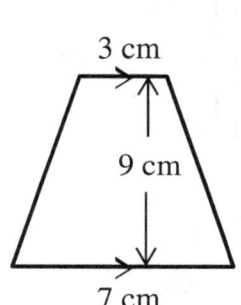

Add the parallel sides $3 + 7 = 10$
Divide by 2
$10 \div 2 = 5$
Multiply by perpendicular height
$5 \times 9 = \mathbf{45 \text{ cm}^2}$

Example 2: Calculate the area of the trapezium below.

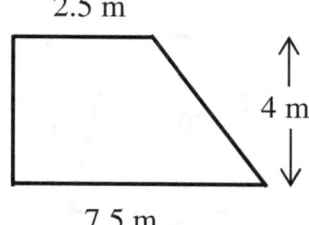

Area $= \frac{1}{2} \times (2.5 + 7.5) \times 4$
$= \frac{1}{2} \times 10 \times 4$
$= \mathbf{20 \text{ m}^2}$

EXERCISE 9D

1) All the lengths are in cm. Work out the area of the trapezia below.

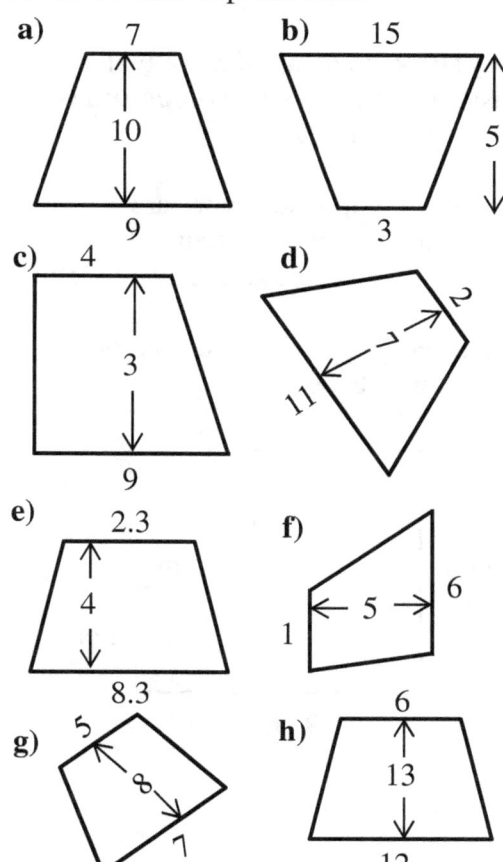

2) Calculate the length of the missing sides in the trapezia below. All the lengths are in *m*.

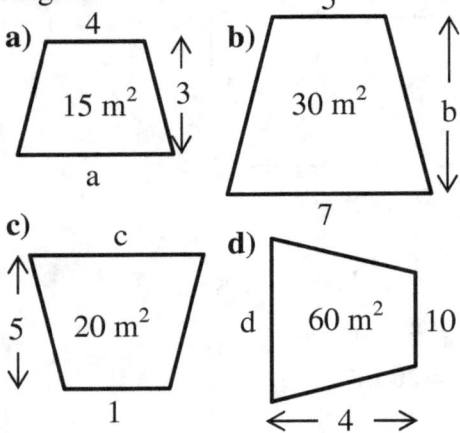

9.3 COMPOUND SHAPES

Any shape made up of more than one basic figure is called a **compound** or **composite** shape.

Areas of compound shapes are calculated by splitting them into its individual shapes and then add together.

Example 1: Calculate the area of the compound shape below.

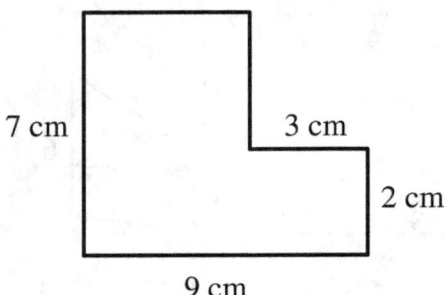

Solution: Split the shape into two rectangles.

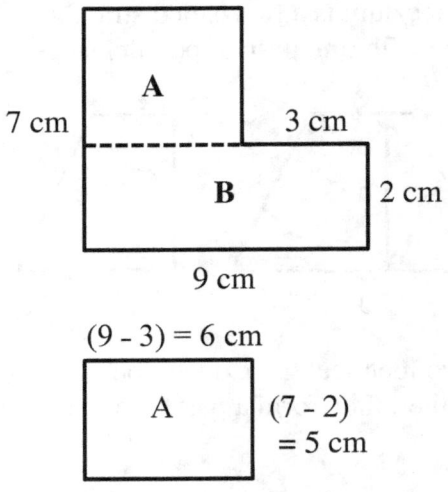

Area of A = 6 × 5 = 30 cm²
Area of B = 9 × 2 = 18 cm²
Total area = 30 + 18 = **48 cm²**

Example 2: Work out the area of the shape.

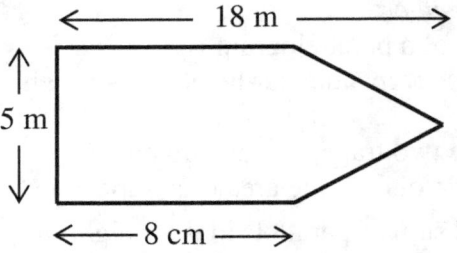

Solution: Split the shape into a rectangle and triangle.

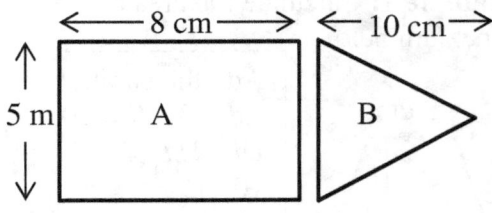

Area of A = 8 × 5 = 40 cm²
Area of B = $\frac{1}{2}$ × 5 × 10 = 25 cm²
Total area = 40 + 25 = **65 cm²**

EXERCISE 9E

1) Calculate the area of the compound shapes. All lengths are in cm.

3) Obiora showed his working out for the area of this shape.

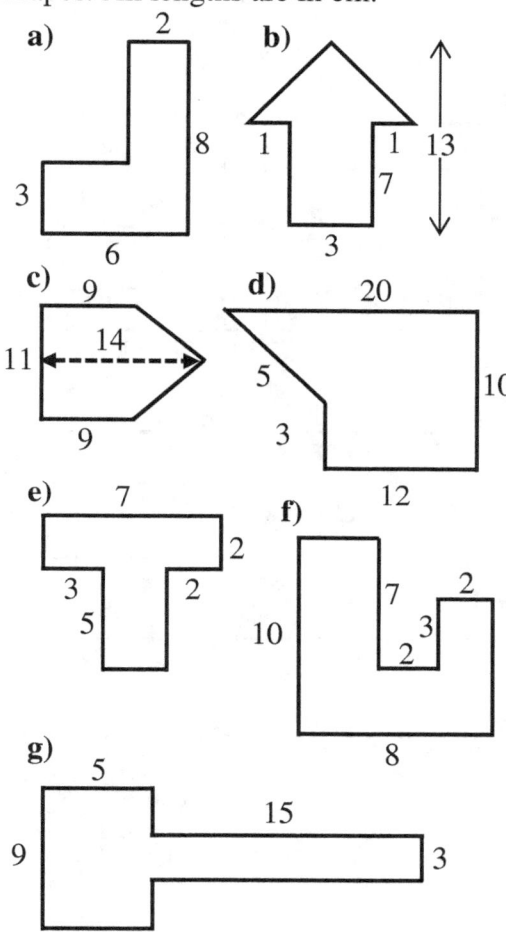

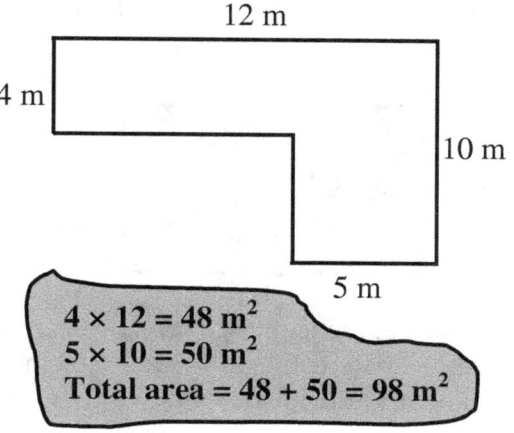

4 × 12 = 48 m²
5 × 10 = 50 m²
Total area = 48 + 50 = 98 m²

Show that Obiora is wrong.

4) Calculate the area of the shapes below. All lengths are in metres.

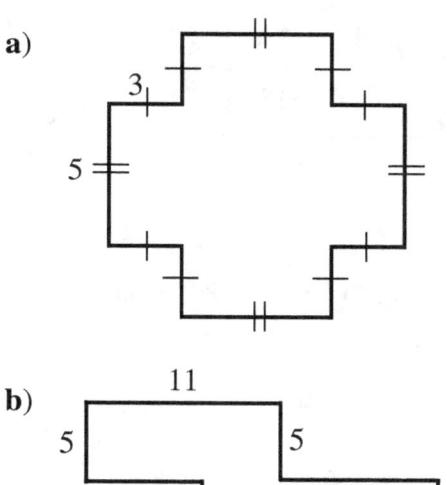

2) The shape below contains two identical triangles at both ends and a rectangle. Calculate the area of the shape. All lengths are in centimetres.

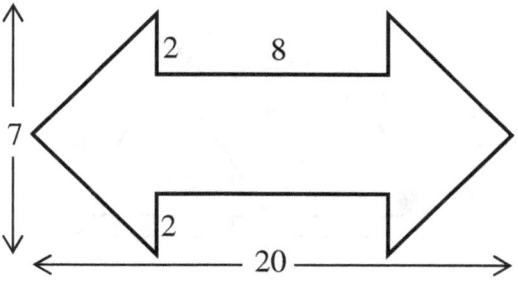

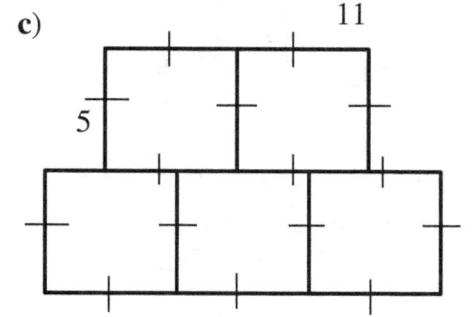

9.4 SHADED AREA

Example 1: Calculate the shaded area.

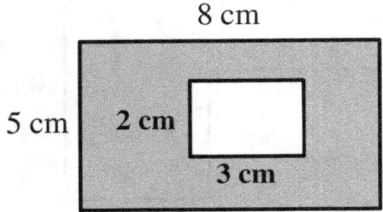

Area of big rectangle = 8 × 5 = 40 cm²
Area of small rectangle = 3 × 2 = 6 cm²
So, are of shaded part = 40 - 6 = 34 **cm²**

Example 2: Calculate the shaded area

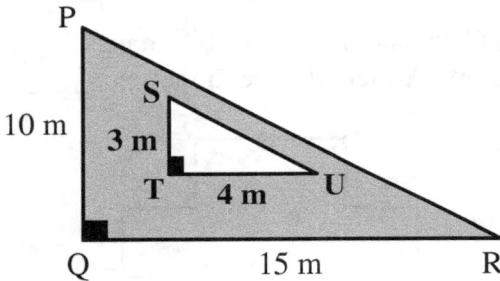

Area of $\triangle$PQR = $\frac{1}{2}$ × 10 × 15 = 75 m²
Area of $\triangle$STU = $\frac{1}{2}$ × 4 × 3 = 6 m²
Shaded area = 75 - 6 = **69 m²**

Example 3: Work out the shaded area.

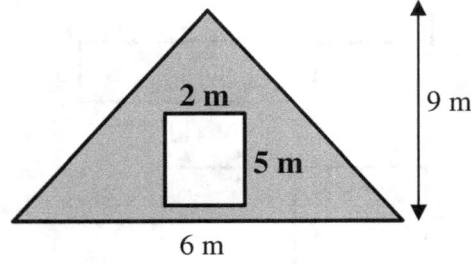

Area of triangle = $\frac{1}{2}$ × 6 × 9 = 27 m²
Area of rectangle = 2 × 5 = 10 m²
Shaded area = 27 - 10 = **17 m²**

EXERCISE 9F

1) Calculate the shaded area of the shapes below. All lengths are in metres.

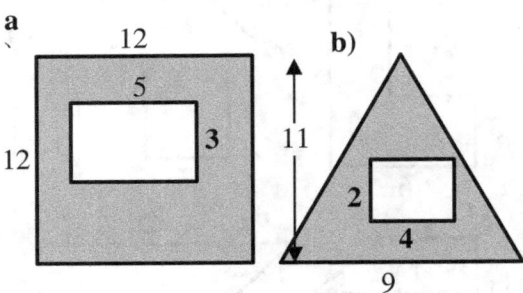

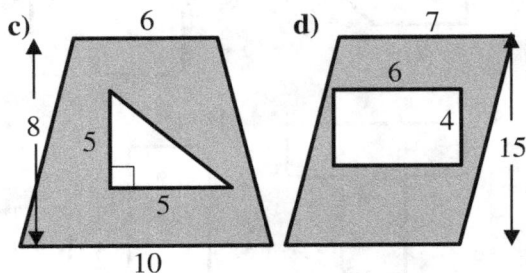

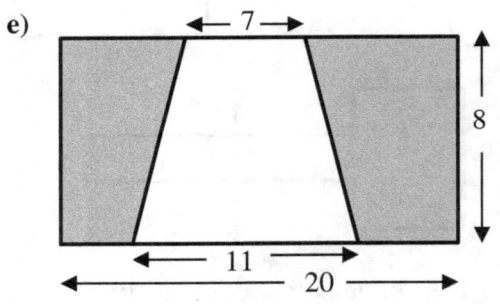

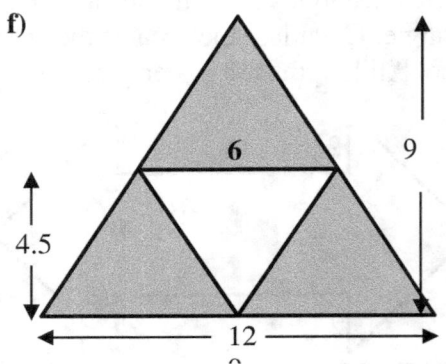

9.5 AREA OF A CIRCLE

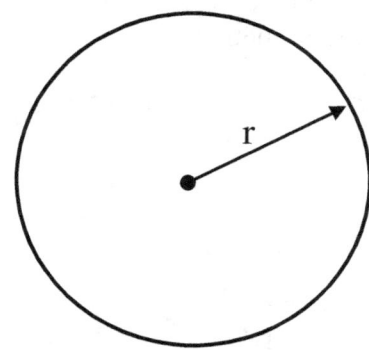

Divide the circle above into different sectors, cut them out and rearrange to form a rectangle. The length of the rectangle will be half the circumference, that is $\frac{1}{2} \times \pi \times$ diameter.
However, diameter = 2 × radius(r)

Length = $\frac{1}{2} \times \pi \times 2 \times r = \pi r$

The shape formed is **close to** a rectangle as shown below.

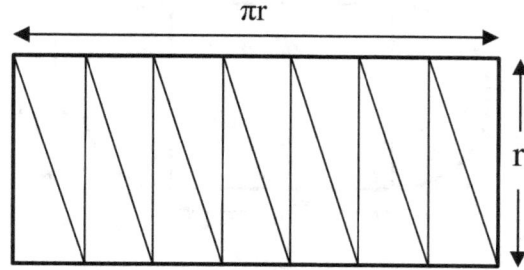

Therefore, the **area of the circle** which is also the area of the rectangle will equal $\pi \times r \times r = \pi r^2$

Area of circle = πr^2
Always remember to use the radius instead of the diameter when working out the area of a circle.

Example 1: A circle has a radius of 7 cm. Calculate the area of the circle. Use π as $\frac{22}{7}$.

Area = $\pi r^2 = \frac{22}{7} \times 7^2$
     = $\frac{22}{7} \times 7 \times 7 =$ **154 cm²**

Example 2: Calculate the area of the circle. Use π as $\frac{22}{7}$ and round to two decimal places.

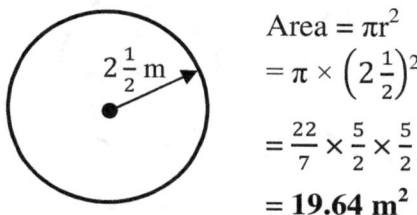

Area = πr^2
     = $\pi \times \left(2\frac{1}{2}\right)^2$
     = $\frac{22}{7} \times \frac{5}{2} \times \frac{5}{2}$
     = **19.64 m²**

Example 3: Work out the area of the circle. Use π as 3.14 and round to one decimal place.

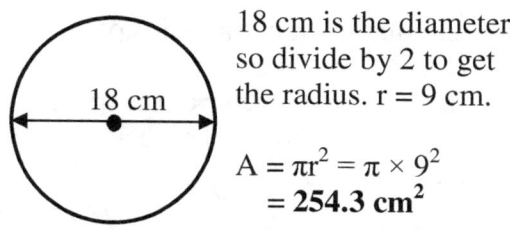

18 cm is the diameter, so divide by 2 to get the radius. r = 9 cm.

A = $\pi r^2 = \pi \times 9^2$
  = **254.3 cm²**

Example 4: Calculate the area of the semicircle. Use π as $\frac{22}{7}$.

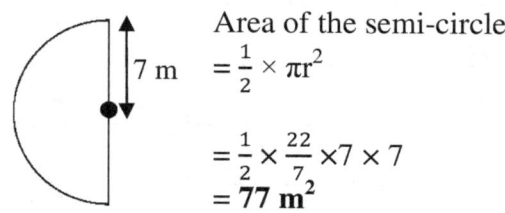

Area of the semi-circle
= $\frac{1}{2} \times \pi r^2$

= $\frac{1}{2} \times \frac{22}{7} \times 7 \times 7$
= **77 m²**

EXERCISE 9G

1) Calculate the area of the shapes below. Use π as $\frac{22}{7}$ and round to one decimal place where possible.

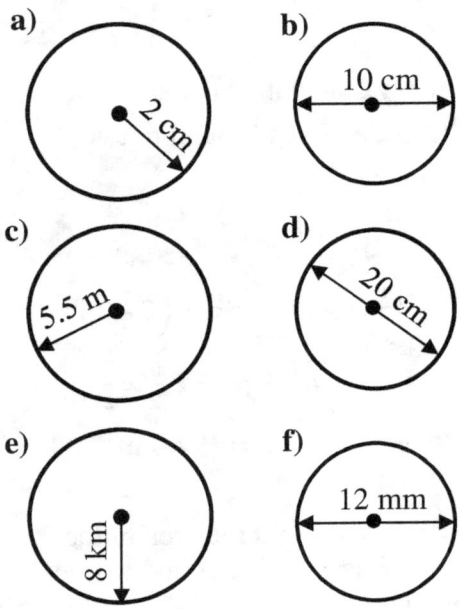

a) 2 cm
b) 10 cm
c) 5.5 m
d) 20 cm
e) 8 km
f) 12 mm

2) A circular track has a radius of 70 m.
a) Calculate the diameter of the track.
b) Calculate the area of the track.
Use π as 3.14.

3) Copy and complete the table below. Use π as $\frac{22}{7}$.

Diameter(m)	Radius(m)	Area(m²)
14		
42		
	21	
	14	

4) Calculate the area of the semicircles and quadrants. Give your answers to one decimal place where possible. All lengths are in metres. Use π as 3.14.

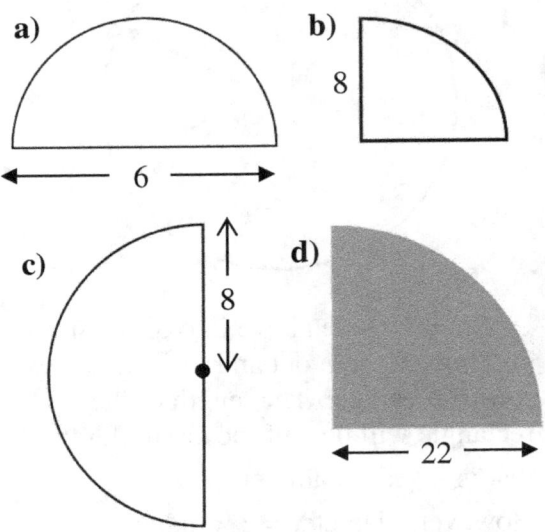

a) 6
b) 8
c) 8
d) 22

5) Calculate the area of the compound shapes made up of semicircles and rectangles. Use π as 3.14.

a)
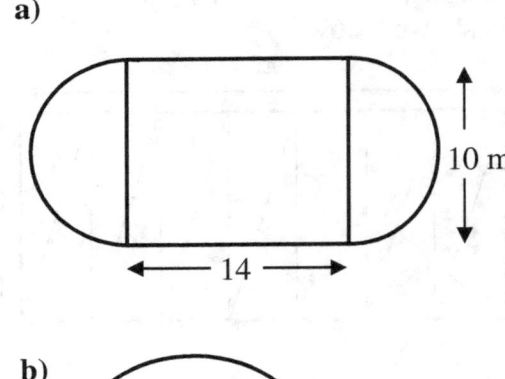
14, 10 m

b)
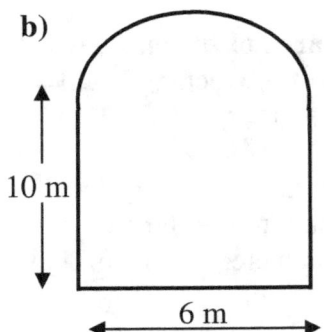
10 m, 6 m

Example 5: Calculate the area of the shaded part. Use π as 3.14.

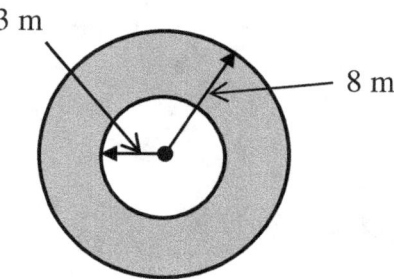

Area of the big circle = πr²
= 3.14 × 8²
= 200.96 m²

Area of the small circle = πr²
= 3.14 × 3²
= 28.26 m²

Area of shaded part = 200.96 - 28.26
= **172.7 m²**

Example 6: A rectangle is inscribed in a circle shown below. Calculate the shaded area to one decimal place. Take π as $\frac{22}{7}$.

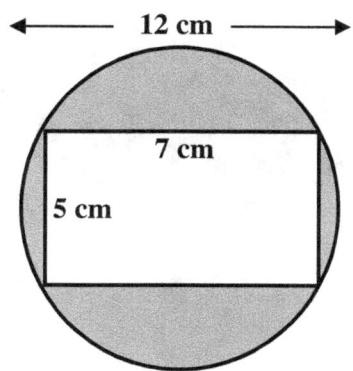

Area of rectangle = 7 × 5 = 35 cm²

Area of circle = πr²
...But diameter = 12 cm, therefore radius = 12 ÷ 2 = 6 cm.

Area = $\frac{22}{7}$ × 6² = $\frac{22}{7}$ × 6 × 6
= 113.1428571 cm²

Shaded area = 113.1428571 - 35
= **78.1 cm²**

EXERCISE 9H

1) Calculate the area of the shaded part of each of the diagrams below. Take π as 3.14 and round your answers to one decimal place where possible.

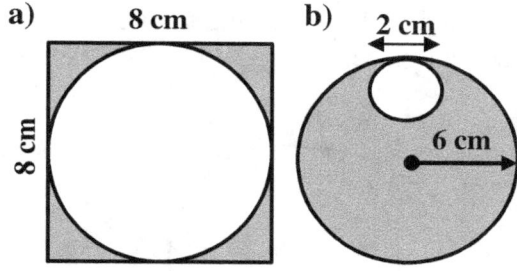

2) Three identical circles are placed in a rectangular box as shown below.

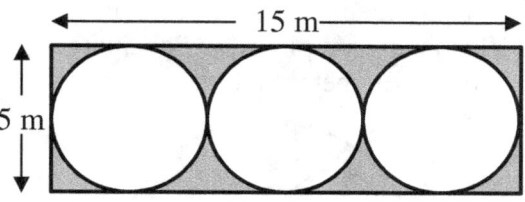

a) Work out the area of a circle.
b) Work out the total area of the three circles.

c) Work out the shaded area.
(Take π as 3.14 and round to two decimal places where possible)

3) Calculate the percentage of the shaded area. Take π as 3.14.

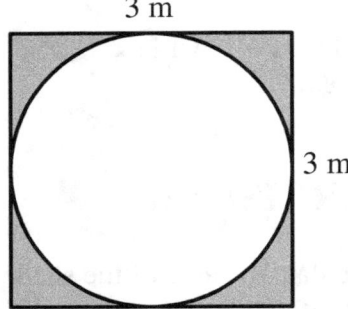

4) A circle of diameter 20 cm fits inside a semi-circle. Calculate the shaded area. Take π as 3.14 and round to one decimal place.

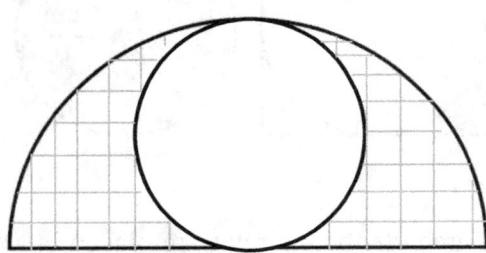

5) Three identical circles of diameter 10 cm each are placed inside a large circle of diameter 30 cm.

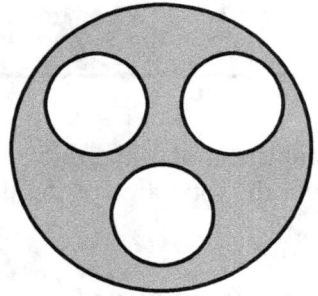

a) Calculate the area of the shaded part.
b) Work out the percentage of the shaded part. Use π as 3.14.

6) Calculate the area of the shaded part. Use π as 3.142.

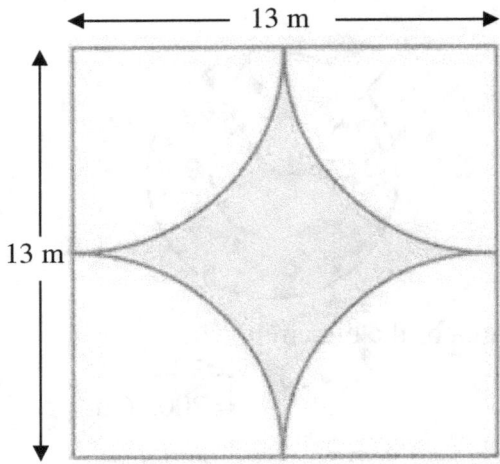

7) You have three semi-circles. The diameter of the large semicircle is 30 m. Work out the area of the shaded part and round your answer to one decimal place.
Use π as 3.142.

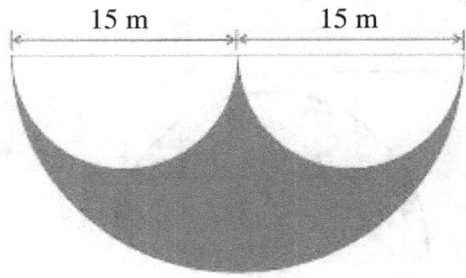

Chapter 9 Review Section
Assessment

1) What number belongs in each box?

a)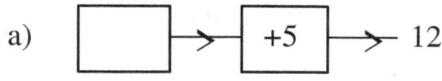

b)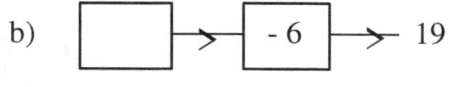

..................2 marks

2) What rule goes in the box?

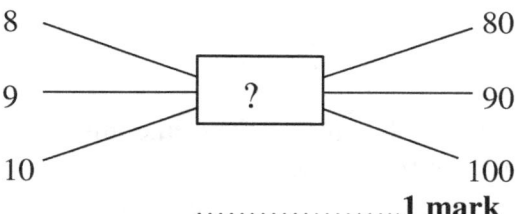

....................1 mark

3) Copy and complete the table for the function machine.

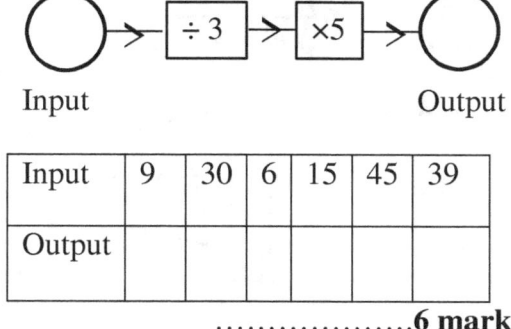

....................6 marks

4) Complete the table for the magic squares below.

a)

8		10
	7	
4		

b)

9		
	8	
5		7

.... 4 marks

5) On the centimetre grid below, calculate the perimeter of the shapes.

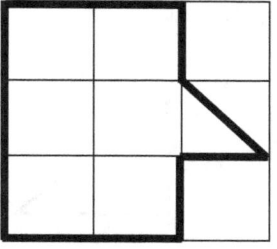

....................1 mark

6) Calculate the perimeter of the object below. Use π as 3.14.

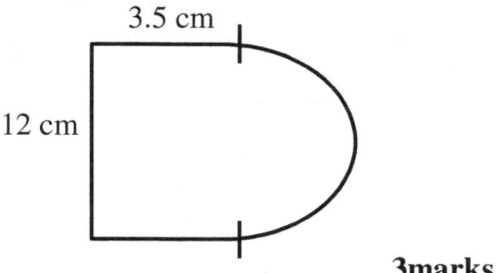

............3marks

7) Calculate the circumference of the circles below. Use π as $\frac{22}{7}$.

a)   b)

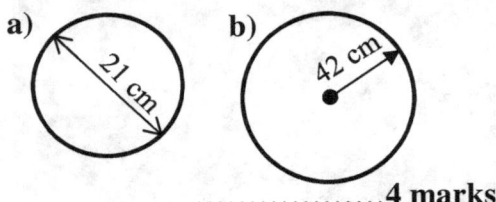

..................4 marks

8) Work out the area of the shaded part.

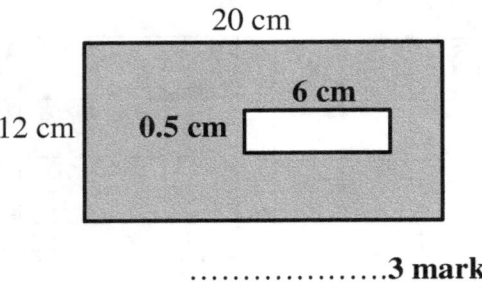

..................3 marks

9) All lengths are in cm. Work out the area of the triangles below.

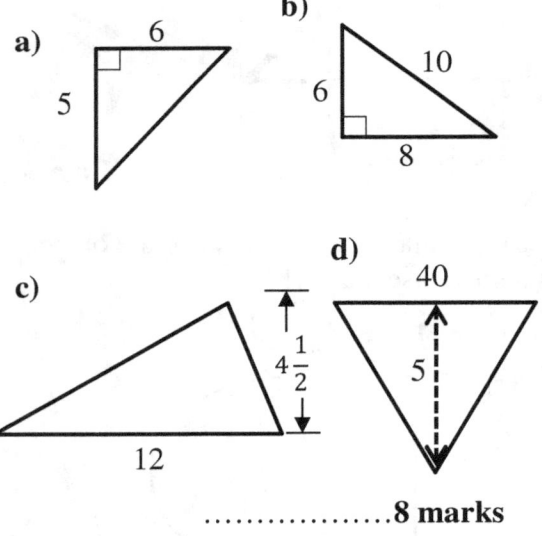

..................8 marks

10) A triangle has an area of 21 cm² with a base length of 7 cm. Calculate the perpendicular height of the triangle.

..................2 marks

11) All lengths are in cm. Work out the area of the parallelograms and triangles below.

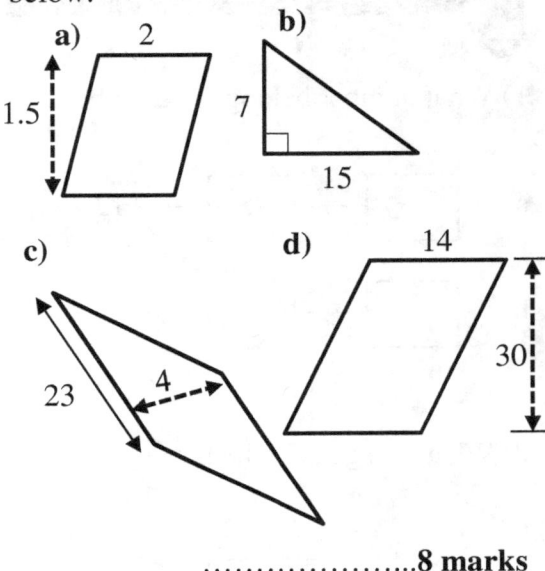

..................8 marks

12) Calculate the length of the missing sides in the trapezia below.

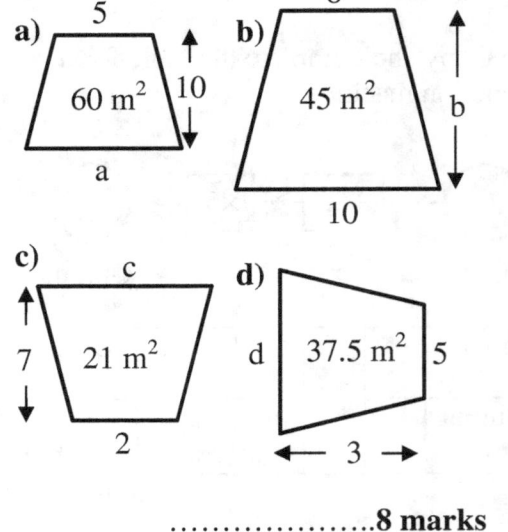

..................8 marks

13) Calculate the area of the compound shapes. All lengths are in cm.

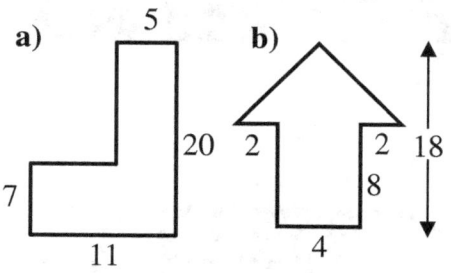

a) b)

…………………..4 marks

14) Calculate the area of the shapes below. Use π as $\frac{22}{7}$ and round to one decimal place where possible.

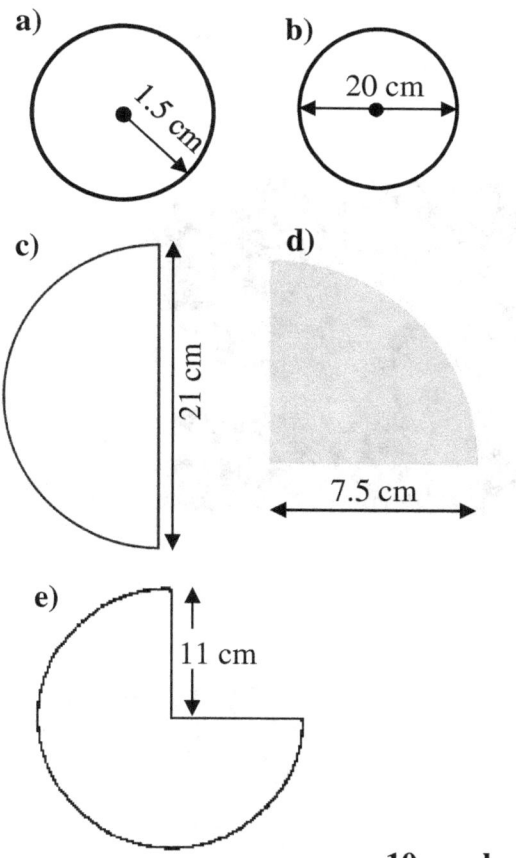

a) b)

c) d)

e)

….10 marks

15) Four identical white squares of length 7 cm each are placed inside a large circle of diameter 25 cm.

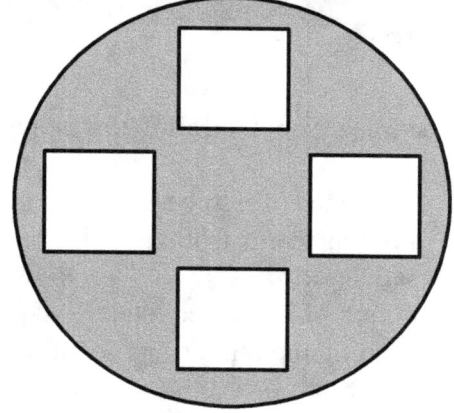

a) Calculate the area of the shaded part.
……………….. 3 marks

b) Work out the percentage of the shaded part. Use π as 3.14.
………………..2 marks

16) An arts theatre has a circular floor as the base. The diameter is 40 metres. A tile specialist charges ₦2 500 per square metre to tile the whole floor. Use π as $\frac{22}{7}$ and work out the cost of tiling the entire floor of the art theatre.
……………….. 4 marks

17) A school track is made up of two straights and two semi-circular ends. Calculate the enclosed area within the track. Take π as 3.14 and round to 2 decimal places.

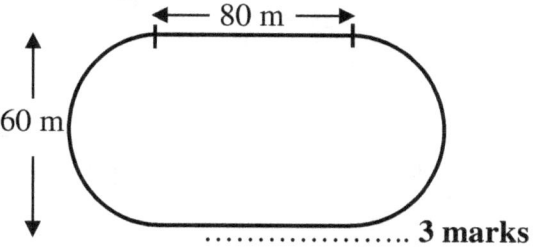

……………….. 3 marks

10 Sequences

This section covers the following topics:

- Sequence of numbers
- Finding terms from the nth term of a sequence
- Graph of linear sequence
- Finding the nth term of a number sequence
- Quadratic sequences

LEARNING OBJECTIVES

By the end of this unit, you should be able to:

a) Produce sequence of numbers
b) Understand terms and nth terms
c) Find a formula for the nth term of a linear sequence
a) Draw graphs to show linear sequences
b) Find quadratic terms and nth terms

KEYWORDS

- Sequence
- Terms
- Nth terms
- Patterns
- Rule
- Quadratic

10.1 LINEAR SEQUENCE

A number **sequence** is a list of numbers that follow a particular rule (pattern). It is linear if it can be represented on a straight line graph. It increases or decreases in equal-sized steps.

Each number in a sequence is called a **term**.

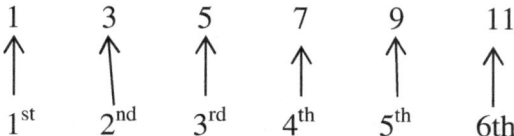

Consecutive terms are terms that are next to each other. They are usually separated by commas. In the example above, 1 and 3, 3 and 5, 5 and 7, 7 and 9, 9 and 11 are consecutive odd numbers.

There are lots of different number patterns. A sequence can continue forever (infinite). For example, 2, 4, 6, 8, 10…

Sequences follow a **rule**.

The sequence 4, 8, 12, 16, 20 will continue if we keep on adding 4.

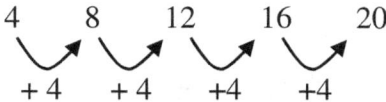

The rule for this sequence is to **add 4** each time. It is often called **term-to-term** rule.
Therefore, 4, 8, 12, 16, 20 is a linear sequence because it has a common difference of 4.

Some common sequences to remember are:

1, 3, 5, 7, 9…odd numbers
1, 4, 9, 16, 25… square numbers
2, 4, 6, 8, 10 … even numbers
10, 100, 1000, 10000… powers of ten
1, 3, 6, 10, 15… triangular numbers
1, 1, 2, 3, 5, 8… Fibonacci sequence

TRIANGULAR NUMBERS

Triangular numbers are formed by using consecutive numbers. The first triangular number is 1.

1
$1 + 2 = \mathbf{3}$
$1 + 2 + 3 = \mathbf{6}$
$1 + 2 + 3 + 4 = \mathbf{10}$
$1 + 2 + 3 + 4 + 5 = \mathbf{15}$
$1 + 2 + 3 + 4 + 5 + 6 = \mathbf{21}$

FIBONACCI NUMBERS (SEQUENCE)

Fibonacci was an Italian mathematician regarded as the "Greatest European mathematician of the middle ages." His full name was Leonardo Pisano.

Source: Wikipedia

Fibonacci numbers are:1,1,2,3,5,8….

Each term in the Fibonacci sequence is the sum of the previous two terms.

1
$1 + 0 = 1$
$1 + 1 = 2$
$2 + 1 = 3$
$3 + 2 = 5$
$5 + 3 = 8$
$8 + 5 = 13$

Example 1: For each of the sequence, work out the missing terms.
    a) 4, 7, 10, 13, ..., ..., ...
    b) 1, ..., 13, ..., ..., 31, 37

Answers:
a) The rule is to add 3 each time, so the missing numbers are 16, 19 and 22.

b) Look at the 6th and 7th terms. The difference between 31 and 37 is 6. The rule is + 6 each time. The missing numbers are: 7, 19, 25

EXERCISE 10A

1) Find the next two terms in each sequence.
a) 4, 5, 6, 7, ___, ___
b) 7, 9, 11, 13, ___, ___
c) 1, 6, 11, 16, ___, ___
d) 4, 7, 13, 25, ___, ___
e) 3, 7, 11, 15, ___, ___
f) 22, 16, 10, 4, ___, ___
g) 5, 6, 8, 11, ___, ___
h) 7, 6.9, 6.8, 6.7, ___, ___
i) 2, 4, 8, 16, ___, ___
j) 1, 4, 8, 13, ___, ___
k) 3, 9, 27, ___, ___
l) $\frac{1}{2}, \frac{2}{6}, \frac{3}{18}, \frac{4}{54}$, ___, ___

2) Write the term-to-term rule for questions 1a, b and c.

3) The term-to-term rule is +7. Write two different sequences that fit the rule.

4) For the sequence 4, 9, 14, 19, write down
a) the term-to-term rule
b) the first term
c) the 10th term

5) You are given the first term and the rule of different sequences. Write down the first four terms of each sequence.

First term	Rule
4	Add 6
47	Add 15
2	Subtract 7
3	Triple
200	Divide by 2

6) Copy and complete the sequences.

$3 \times 88 = 264$
$4 \times \square = 352$
$\square \times 88 = 440$
$6 \times \square = 528$
$7 \times 88 = \square$
.   .   .
.   .   .
$19 \times 88 = \square$

7) Find i) the rule ii) the missing numbers for each sequence

a) 7, 15, 31, 63, ___
b) 225, 180, 135, ___, ___
c) ___, 80, 77, 74, 71
d) 0.3, ___, 30, 300, ___
e) 8, 9, 11, 14, ___

8) Look at the sequence below.
$5^2 =$ 25
$55^2 =$ 3025
$555^2 =$ 308025
$5555^2 =$ 30858025

What is the value of 55555^2?

TERM NUMBER FROM A FORMULA

From the sequence, 4, 7, 10, 13, 16, a formula can be written as $3 \times n + 1$ or $3n + 1$ where n is the number of terms. Using the formula $3n + 1$, you could find any number of terms. For example, the 2000th term will be $3 \times 2000 + 1 =$ **6001**

Example 2: Write down the first three terms of the formula (nth term) $8n + 5$.

First term is when n = 1,
$(8 \times 1) + 5 = \mathbf{13}$.
Second term is when n = 2,
$(8 \times 2) + 5 = \mathbf{21}$.
Third term is when n = 3,
$(8 \times 3) + 5 = \mathbf{29}$
Therefore, the first three terms are: 13, 21 and 29

Example 3: Find the first three terms of the formula 5n.
First term: $5 \times 1 = \mathbf{5}$, second term: $5 \times 2 = \mathbf{10}$, third term: $5 \times 3 = \mathbf{15}$

Example 4: Find the 20th term of the nth term $2n - 3$.
20th term = $(2 \times 20) - 3 = \mathbf{37}$

Example 5: Write down the first five terms of the nth term n - 5

First term: $1 - 5 = \mathbf{-4}$
Second term: $2 - 5 = \mathbf{-3}$
Third term: $3 - 5 = \mathbf{-2}$
Fourth term: $4 - 5 = \mathbf{-1}$
Fifth term: $5 - 5 = \mathbf{0}$
The first 5 terms are: -4, -3, -2, -1, 0

Example 6: Find the 10th term of the nth term formula, $3n^2$.
Note: $3n^2 = 3 \times n^2$. Therefore, the 10th term will be $3 \times 10^2 = 3 \times 100 = \mathbf{300}$

Example 7: Write down the first three terms of the nth term $4n^2 + 1$.

1st term: $4 \times (1)^2 + 1 = 4 \times 1 + 1 = \mathbf{5}$
2nd term: $4 \times (2)^2 + 1 = 4 \times 4 + 1 = \mathbf{17}$
3rd term: $4 \times (3)^2 + 1 = 4 \times 9 + 1 = \mathbf{37}$

A common mistake would be to multiply 4 by n and then square it. It is wrong.
Example: To find the 1st term of $4n^2 + 1$, a common mistake will be $(4 \times 1)^2 + 1 = 4^2 + 1 = 17$. ✘
This is wrong.

Example 8: For the sequence 11, 14, 17, 20, write down a) $T_{(1)}$, b) $T_{(2)}$, c) $T_{(10)}$

Solution: $T_{(1)}$ means the 1st term, $T_{(2)}$ means the 2nd term and $T_{(10)}$ means the 10th term.

$T_{(1)} = 11$
$T_{(2)} = 14$
$T_{(10)} = 38$…….from continuing the sequence.

EXERCISE 10B

1) Write down the first **three** terms of the sequences produced by these nth term formulae.

a) 2n
b) 14n
c) 2n + 2
d) 3n – 1
e) 4n + 1
f) n^2
g) 3 – n
h) 4 + 2n
i) 10 – 5n
j) 8 + 3n
k) $n^2 + 7$
l) $9 + n^2$
m) 20n - 3
n) $2n^2$
o) $3n^2 - 1$
p) $4n^2 + 10$
q) 13 – 2n
r) $\dfrac{1}{n^2}$

2) Write down the first **five** terms of the sequences below.

a) First number is 40, the rule is add 4
b) First number is 7, the rule is add 5
c) First number is 13, the rule is minus 2
d) First number is 2, rule is subtract 3
e) First number is -7, the rule is add 9

3) The third term of a sequence is 16, fourth term, 22 and fifth term, 28. Find the first and second terms.

4) For the sequence -14, -9, -4, 1…
a) Find the term-to-term rule
b) Find the next term in the sequence
c) Find the 100th term in the sequence.

5) For the sequence below
  7, 11, 15, 19, 23…,
Write down
a) $T_{(1)}$ b) $T_{(2)}$ c) $T_{(97)}$

6) Each number in this sequence is -3 times the previous number.
  -4, 12, -36, …
Write down
a) the 4th term b) the 6th term

7) Fill in the missing numbers

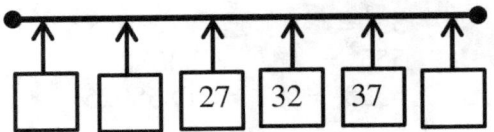

8) Write down the first three terms of the sequences where $T_{(n)}$ is:

a) 2n + 11
b) n – 3
c) n^2
d) $3n^2 + 3n$

SEQUENCES AND GRAPHS

Number patterns can be drawn on a graph. For example, the number pattern 3, 6, 9, 12, 15 can be shown on a graph as follows.

Term	1	2	3	4	5
Magnitude	3	6	9	12	15

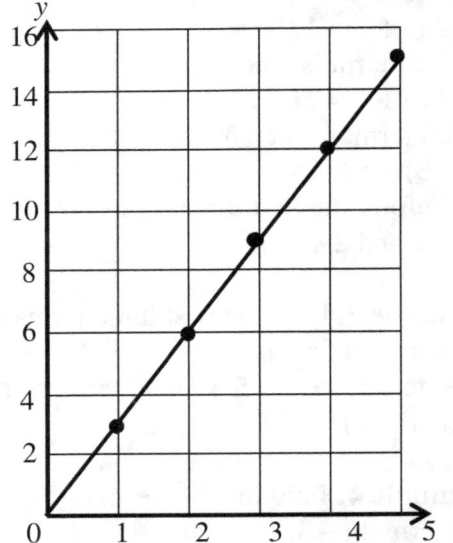

The graph of the sequence is a straight line indicating linear sequence.

10.2 NTH TERM FORMULA

For most mathematical sequences, there are formulae (nth terms). The nth term is very important in predicting the number of terms in a sequence or the term number.

A general term in the sequence is the **nth term**, where *n* stands for any number of terms.

For a linear sequence, the difference is constant. The sequence 4, 6, 8, 10 …have a constant difference of 2, all through. The sequence is linear. To find the nth term of a linear sequence, try to memorise **DNO**.

D = difference between the terms
N = Number of terms
O = **Zero** term
(Term before the 1st term)

Example 1: Find the nth term for the sequence 9, 14, 19, 24, 29…

Solution:

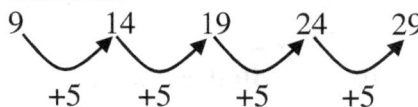

Difference between the terms is 5.
N, which is the number of terms, is written beside the difference to give 5n.
O, which is the zero term is the number before 9.
So we work backwards to get the zero term. 9 – 5 = **4**
Our sequence would look like this:

4, 9, 14 19, 24, 29
↗
Zero term
Using DNO
↓ ↓ ↓
5 n +4

The nth term is now **5n + 4** ✓

CHECK YOUR ANSWER
Using the nth term 5n + 4 obtained, find the first 2 or three terms. If they equal the original sequence, then the nth term is correct.
1st term: $5 \times 1 + 4 = 9$
2nd term: $5 \times 2 + 4 = 14$
3rd term: $5 \times 3 + 4 = 19$

Since the numbers obtained are the original numbers in the sequence, the nth term 5n + 4 is correct.

Example 2: Find the nth term formula for the sequence 2, 5, 8, 11, 14,..
Solution
Difference = 3, which implies a 3n term
Zero term = 2 – 3 = -1.
Nth term using DNO = **3n - 1** ✓

Example 3: Find the nth term formula for the sequence 20, 18, 16, 14, 12…
Solution
Difference = 18 – 20 = -2
which implies -2n term
Zero term = 20 – (-2) = 22
Nth term = **-2n + 22** ✓

Example 4: Find the nth term of the sequence $\frac{5}{7}, \frac{6}{9}, \frac{7}{11}, \frac{8}{13}, \ldots$
For numerator, nth term = n + 4
For denominator, nth term = 2n + 5
Nth term = $\frac{n+4}{2n+5}$ ✓

EXERCISE 10 C

1) Find the nth term of each of the sequences below.
a) 5, 7, 9, 11, 13…
b) 11, 14, 17, 20, 23…
c) 4, 5, 6, 7, 8, 9 …
d) 30, 25, 20, 15, 10 …
e) 1, 5, 9, 13, 17 …
f) 9, 12, 15, 18, 21 …
g) 7, 14, 21, 28, 35 …
h) 0, 4, 8, 12, 16 …
i) -1, 1, 3, 5, 7 …
j) 17, 14, 11, 8, 5 …
k) 500, 550, 600, 650, 700 …

2) Copy and complete the mapping diagram below.

Term number (n)	4n	Term
1	4	7
2	8	11
3		
4		19
70		

3) Write down each sequence and select the correct expression for the nth term.

a) 4, 6, 8, 10, 12…   d) 2, 5, 8, 11, 14…
b) 33, 30, 27, 24…   e) 3, 4, 5, 6…
c) 2, 4, 6, 8…   f) $1^2, 2^2, 3^2, 4^2$…

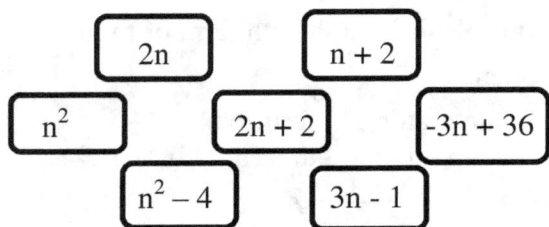

4) This sequence of patterns is made from regular pentagons with sides of 1 centimetre.

 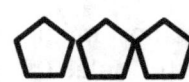

Pattern 1   Pattern 2   Pattern 3

a) What is the perimeter of pattern number 3?
b) Draw pattern numbers 4 and 5
c) Copy and complete the table below.

Pattern number	1	2	3	4	5	6
Perimeter (cm)						

d) Find a formula for the perimeter of the nth pattern.
e) Find the perimeter of the 100th pattern.
f) Which pattern number would have a perimeter of 3150?

5) Rods are placed to form a pattern.

Pattern 1   Pattern 2   Pattern 3

a) How many rods will there be in pattern number 5?
b) How many rods are needed for the nth pattern?
c) What pattern number will have 59 rods?
d) Copy and complete the table below.

Pattern number	1	2	3	50
Number of rods				

10.3 QUADRATIC SEQUENCES

In a linear sequence, the first difference is constant whereas, in a **quadratic sequence**, the second difference is constant (the same).

Nth Term formula for a quadratic sequence

Example 1: Find the nth term for 1, 4, 9, 16, 25

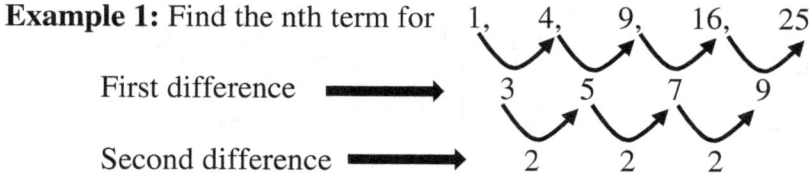

First difference → 3, 5, 7, 9

Second difference → 2, 2, 2

The sequence 1, 4, 9, 16, 25 is **quadratic** because the second difference is the same.

Once the second difference is constant, we know that the nth term formula must contain n^2. To know the coefficient of n^2 (the number before n^2), we divide the constant by **2**. $2 \div 2 = 1$. It follows that the nth term formula for the quadratic sequence must start with $1n^2$. However, in algebra, it is not the convention to write the number, 1, in front of a letter. Therefore, the nth term of 1, 4, 9, 16, 25 must start with n^2.

Now, make a table

n	n^2	Sequence	Difference
1	1	1	0
2	4	4	0
3	9	9	0
4	16	16	0
5	25	25	0

Since the difference from the table is the same, the nth term for 1, 4, 9, 16, 25 = n^2 ✓

Example 2: Find the nth term for 5, 8, 13, 20, 29..

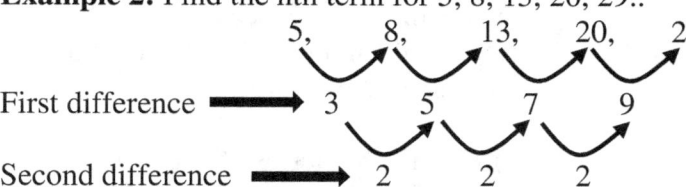

First difference → 3, 5, 7, 9

Second difference → 2, 2, 2

$2 \div 2 = 1$, therefore the nth term must start with n^2.

n	n^2	Sequence	Difference
1	1	5	+4
2	4	8	+4
3	9	13	+4
4	16	20	+4
5	25	29	+4

Since the difference from the table is the same, the nth term of 5, 8, 13, 20, 29 = $n^2 + 4$ ✓

Example 3: Find the nth term of the sequence 15, 25, 39, 57,....

First difference = 10, 14, 18
Second difference = 4, 4, 4

$4 \div 2 = 2$, therefore, the nth term must start with $2n^2$. Next, draw the table.
Remember, $2n^2 = \mathbf{2 \times n^2}$.
Find n^2 first and then multiply by 2.

n	$2n^2$	Sequence	Difference
1	2	15	13
2	8	25	17
3	18	39	21
4	32	57	25

As you can see, the difference is **not** the same from the table. We, therefore, find the linear sequence of 13, 17, 21, 25 because the difference is constant. Refer to section 10.2 above. The nth term for 13, 17, 21, 25 is $4n + 9$.
Therefore, the nth term for 15, 25, 39, 57,.. is $\mathbf{2n^2 + 4n + 9}$ ✓

Check the result: $2n^2 + 4n + 9$
1st term = $2(1)^2 + 4(1) + 9 = 15$
2nd term = $2(2)^2 + 4(2) + 9 = 25$
3rd term = $2(3)^2 + 4(3) + 9 = 39$
4th term = $2(4)^2 + 4(4) + 9 = 57$
Since the terms are the same with the original sequence, the nth term is correct.

Points to remember:
1) If the second difference is 2, the nth term must start with n^2.
2) If the second difference is 4, the nth term must start with $2n^2$.
3) If the second difference is 6, the nth term must start with $3n^2$.

Example 4: Find the nth term of the sequence 4, 22, 52, 94

Fist difference = 18, 30, 42
Second difference = 12, 12, 12
$12 \div 2 = 6$, therefore, the nth term must start with $6n^2$.

N	$6n^2$	Sequence	Difference
1	6	4	-2
2	24	22	-2
3	54	52	-2
4	96	94	-2

Since the difference from the table is constant (-2), the nth term of 4, 22, 52, 94 is $\mathbf{6n^2 - 2}$ ✓

EXERCISE 10 D
1) Look at the sequences below:
a) 1,   3,   5,   7,....
b) 11, 9,   7,   5, ...
c) 7,   17, 31, 49, ...
d) 15, 25, 39, 57, ...
e) 20, 30, 40, 50, ...
f) 33, 47, 63, 81, ...

i) Which of these sequences are linear?
ii) Which of these sequences is quadratic?
iii) Write down the next two terms of the quadratic sequences.
iv) Find the nth terms of the quadratic sequences above.

2) Find the nth terms of each sequence.
a) 2,5,10,17,26   f) 4,11,22,37,56
b) 6,9,14,21,30   g) 4,13,26,43
c) 3,9,19,33,51   h) 3,13,29,51
d) 0,6,16,30,48   i) 3,10,21,36
e) 6,15,30,51,78   j) 6,24,54,96

3) A quadratic sequence begins with 6, 9, 14, 21,….. Find the 100th term.

4) The fourth, fifth and sixth terms of a quadratic sequence are 30, 41 and 54. Find the first, second and third terms of the sequence.

5) Squares are arranged to form a sequence of rectangular patterns as shown below.

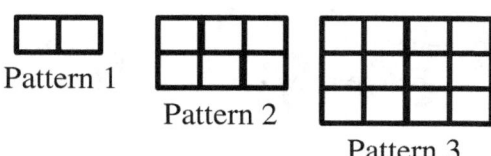
Pattern 1   Pattern 2   Pattern 3

a) How many squares will be needed in pattern 4?

b) Write an expression in terms of n for the number of squares required to form the nth pattern.

c) How many squares will there be in pattern number 300?

6) Triangular numbers are used to make the patterns below.

a) Draw the next pattern.

b) If represents a circle, find the number of circles in the nth term of the sequence.

c) How many circles will be in the 100th pattern?

TERMS FROM QUADRATIC NTH TERMS

Example 1: Write down the first three terms of the nth term $n^2 + 1$.

1st term is when n = 1: $(1)^2 + 1 = \mathbf{2}$
2nd term is when n = 2: $(2)^2 + 1 = \mathbf{5}$
3rd term is when n = 3: $(3)^2 + 1 = \mathbf{10}$
The first three terms are: 2, 5 and 10.

Example 2: Write down the first five terms of $3n^2 - 5$

This is $(3 \times n^2) - 5$
1st term = $(3 \times (1)^2) - 5$
        = $(3 \times 1) - 5 = 3 - 5 = \mathbf{-2}$
2nd term = $3 \times (2)^2 - 5 = 3 \times 4 - 5 = \mathbf{7}$
3rd term = $3 \times (3)^2 - 5 = 3 \times 9 - 5 = \mathbf{22}$
4th term = $3 \times (4)^2 - 5 = 3 \times 16 - 5 = \mathbf{43}$
5th term = $3 \times (5)^2 - 5 = 3 \times 25 - 5 = \mathbf{70}$

Example 3: Find the 50th term of the nth term $2n^2 + n$.

Replace n with 50 and work out the value of the expression.
$2 \times (50)^2 + 50 = 2 \times 2500 + 50 = \mathbf{5050}$

EXERCISE 10E

1) Write down the first three terms of
a) $n^2 + 3$ f) $3n^3 + 2n$
b) $n^2 - 1$ g) $2n^2 + 2$
c) $n^2 + 10$ h) $2n^2 - 6$
d) $4 + n^2$ i) $2n^2 + n + 2$
e) $3n^2 + 1$ j) $3n^2 + 3n - 3$

2) Write down the 70th term for
a) $2n^2 + 9$ b) $3n^2 - 3n + 2$

Chapter 10 Review Section
Assessment

1a) What is the next number in the sequence 4, 7, 10, 13, 16.. ? 1mark
   b) One number in the sequence is k.
      i) Write in terms of k, the next number in the sequence..........................1mark
      ii) Write in terms of k, the number in the sequence before k. ....1 mark

2) A sequence begins 6, 15, 30, 51, 78,...
   a) Write an expression, in terms of n, for the nth term of this sequence..........3 marks
   b) Work out the 200th term.. 2 marks

3) A sequence begins with 4, 13,....... The next term in the sequence is found by using the rule: **multiply the previous number by 3 and add 1.** Use this rule to find the next four numbers in the sequence...2 marks

4) Write down i) the nth term ii) the 20th term of the sequences below.

a) 5, 8, 11, 14, 17, ...2 marks
b) 70, 60, 50, 40, 30, ...2 marks
c) 0, 6, 16, 30, 48, ...3 marks
d) 31, 34, 39, 46, ...3 marks
e) $\frac{3}{5}, \frac{4}{7}, \frac{5}{9}, \frac{6}{11},$ ....3 marks
f) (1 × 3), (2 × 4), (3 × 5), (4 × 6), ...2 marks

5) Look at patterns 1 and 2. Find the values of *c* and *d*.
    Pattern 1:   4   9   16   25   d

    Pattern 2:   5   7   9   c   ....2 marks

6) The term of a particular sequence is given by $\mu n = 3n^2 - 3$.
a) Write down the first five terms.   ....5 marks
b) Work out the 25th term of the sequence.   ....2 marks

11 Ratio & Proportion

This section covers the following topics:

- Simplifying ratios
- Dividing/sharing ratios
- Simple proportion
- Maps, scales and ratios

LEARNING OBJECTIVES

By the end of this unit, you should be able to:

a) Simplify ratios
b) Divide a quantity in a given ratio
c) Solve word problems involving ratio and proportion
d) Understand direct and indirect proportions
e) Interpret and solve mathematical problems involving maps

KEYWORDS

- Ratio
- Share/divide
- Simplify
- Proportion
- Rate

11.1 RATIOS

Ratio is used to compare quantities.

There are 9 shaded squares and 26 plain squares. The ratio of shaded squares to plain squares can be written as **9:26**. Ratios compare the sizes of parts to each other.

Example 1: What is the ratio of shaded triangles to plain triangles in the diagram below?

Shaded: plain
5: 4
Notice that the order is 5 and 4 and **not** 4 and 5.

EQUIVALENT RATIOS

Equivalent ratios can be obtained by multiplying or dividing numbers in a ratio by the same number.

4:7 is equivalent to 8:14 because 4 and 7 are multiplied by 2.
Also, 15:10 is equivalent to 3:2 because 15 and 10 are divided by 5.

SIMPLIFYING RATIOS

The ratios 4:5 and 3:4 are in their simplest forms or terms. This is because there is no whole number which will divide into both sides **exactly,** without any remainder(s).

The easiest way to simplify a ratio is by dividing each part by the highest common factor.

Example 2: Simplify the following ratios to their lowest terms.
a) 4:10   b) 24:16   c) 15:5:35

a)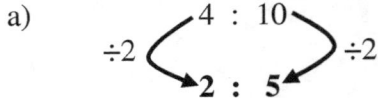

b)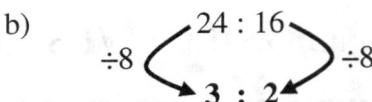

c)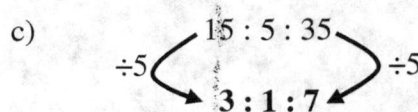

Note: Ratios **must** be simplified in the **same unit** even if they are expressed in different units.

Example 3: Simplify 40p: £1
The units are not the same, so change £1 to 100p since £1 = 100 pence.

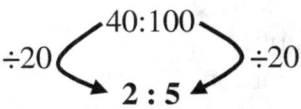

Example 4: Simplify to its lowest form
4000 g: 2 kg: 1 tonne

The ratios have different units so we must convert to the same unit before simplifying.

Convert all to kg
1000g = 1 kg, therefore 4000 g = *4 kg*
1 tonne = *1000 kg*

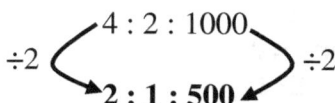

Example 5: Express 35:105 as a fraction in its lowest term.
$35:105 = \frac{35}{105} = \frac{1}{3}$

Example 6: Simplify 0.6:3
Change the decimal to a whole number to make the calculations easier. To achieve this, multiply both sides by 10.

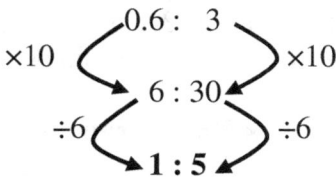

Example 7: Simplify $\frac{3}{5}$: 3
Change the fraction $\frac{3}{5}$ to a whole number by multiplying both sides by 5.

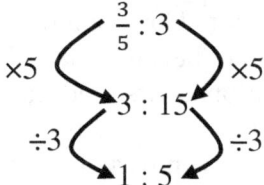

Therefore, $\frac{3}{5}$:3 in its simplest form is **1:5**.

Example 8: In a studio, the ratio of women to men is 4:7. There are 12 women, how many men are in the studio?

Solution: Women: Men
4 : 7
12: **?**
Using the knowledge of equivalent ratios,

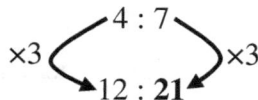

There are 21 men in the studio.

EXERCISE 11A

1) Write down the ratio of shaded to unshaded shapes in each diagram below. Simplify where possible.

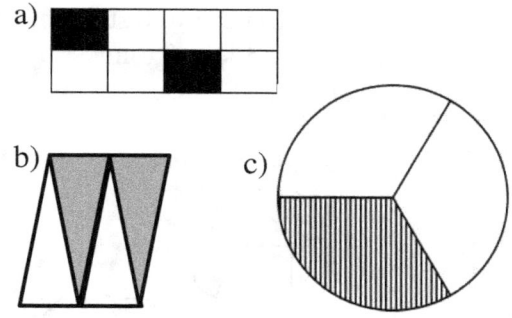

2) Express each of the ratios below in its simplest form.

a) 4:8   g) 38:60   m) 2:8:20
b) 3:6   h) 12:36   n) 100:75:25
c) 8:4   i) 9:6   o) 7:14:49
d) 6:3   j) 12:18   p) 3:9:3
e) 26:13   k) 14:22   q) 4:6:12
f) 5:40   l) 40:48   r) 60:20:30

3) Write each ratio in its simplest form.
   a) 1.2 : 1.8
   b) 1.2 : 2
   c) 4.9 : 9.8
   d) 0.5 : 1.5
   e) 2.7 : 10.8
   f) $\frac{1}{2}$: 5
   g) $\frac{3}{4}$: 3
   h) $1\frac{1}{2}$: 3

4) In a shop, there are 84 chairs and 14 tables. Find the ratio of chairs to tables in their lowest term.

5) In a class of 50 students, there are 14 boys. What is the ratio of girl to boys in their lowest form?

6) In a box, the ratio of blue pens to red pens to black pens is 4:3:5. If there are seventy black pens, how many red pens and blue pens are there in the box?

7) In a class of 33 pupils, 17 are girls. What fraction are boys?

8) Write each ratio in its simplest form.
   a) 4 kg : 2000 g   c) 3 kg : 6 tonnes
   b) 140 cm : 7 m   d) £8 : 40p

9) Fill in the missing numbers.

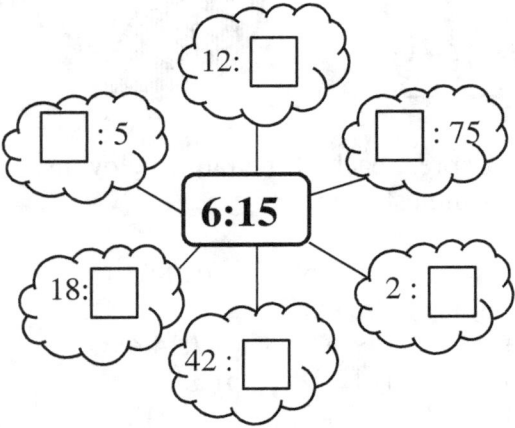

DIVIDING/SHARING RATIOS

Example 1: Divide £20 in the ratio 2:3.

We could write the ratio as
2 parts**:** 3parts since ratios are made up of parts.
Total number of parts = 2 + 3 = 5 parts.
Each part = 20 ÷ 5= 4 parts.

Therefore, 2 parts = 2 × 4 = **£8** ✓
3 parts = 3 × 4 = **£12** ✓

Always check that the parts add up to the total amount. £8 + £12 = £20

Example 2: 75 oranges are shared between Isabel, Edward and Angela in the ratio 3:5:7. How many oranges will each person get?

Solution:
Add up all the ratios: 3 + 5 + 7 = 15
Each share = 75 ÷ 15 = **5** oranges

Isabel gets   3 × 5 = **15 oranges**
Edward gets 5 × 5 = **25 oranges** and
Angela gets   7 × 5 = **35 oranges**

Check: 15 + 25 + 35 = 75

Also, remember that the order of the ratios matters a lot. Three shares/parts **must** be for Isabel because Isabel's name was mentioned first and the first ratio is 3. Likewise, five shares/parts are for Edward and seven parts for Angela.

EXERCISE 11B

1) Share £36 in the ratio
a) 1:2   b) 2:7   c) 5:7

2) Divide £420 in the ratio
a) 1:6   b) 2:3   c) 4:3

3) Joseph and Mark shared £12000 in the ratio 7:3.
a) How much did Mark receive?
b) How much more is Joseph's share than Marks share?
c) What is Joseph's percentage share?

4) The amount £880 is divided in the ratio 4:5:2. What is the difference between the largest and the smallest share?

5) A quadrilateral has angles which are in the ratio 5:7:2:4

a) Find all the angles in the quadrilateral
b) What is the difference between the largest angle and the smallest angle?

6) Two equilateral triangles are shown below with their perimeters, P. What is the ratio of the lengths of their sides?

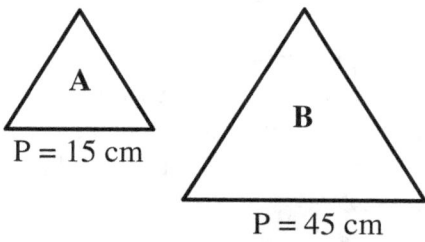

7) x and y are two numbers and their sum is 35. However, x is 9 less than y. Find the ratio of $x:y$.

8) Look at the spider diagram below. Divide the amount given in the ratios provided.

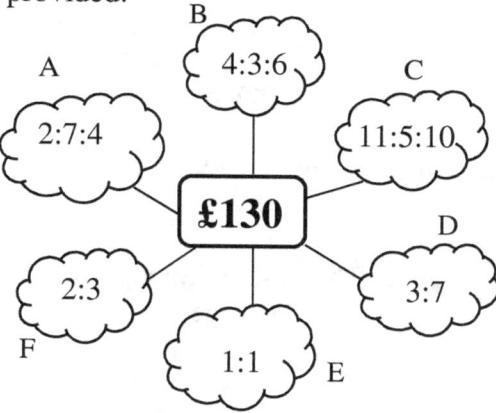

11.2 PROPORTION

Proportion is a bit different from the ratio. Proportion compares the size of a part to the size of a whole.

Example 1: What proportion of these squares is shaded?

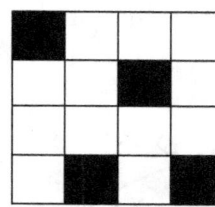

Answer
$\frac{4}{16} = \frac{1}{4}$ or 0.25

Proportion can be expressed as fraction, decimal or percentage.

DIRECT PROPORTION

If two quantities increase or decrease at the same rate, they are **directly proportional**. A typical example is petrol consumption in a car. The more petrol, the more the distance travelled. Also, the less petrol in a car, the less distance travelled.

Example 1: The cost of 7 books is £24.50. Find the cost of a book.

$$\div 7 \begin{pmatrix} 7 \text{ books} = £24.50 \\ 1 \text{ book} = \mathbf{£3.50} \end{pmatrix} \div 7$$

The cost of a book is £3.50

Example 2: The cost of 5 oranges is £2. Find the cost of 15 oranges.

$$\times 3 \begin{pmatrix} 5 \text{ oranges} = £2 \\ 15 \text{ oranges} = \mathbf{£6} \end{pmatrix} \times 3$$

The cost of 15 oranges is £6.

Example 3: If three paintings cost £300, find the cost of 11 paintings.

Method 1: <u>Unitary method</u>
The unitary method requires finding the cost of **one item** and then multiplying by the quantity required.

$$\begin{array}{c} \div 3 \\ \times 11 \end{array} \begin{pmatrix} 3 \text{ paintings} = £300 \\ 1 \text{ painting} = \mathbf{£100} \\ 11 \text{ paintings} = \mathbf{£1\,100} \end{pmatrix} \begin{array}{c} \div 3 \\ \times 11 \end{array}$$

Therefore, 11 paintings will cost **£1100**.

Method 2: Using fractions.
3 paintings cost £300
11 paintings will cost $\frac{11}{3} \times £300$
= **£1 100**

EXERCISE 11C

1) If ten computers cost £4 500,
a) find the cost of one computer.
b) find the cost of 12 computers.

2) Four books cost £20. Find the cost of 9 books.

3) 5 red pens cost ₦220. Find the cost of 3 red pens.

4) The car boot of a car contains one tyre, two jacks, five spanners and two buckets. Find the proportion of
a) Jacks   c) Buckets
b) Spanners d) Equipment that are not spanners.

5) Yvonne has £5000. She spends $\frac{2}{5}$ of her money and saves the rest.
a) What proportion did he save?
b) How much did he spend?

6) Seven second-hand dining tables cost £350. How much will eight of such dining tables cost?

7) At a seminar, there are 69 delegates. There are twice as many women as men.
a) What proportion are men?
b) What proportion are women?
c) How many men are in the seminar?
d) How many women are there?

INVERSE PROPORTION

Two quantities are inversely proportional if an increase or decrease in one quantity produces a decrease or increase in the second quantity.

Example 1:

It took ten days for three men to complete the building of a conservatory. How long would it take 15 men to complete it?

Solution:
One man will take $3 \times 10 = 30$ days to complete the conservatory.
Therefore,
15 men will take $(30 \div 15) =$ **2 days**.

EXERCISE 11D

1) 5 women can clean a shop in 7 hours. How many hours would it take seven women to clean the shop?

2) 20 cows can feed in a field for six days. How long would it take to feed
a) 3 cows b) w cows?

3) 12 people take 4 hours to cut grasses in a field. How long will it take eight people to do the work?

4) 60% of biology books in a cabinet is red. The rest are blue.
a) What is the ratio of red books to blue books?
b) If the numbers of blue books are 18, how many red books are in the cabinet?

RATE

A ratio which involves two quantities that are different is known as **rate**. Assuming a parent gives a child £500 every week as pocket money; this is an example of rate. It could be written as £500 per week. This tells us the amount of money spent every week by the child. Other examples of rate are metre per second (m/s), km per litre (km/l) and so on.

Example 1: Joe ran 120m in 15 seconds.
a) Find the rate in metre per second.
b) Find the rate in metre per minute.

a) Rate $= \dfrac{120\ m}{15\ s} =$ **8 m/s**

b) Rate $= \dfrac{120\ metre}{15\ sec}$

....but 60 seconds = 1 minute
15 seconds = 0.25 or $\dfrac{1}{4}$ minute

Therefore, rate $= \dfrac{120\ m}{0.25\ m} =$ **480 m/min**

EXERCISE 11E

1) Emily gets paid £150 for 10 hours of work. What is the rate of pay per hour?

2) Joseph ran 200 m in 20 seconds. Find the rate in a) metres per second, b) metres per minute.

3) Jude pays £9600 per annum to rent a shop. What would be the monthly rate?

11.3 MAP, SCALES AND RATIO

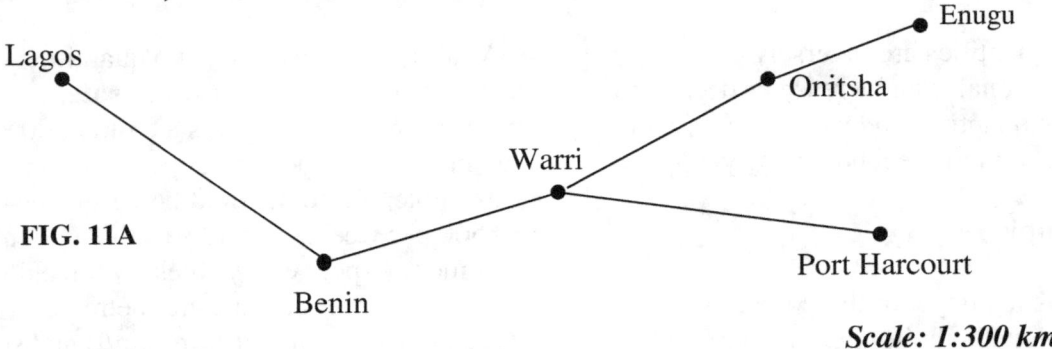

FIG. 11A

Scale: 1:300 km

Example 1: Work out the actual distance between Lagos and Benin.

1 cm on the map = 300 km on land.

By measurement, the distance from Lagos to Benin is 4 cm.

Using the scale, 4 cm = 4 × 300 = **1200 km**

EXERCISE 11F

Use Fig. 11A for questions 1 and 4.

1) Benin and Port Harcourt appears 7cm apart. Find the actual distance between the two towns.

2) Find the actual distance between Benin and Enugu which appears 8 cm apart. Write your answer **in metres**.

3) Find the actual distance between Warri and Port Harcourt.

4) Find the actual distance between Lagos and Enugu.

5) A rectangular museum has a length of 5 cm and a width of 3 cm. If the scale on a map is 1: 2000, calculate a) the real life dimensions of the museum b) the area of the museum in real life.

6) 3.5 km is the distance between two points. How far apart will the two points be on **the map** with a scale of 1:50 000?

12 Brackets & Factorisation

This section covers the following topics:

- Expanding brackets
- Factorisation
- Substitution
- Algebraic Fractions

LEARNING OBJECTIVES

By the end of this unit, you should be able to:

a) Expand single and double brackets
b) Factorise expressions
c) Solve equations involving brackets and fractions
d) Form and solve equations
e) Understand additive inverse
f) Find HCF AND LCM of algebraic fractions
g) Simplify algebraic fractions

KEYWORDS

- Expand
- Factorise
- Substitute
- Algebraic fractions
- Algebraic expression
- Terms

12.1 EXPANDING BRACKETS

Removing or expanding a bracket means to **multiply** each term in the bracket by the term outside the bracket.

Example 1: Simplify/Expand $2(3 + 5)$

Multiply 2 by 3 and then 2 by 5 and add them together.
$2 \times 3 = 6$ and $2 \times 5 = 10$, therefore, $6 + 10 = $ **16**

Alternatively, using BIDMAS perform the operation in the bracket first and multiply by 2. $3 + 5 = 8$. $8 \times 2 = $ **16**

Example 2: Expand the following:
a) $5(n + 7)$, b) $3(n - 10)$, c) $(4 + n)3$
d) $7(5p - 3)$, e) $w(w + 2)$, f) $3c(4c - 10)$
g) $5(x + y - z)$
Solutions:
a) $5(n + 7)$
↑
(Invisible multiplication sign)

There is an invisible multiplication (×) sign between 5 and the bracket. This means $5 \times (n + 7)$
5 is multiplying **everything** in the bracket, not just n.

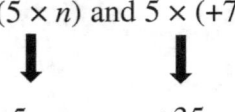
$5(n + 7)$

$= (5 \times n)$ and $5 \times (+7)$
↓　　　　　↓
$= 5n$　　　$+35$
$= $ **5n + 35**

b) $3(n - 10)$
$= (3 \times n)$ and (3×-10)
↓　　　　　↓
$3n$　　　$- 30$
$= $ **3n - 30**

c) $(4 + n)3$
This is the same as $(4 + n) \times 3$ or $3 \times (4 + n)$
$= 3 \times 4$ and $3 \times (+n)$
↓　　　　　↓
12　　　$+3n$
$= $ **12 + 3n**

d) $7(5p - 3)$
$= 7 \times 5p$ and $7 \times (-3)$
$= $ **35p - 21**

e) $w(w + 2)$
$= w \times w$ and $w \times (+2)$
$= $ **$w^2 + 2w$**

f) $3c(4c - 10)$
$= 3c \times 4c$ and $3c \times (-10)$
$= $ **$12c^2 - 30c$**

g) $5(x + y - z)$
$= 5 \times x$ and $5 \times y$ and $5 \times (-z)$
$= $ **5x + 5y − 5z**

EXERCISE 12A

1) Expand the following:
a) $2(x + 3)$　　l) $a(a + 3)$
b) $3(x + 4)$　　m) $w(w + 7)$
c) $4(x + 2)$　　n) $3x(x + 5)$
d) $5(x - 3)$　　o) $m(m + 9)$
e) $7(x - 4)$　　p) $6(7 - 3c)$
f) $10(4 - x)$　　q) $4d(d + 2)$
g) $8(3 + x)$　　r) $a(7 - a)$
h) $12(x + 5)$　　s) $2(x + 2c - 9)$
i) $8(x - 10)$　　t) $0.5x(x + 4)$
j) $99(x + 10)$　　u) $\frac{1}{4}(12x - 36)$
k) $12(5 - x)$　　v) $0.9(0.2c + 3)$

EXPANDING WITH NEGATIVE NUMBERS AND TERMS

Always be careful when multiplying a negative number or term outside a bracket.

Recollect: **+ + = +** | **Like** signs will give positive
- - = +

+ - = - | **Unlike** signs will give negative
- + = -

Example: Expand the brackets.
a) -3(n + 5) **b)** -6(n – 3) **c)** -4(5n + 3)
d) -2x (-3x – 5y)

a) -3(n + 5)
= -3 × n and -3 × (+5)
  ↓   ↓
  -3n   -15
= **-3n - 15**

b) -6(n – 3)
= -6 × n and -6 × (-3)
= **-6n + 18**

c) -4(5n + 3)
= -4 × 5n and -4 × (+3)
= **-20n - 12**

d) -2x (-3x – 5y)
= -2x × -3x and -2x × -5y
    ↓   ↓
= 6x^2 and (+10xy)

= **6x^2 + 10xy**

EXERCISE 12B

1) Expand each bracket.

a) -2(x + 2)
b) -8(x + 3)
c) -3(x + 10)
d) -5(x + 2)
e) -9(x + 4)
f) –(x + 2)
g) –4(x + 2)
h) -7(x + 7)
i) -10(x + 6)
j) -20(x + 5)
k) -6(7 + x)
l) -3(x – 4)
m) -9(x – 1)
n) -4(x – 2)
o) -8(x – 5)
p) -7(5 – x)
q) –a(a + 3)
r) –w(w + 10)
s) -6w(6w^2 + 5w)
t) –v(3v – 2)
u) -4w^2(7w + 6)
v) –a(a + 4n - 2)

EXPANDING AND SIMPLIFYING BRACKETS

To expand and simplify simply means to multiply (remove) bracket and collecting like terms. Each bracket is treated separately and then combined by collecting like terms.

Example: Remove the brackets and simplify.

a) 3(a + 2) + 4(a + 1)
b) 4(a + 5) + 3(a - 2)
c) 6y - 4(y + 2) - 3
d) 5(y - 3) - (2y - 7)
e) -2(a + 7) + 3(2a - 5)
f) 13(2k - 2) - 4k

a) 3(a + 2)   +   4(a + 1)
= 3 × a + 3 × 2   +   4 × a + 4 × 1
= 3a + 6   +   4a + 4
Collecting like terms gives **7a + 10**

b) 4(a + 5) + 3(a – 2)
= 4a + 20 + 3a - 6
= **7a + 14**

c) 6y - 4(y + 2) - 3
= 6y - 4y - 8 - 3
= **2y - 11**

d) 5(y - 3) - (2y - 7)
= 5y - 15 - 2y + 7
= **3y - 8**

e) -2(a + 7) + 3(2a - 5)
= -2a - 14 + 6a - 15
= **4a – 29**

f) 13(2k - 2) - 4k
= 26k - 26 - 4k
(Do not multiply 13 and 4k)
= **22k - 26**

EXERCISE 12C

1) Expand and simplify.

a) 2(x + 3) + 3(x + 1)
b) 3(x + 5) + 6(x + 2)
c) 4(x + 7) + 8(x + 3)
d) (x + 4) + (x + 4)
e) 5(x + 3) – 2(x + 2)
f) 2(x + 4) – 3(x + 5)
g) 3(x – 2) – 2(x – 5)
h) 2(x + 3) + 3x
i) 8(2y – 2) – 4y
j) 7y – y(3 – y)
k) 5(6t + 2) - 8
l) 3(x + 3) + 3(x + 3)
m) 4(4d – 2) + 6(4d – 3)
n) 5d + 6(d – 1) + 7
o) x(x – 1) + x(x – 2)
p) -3d (d + 4) – d(d – 3)
q) n(n – 7) + 4n(n – 2)
r) -5(7w + 2) -9(7w – 4)
s) -2s(3s – 6) – (7s – 9) + 50

WORDED PROBLEMS

Example 1: Write an expression for the perimeter of the rectangle below.

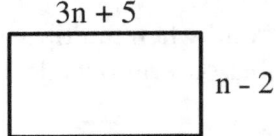

Add all the sides.
3n + 5 + 3n + 5 + n - 2 + n - 2
Collect like terms
= 3n + 3n + n + n + 5 + 5 - 2 - 2
= **8n + 6**

EXERCISE 12D

1) A rectangular swimming pool has length (5x – 3) m and width (x + 7) m. Write an expression for the pool's perimeter.

2) Three shapes W, X and Y have lengths c, c + 7 and c + 11 respectively.

In the diagram below, express the length, **a** in terms of **c**. Simplify your answers where possible.

a)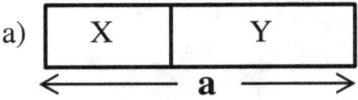

b)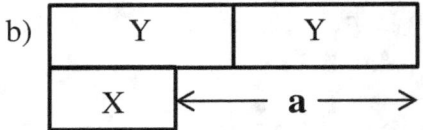

c)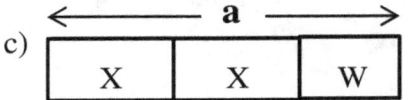

3) A cuboid measures **f** cm by **f** cm by (**f** − 2) cm as shown.

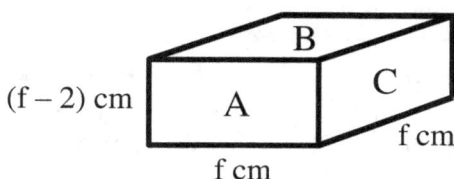

Write an expression for the area of
a) face A
b) face B.
c) Write an expression for the total surface area of the cuboid.
d) Write an expression for the volume of the cuboid.

4) The top of a luxurious dining table is rectangular which consists of 30 small rectangular pieces as shown below.

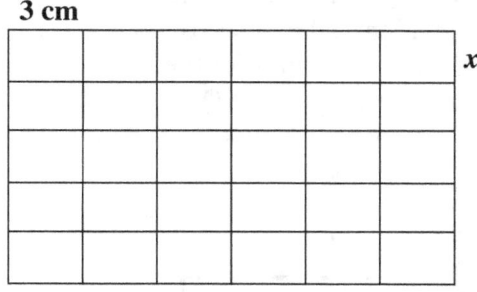

a) Write an expression for the area of a small rectangular piece.
b) Write an expression for the perimeter of a small rectangular piece.
c) Write an expression for the perimeter of the table top.
d) Write an expression for the area of the table top.
e) What is the difference between the perimeter and area of the table top?
f) If 7 rectangular pieces are removed from the table top, what is the area of the remaining table top?

12.2 EXPANDING DOUBLE BRACKETS

Example 1: Expand $(n + 1)(n + 2)$

Method 1: Grid method

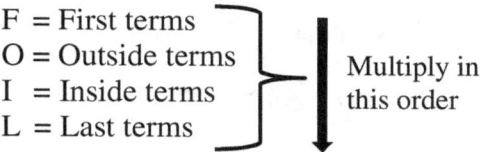

Therefore, the area of the big rectangle
$= n^2 + n + 2n + 2 = \mathbf{n^2 + 3n + 2}$

Method 2: FOIL Method

F = First terms
O = Outside terms
I = Inside terms
L = Last terms

Multiply in this order

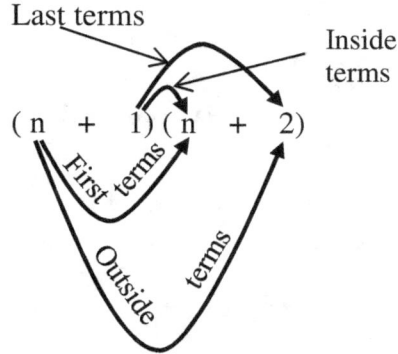

$(n + 1)(n + 2)$
$n \times n = \mathbf{n^2}, \qquad 1 \times n = \mathbf{n},$
$n \times 2 = \mathbf{2n} \qquad 1 \times 2 = \mathbf{2}$
$n^2 + 2n + n + 2 = \mathbf{n^2 + 3n + 2}$
Therefore, $(n+1)(n+2) = \mathbf{n^2 + 3n + 2}$

Example 2: Expand $(x - 2)(x + 4)$
You may use any method. Using the FOIL method,

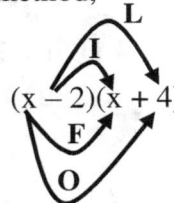

First: $x \times x = x^2$
Outside: $x \times 4 = 4x$
Inside: $-2 \times x = -2x$
Last: $-2 \times (+4) = -8$

Therefore,
$x^2 + 4x - 2x - 8$
$= \mathbf{x^2 + 2x - 8}$

Example 3: Expand $(n - 4)(n - 3)$

F: $n \times n = n^2$
O: $n \times (-3) = -3n$
I: $-4 \times n = -4n$
L: $-4 \times (-3) = 12$

Therefore,
$n^2 - 3n - 4n + 12$
$= \mathbf{n^2 - 7n + 12}$

Example 4: Expand $(2x - 5)^2$

$(2x - 5)^2 = (2x - 5)(2x - 5)$

Using the foil method,

F: $2x \times 2x = 4x^2$
O: $2x \times (-5) = -10x$
I: $-5 \times 2x = -10x$
L: $-5 \times (-5) = 25$

Therefore,
$4x^2 - 10x - 10x + 25$
$= \mathbf{4x^2 - 20x + 25}$

Example 5: Work out the area of the rectangle in terms of w.

    2w + 1
3 - w [   ]

Area $= (2w + 1) \times (3 - w)$
$= (2w + 1)(3 - w)$
Using Foil method
$= 6w - 2w^2 + 3 - w$
$= \mathbf{-2w^2 + 5w + 3}$

EXERCISE 12 E

1) Expand and simplify.
a) $(x + 1)(x + 3)$ f) $(x + 2)(x - 1)$
b) $(x + 2)(x + 5)$ g) $(x + 3)(x - 7)$
c) $(x + 3)(x + 7)$ h) $(w - 5)(w - 1)$
d) $(n + 4)(n + 4)$ i) $(w - 10)(w + 1)$
e) $(n + 6)(n + 9)$ j) $(c + 9)(c - 12)$

2) Remove the brackets and simplify fully.
a) $(6m + 3)(2m + 5)$
b) $2(3x + y)(x + 3y)$
c) $(x - 8)^2$
d) $(a + b)^2$
e) $(7c - 5)(4c - 3)$
f) $w(w + 2)(3w - 8)$

3) In the figure below, slabs A, B, C and D are identical and surround a pond. Find an expression for the:
a) The perimeter of PQRS
b) Area of PQRS
c) Area of the pond.

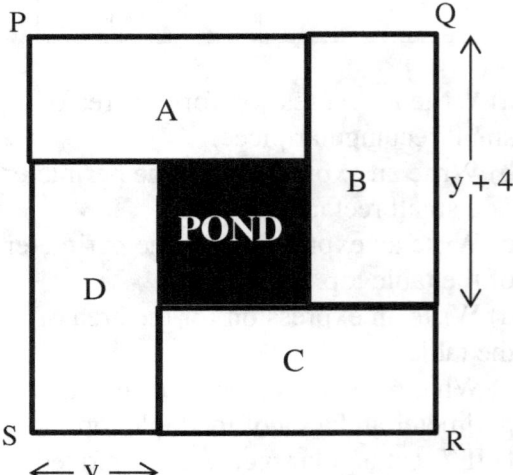

12.3 FACTORISATION

Factorisation is the opposite of expanding brackets. Consider the expression 3y + 15. The terms 3y and 15 have a common factor. The common factor is 3. We may write the expression using a bracket.

Example 1: Factorise 3y + 15

The highest common factor (HCF) of 3y and 15 is 3. Write 3 and then put a bracket like 3(   ).
Now, divide each term by 3.

$$3\left[\frac{3y}{3} + \frac{15}{3}\right] = 3(y + 5) \checkmark$$

Check that you have factorised properly by expanding the bracket. If the original expression is obtained, then it is correct.
(3 × y) + (3 × 5) = 3y + 15. This is the same as the original expression.

Example 2: Factorise $5w - 10$
HCF = 5 and it comes outside the bracket.
Therefore, $5\left[\frac{5w}{5} - \frac{10}{5}\right] = 5(w - 2) \checkmark$

Example 3: Factorise $12m + 18$
HCF = 6.
Therefore, $6\left[\frac{12m}{6} + \frac{18}{6}\right] = 6(2m + 3) \checkmark$

Example 4: Factorise $2x^2 + 4xy$
HCF = $2x$.
Therefore, $2x(x + 2y) \checkmark$

Example 5:
Factorise $20x^2y + 8xy^2 + 6xy$

Solution
HCF of 20, 8 and 6 = 2
HCF of x^2, x and x = x
HCF of y, y^2 and y = y

Therefore, the overall HCF = $2xy$

$$2xy\left[\frac{20x^2y}{2xy} + \frac{8xy^2}{2xy} + \frac{6xy}{2xy}\right]$$

= **$2xy(10x + 4y + 3)$** $\checkmark$

EXERCISE 12 F

1) Factorise the following expressions.

a) 4x + 16
b) 6x + 9
c) 12x + 16
d) 5a + 10
e) 9x + 90
f) 6x – 3
g) 9x – 18
h) 5x – 30
i) 30x – 45
j) 7m - 49
k) 28n + 21
l) 7x – 14b + 21c
m) 4n – 6ny
n) 12x + 14
o) 16y - 16
p) abcd - bcd
q) 5m + 20n - 5
r) 75k – 450

EXERCISE 12 G

1) Factorise the expressions below.

a) $x^2 + 5x$
b) $4t^2 – 16t$
c) $bc^2 – dc$
d) $7p^2 + 28p$
e) $c^2d + cd^2$
f) $f^3 – 2f$
g) ps + qs – prs
h) $11x^2 + x$
i) $d^2f^2 - mnd^2f^2 + df$
j) $3x^2 – 9x + 27x^3$
k) $14y – 7y^2 + 28y^3$
l) cd + ce + fd + fe
m) x(b + c) + 3(b + c)
n) $-30y^2 - 9$

12.4 ALGEBRAIC FACTORS

Examples of algebraic terms are 4x, 3xy, $12x^2y$.....

The factors of 8 are 1, 2, 4 and 8. The factors of 20 are 1, 2, 4, 5, 10 and 20. The number, 1, has one factor, 1.

Other than the number 1, all whole numbers have two or more factors. Numbers that have only two factors are called prime numbers.

Example 1: Write the factors of 10x.
10x: 1, 2, 5, 10, x, 2x, 5x, 10x

Example 2: Write down the factors of $4cd^2$.
$4cd^2$: 1, 2, 4, c, 2c, 4c, d, 2d, 4d, d^2, $2d^2$, $4d^2$, cd, 2cd, 4cd, cd^2, $2cd^2$, $4cd^2$

EXERCISE 12 H

1) Write down all the factors of

a) 4c
b) 12t
c) ac
d) 7xy
e) y^2
f) 6xy
g) $10x^2$
h) 35ef

12.5 HCF OF ALGEBRAIC EXPRESSIONS

First, work out the prime factors (using factor tree) of the numerical terms and then find the factors of the letters.

Example 1: Find the HCF of 18xy and $40x^2y$.

Prime factors of 18 from factor tree are $2 \times 3 \times 3$.
Therefore, $18xy = 2 \times 3 \times 3 \times x \times y$

Prime factors of 40 from factor tree are $2 \times 2 \times 2 \times 5$.
Therefore,
$40x^2y = 2 \times 2 \times 2 \times 5 \times x \times x \times y$

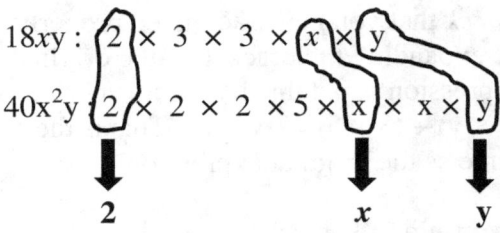

The HCF of 18xy and $40x^2y$ is $2 \times x \times y$ = **2xy** ✓

Example 2: Find the HCF of abc and bcd.

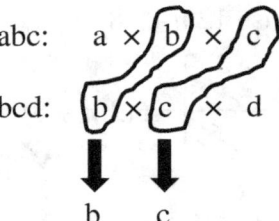

HCF of abc and bcd = $b \times c$ = **bc** ✓

EXERCISE 12 I

1) Find the HCF of the following algebraic expressions.

a) pqr and qrs
b) 3mn and 7gn
c) $10c^2$ and $25c^2$
d) $30m^2n$ and 60mn
e) ab and cb
f) $25de^2$ and 5cf
g) 6jk and 6jm
h) n^2 and 7n
i) 5ab and $15a^2b$
j) $6n^2y$ and $24ny^2$

12.6 LCM OF ALGEBRAIC EXPRESSIONS

First, work out the HCF using the method described in section 12.5 and multiply the HCF together with the leftovers to give the LCM.

Example 1: Find the LCM of 18xy and $40x^2y$.

Prime factors of 18 from factor tree are $2 \times 3 \times 3$.
Therefore, $18xy = 2 \times 3 \times 3 \times x \times y$

Prime factors of 40 from factor tree are $2 \times 2 \times 2 \times 5$.
Therefore,
$40x^2y = 2 \times 2 \times 2 \times 5 \times x \times x \times y$

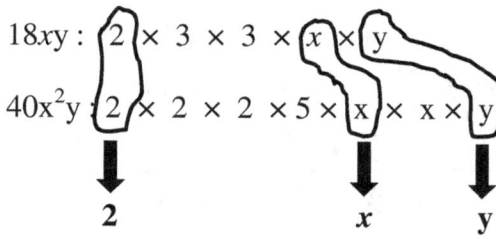

The HCF = $2 \times x \times y = 2xy$

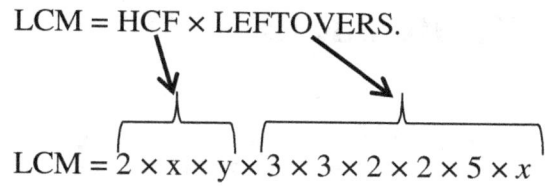

LCM = $2 \times x \times y \times 3 \times 3 \times 2 \times 2 \times 5 \times x$

LCM = **$360x^2y$** ✓

Example 2: Find the LCM of abc and bcd.

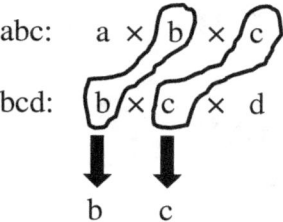

HCF of abc and bcd = $b \times c$ = **bc**

Therefore,
LCM = HCF × LEFTOVERS
= $b \times c \times a \times d$ = **abcd** ✓

USING VENN DIAGRAM
Refer to Example 1 above.

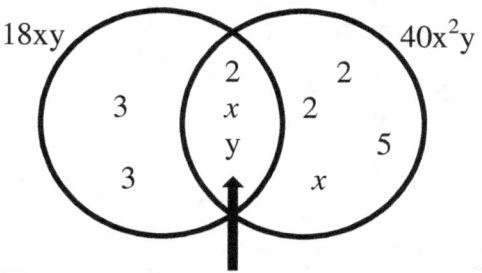

The overlapping section gives the HCF which is $2 \times x \times y = 2xy$.

LCM = HCF × LEFTOVERS
    = $2 \times x \times xy \times 3 \times 3 \times 2 \times 2 \times 5 \times x$
    = **$360x^2y$** ✓

EXERCISE 12 J

1) Work out the LCM of
a) 4p and 6p
b) c and d
c) 3t and 5u
d) n^2 and 7n
e) 8cd and $4cd^2$
f) $5p^2q$ and $30pq^2$
g) abc and $10b^2$
h) mn and an
i) 15mn and $7n^2$

12.7 ALGEBRAIC FRACTIONS

Algebraic fractions are worked out the same way normal fractions are calculated. They follow the same principle(s).

MULTIPLYING ALGEBRAIC FRACTIONS

Remember: $\frac{1}{8} \times \frac{2}{3} = \frac{1 \times 2}{8 \times 3} = \frac{2}{24} = \frac{1}{12}$

Example 1: Simplify $\frac{f}{g} \times \frac{h}{t}$

$= \frac{f \times h}{g \times t} = \frac{fh}{gt}$ ✓

Example 2: Simplify $\frac{20x}{5xy}$

$\frac{20x}{5xy} = \frac{\cancel{20}^4 \times \cancel{x}}{\cancel{5} \times \cancel{x} \times y}$

$= \frac{4}{y}$ ✓

Example 3: Simplify $\frac{8c}{3g} \times \frac{d}{3h} \times \frac{f}{k}$

Numerator: $8c \times d \times f = 8cdf$
Denominator: $3g \times 3h \times k = 9ghk$

Therefore: $\frac{8cdf}{9ghk}$ ✓

Example 4: Simplify $\frac{16w}{4} \times \frac{7}{w^2}$

$= \frac{\cancel{16}^4 \times \cancel{w}}{\cancel{4}} \times \frac{7}{\cancel{w} \times w}$

$= \frac{4 \times 7}{w} = \frac{28}{w}$ ✓

DIVIDING ALGEBRAIC FRACTIONS

Remember: Keep, Change, Flip

$\frac{3}{4} \div \frac{1}{2} = \frac{3}{4} \times \frac{2}{1} = \frac{3 \times 2}{4 \times 1} = \frac{6}{4} = 1\frac{1}{2}$

Example 5: Simplify $\frac{11}{x} \div \frac{2}{x}$

$\frac{11}{x} \times \frac{x}{2} = \frac{11 \times \cancel{x}}{\cancel{x} \times 2} = \frac{11}{2} = 5\frac{1}{2}$ ✓

Example 6: Simplify $\frac{y^2 + y}{y^2 - y}$

Factorise numerator and denominator.

$\frac{y^2 + y}{y^2 - y} = \frac{\cancel{y}(y+1)}{\cancel{y}(y-1)} = \frac{y+1}{y-1}$ ✓

EXERCISE 12 K

Simplify the following.

1) $\dfrac{10t}{5}$

2) $\dfrac{8x}{4y}$

3) $\dfrac{2x}{7x}$

4) $\dfrac{2ac}{10a}$

5) $\dfrac{26abc}{13a}$

6) $\dfrac{7ab}{4} \times \dfrac{2}{b}$

7) $\dfrac{16xy}{3} \times \dfrac{9}{2x}$

8) $\dfrac{d}{e} \times \dfrac{f}{g}$

9) $\dfrac{7pqr}{14qr}$

10) $\dfrac{5p}{50pq} \times \dfrac{75q}{ps}$

Simplify and following

11) $\dfrac{7x^2}{x}$

12) $\dfrac{5pq^2}{25pq}$

13) $\dfrac{3ab^2}{27a^2b}$

14) $\dfrac{10st}{3s^2t}$

15) $\dfrac{x^2}{3x}$

16) $\dfrac{42c^2}{7c^2}$

17) $\dfrac{3w^2}{4} \times \dfrac{2}{3y}$

18) $\dfrac{c}{8} \times \dfrac{d^2}{c^2}$

19) $\dfrac{6p}{a^2} \times \dfrac{a}{3p^2}$

20) $\dfrac{cd^2}{de^2} \div \dfrac{de}{4d}$

21) $\dfrac{18y^2}{8w^2} \div \dfrac{y^3}{w^3}$

22) $\dfrac{g^2 + g}{g^2 - g}$

23) $\dfrac{y^2}{y + 2y} \div \dfrac{y}{y + 2}$

24) $\dfrac{(3x)^2}{3x}$

ADDING / SUBTRACTING ALGEBRAIC FRACTIONS

Adding and subtracting fractions follow the same rule as arithmetic fractions.

Points to remember
1) Before adding or subtracting fractions, the denominators must be the same.
2) Do to the numerator what you did to the denominator.
3) Add or subtract the fraction.

Example 1: Simplify $\dfrac{x}{3} + \dfrac{y}{4}$

Solution: Make the denominators the same.
$$\dfrac{x \times 4}{3 \times 4} + \dfrac{y \times 3}{4 \times 3} = \dfrac{4x}{12} + \dfrac{3y}{12} = \dfrac{4x + 3y}{12} \checkmark$$

*Note: Since 4x and 3y are not like terms, **do not** add them.*

Example 2: Simplify $\dfrac{2c}{3} - \dfrac{c}{5}$

$$\dfrac{2c \times 5}{3 \times 5} - \dfrac{c \times 3}{5 \times 3} = \dfrac{10c}{15} - \dfrac{3c}{15} = \dfrac{7c}{15} \checkmark$$

Example 3: Simplify $\dfrac{2n-4}{8} + \dfrac{n-2}{4}$

$$\dfrac{(2n-4) \times 1}{8 \times 1} + \dfrac{(n-2) \times 2}{4 \times 2}$$

$$= \dfrac{2n-4}{8} + \dfrac{2n-4}{8} = \dfrac{2n-4+2n-4}{8}$$

$$= \dfrac{4n-8}{8} \quad \text{(Now factorise the numerator)}$$

$$= \dfrac{4(n-2)}{8} = \dfrac{\cancel{4}(n-2)}{\cancel{8}_2} = \dfrac{n-2}{2} \checkmark$$

Example 4: Simplify $\dfrac{3}{y} + \dfrac{4}{2y} + \dfrac{5}{3y}$

Make the denominators the same.

$\dfrac{3 \times 6}{y \times 6} + \dfrac{4 \times 3}{2y \times 3} + \dfrac{5 \times 2}{3y \times 2}$

$= \dfrac{18}{6y} + \dfrac{12}{6y} + \dfrac{10}{6y} = \dfrac{18 + 12 + 10}{6y}$

$= \dfrac{40}{6y} = \dfrac{20}{3y}$ ✓

EXERCISE 12 L

Simplify the following:

1) $\dfrac{3x}{7} + \dfrac{x}{7}$ 11) $\dfrac{c-1}{5} + \dfrac{c+1}{2}$

2) $\dfrac{5n}{6} + \dfrac{2n}{6}$ 12) $\dfrac{b}{5} + \dfrac{b-3}{3}$

3) $\dfrac{7x}{7} - \dfrac{4x}{7}$ 13) $4d - \dfrac{5f}{7g}$

4) $\dfrac{1}{y} + \dfrac{n}{y}$ 14) $8u + \dfrac{7u+v}{9}$

5) $\dfrac{2xy}{4w} + \dfrac{3xy}{4w}$ 15) $\dfrac{k}{3} - \dfrac{2k}{9}$

6) $\dfrac{2c}{3} - \dfrac{d}{6}$ 16) $\dfrac{2x-4}{8} + \dfrac{x-2}{4}$

7) $\dfrac{3x}{u} + \dfrac{t}{4}$ 17) $\dfrac{(4x+7)}{7} - \dfrac{(x-4)}{7}$

8) $\dfrac{5}{y} + \dfrac{7}{2y} + \dfrac{9}{3y}$

9) $\dfrac{x}{10} + \dfrac{3}{c}$

10) $\dfrac{5y}{7} - \dfrac{x}{3}$

Chapters 11 & 12 Review Sections
Assessment

1) Write each ratio in its simplest form.
a) 2.2 : 2
b) 0.5 : 3.5
c) $\frac{1}{2}$: 7
d) $1\frac{1}{2}$: 6
……..**4 marks**

2) In a class of 70 students, there are 14 boys. What is the ratio of girl to boys in their lowest form? …………..**2 marks**

3) Fill in the missing numbers.

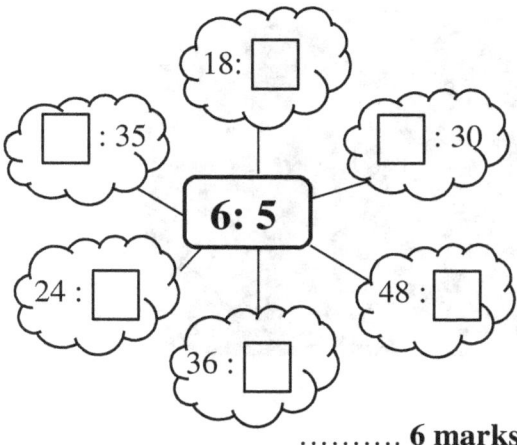

………. **6 marks**

4) 75 oranges are shared between Isabel, Edward and Angela in the ratio 3:5:7. How many oranges will each person get?
…………..**3 marks**

5) Four books cost £20. Find the cost of 9 books. ……….**2 marks**

6) Expand each bracket.
a) -4(x + 5)   b) -3(x – 4)
c) -3s(3s – 4) – (7s – 9) + 7
………...**6 Marks**

7) Factorise the expressions below.
a) $w^2d + wd^2$ c) $16ry^2 + 8r^2y - 4ry$
b) a(b + c) + 3(b + c)   ……….**6 marks**

8) Find the HCF and LCM of the following algebraic expressions.
a) $15w^2x$ and 30wx
b) $5n^2y$ and $25ny^2$ …………….**3 marks**

9) Simplify the following.
a) $\frac{54abc}{9a}$ b) $\frac{15p}{30pq} \times \frac{q}{ps}$
………..**4 marks**

10) Simplify the following:
a) $\frac{5x}{9} + \frac{x}{9}$
b) $\frac{3n}{3} + \frac{2n}{3}$
c) $\frac{3c}{3} - \frac{d}{7}$
………. **6 marks**

13 Probability

This section covers the following topics:

- Simple probability
- Probability scales
- Mutually exclusive events
- Tree diagrams
- Experimental probability

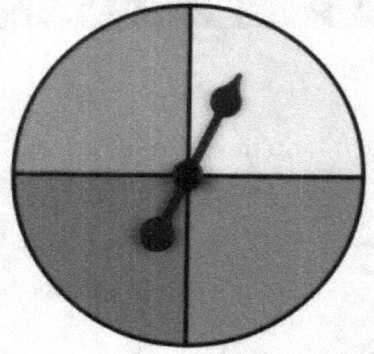

LEARNING OBJECTIVES

By the end of this unit, you should be able to:

a) Find probabilities between 1 and 0
b) Calculate probabilities of mutually exclusive events
c) List all possible outcomes of a combined event
d) Calculate theoretical probability
e) Calculate experimental probability
f) Use tree diagrams to calculate probabilities

KEYWORDS
- Probability
- Events
- Probability scale
- Chance
- Outcomes
- Tree diagram

PROBABILITY SCALE

Probabilities describe how likely or unlikely is that something will happen.

The probability of an event happening is between **impossible** and **certain** as seen on a **probability scale** below.

Example 1
Consider the following numbers below. One is chosen at random.

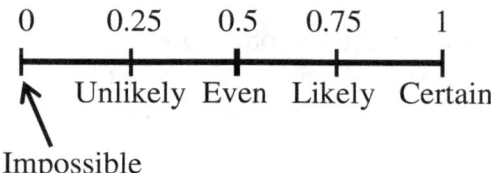

Mark each statement below in the correct place on the scale below.

A – The card is 7 or less………certain
B – The card is a 3 ……….impossible
C – The card is a 7 ………….unlikely

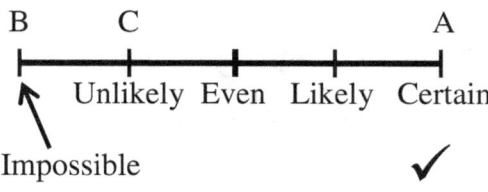

Example 2: A 6-sided die is rolled. Find the probability of getting a

a) 1   b) 3   c) 6   d) 7

Solution
a) Probability of getting a 1 is unlikely, but in terms of numbers, it is $\frac{1}{6}$ because there are 6 outcomes and only one number 1.
b) Same reason as in **a** above, so $\frac{1}{6}$
c) Same reason as in **a** above, so $\frac{1}{6}$
d) Impossible, so probability is zero (0)

Example 3: A six-sided die is rolled 60 times, how many fours would you expect to get?

Solution: The probability of getting a four is $\frac{1}{6}$. To get *a four* in 60 throws would be $\frac{1}{6} \times 60 = 10$.

Example 4: When you roll a non-biased six-sided die, which number(s) are you most likely to get?

Answer: You are **equally likely** to get any of the six numbers from the die.

Example 5: The cards below are placed in a bag and one selected at random.

a) What is the probability of getting the letter V? ………………$\frac{1}{5}$
b) What is the probability of getting the letter E? ………………$\frac{2}{5}$

EXERCISE 13A

1) By using one of the following descriptions; *Impossible, Very Unlikely, Unlikely, Even chance, likely, Very likely, Certain*, describe each one of the statements **a** to **e**.

a) A new baby is a boy   b) Tomorrow is Saturday   c) Your teacher will win the lottery
d) You will live to 300 years   e) 3 × 4 = 12

2) Match each event (P, Q, R, S, T) to the correct description on the probability scale.

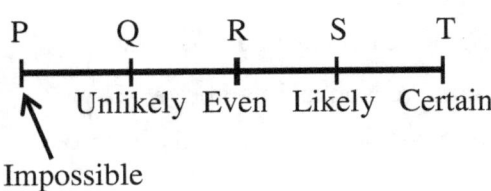

a) It will snow in Britain tomorrow
b) You roll normal die and get a score of 9
c) You roll a fair, six-sided die and get an odd number
d) Tomorrow will surely come
e) You roll a fair, 5-sided die and get an odd number

3) Thirteen cards are placed in a bag and one selected at random.

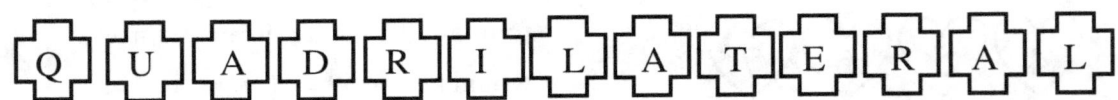

What is the probability of selecting the letters a) U   b) A   c) E   and   d) L?
e) Which **letters** are twice as likely to be on the selected card as the letter Q?
f) Which letter are three times as likely to be on the selected card as the letter D?

4) How many twos would you expect to get if you rolled a fair six-sided die:
a) 6 times,   b) 18 times   c) 114 times   d) 480 times?

5) Describe an event that is: a) impossible   b) certain   and   c) likely to happen.

6) Match the letters P, Q, R, S and T to the correct probability.

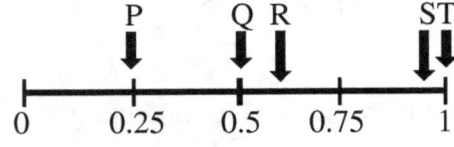

a) $\frac{3}{5}$ b) 50%   c) $\frac{1}{4}$ d) 100%   e) $\frac{9}{10}$

7) Copy the shapes below and shade so that the probability of landing on a shaded section is i) $\frac{1}{4}$ ii) $\frac{1}{2}$ **a)** 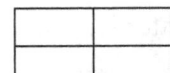   **b)**

CALCULATING PROBABILITIES

If all the possible outcomes are equally likely, probabilities can be worked out using

$$\text{Probability} = \frac{\text{number of ways events can happen}}{\text{total number of possible outcomes}}$$

Example 1: A bag contains five red counters, four blue counters and six yellow counters, numbered 1 to 15. If the counters are selected at random, work out the probabilities that
a) the counter is red   b) the counter is yellow   c) the counter is the number 7
Solutions:
a) There are 5 red counters, so probability $= \frac{5}{15} = \frac{1}{3}$
b) There are 6 yellow counters, so probability $= \frac{6}{15} = \frac{2}{5}$
c) The number 7 is only 1. Therefore the probability is $\frac{1}{15}$

EXERCISE 13B

1) If a card of 52 cards is cut, what is the probability that the card is a) a club b) red c) two d) ace of hearts?

2) One letter is selected at random from the word PERCENTAGE. What is the probability that the letter is
a) a P   b) an E   c) a vowel   d) a T or a G?

3) The spinner is fair and spun once. What is the chance of spinning

a) a 1
b) a 4
c) a 2
d) a prime number
e) an even number
f) a triangle number
g) a square number
h) a cube number
i) a 3 or a 4?

4) Twenty cards are numbered from 1 to 20 and placed in a bag. A card is selected at random from the bag. Find the probability of selecting each of the following.

a) an even number
b) a square number
c) card 7
d) card 6 or 13
e) a number more than 10
f) a factor of 4
g) a multiple of 5
h) a prime number

5) At a musical concert, 200 raffle tickets are sold. Each ticket is purchased by a different person and the winning number drawn randomly. Steve was the first to buy a ticket.
a) What is the probability that Steve wins the prize?
b) What is the probability that the 27th person to buy a ticket wins the prize?

MUTUALLY EXCLUSIVE EVENTS

When events **cannot** happen at the same time, they are said to be **mutually exclusive**.

Driving North and driving south are mutually exclusive as you cannot do both at the same time. You either go north or south.
Heads and tails are mutually exclusive when a coin is tossed. You cannot get a head and a tail at the same time.

INDICATORS FOR MUTUALLY EXCLUSIVE EVENTS

1) Probabilities of mutually exclusive events covering all the possible outcomes will always add up to 1.

2) The probability of something **not happening** = 1 − Probability of something happening.

Example 1: The probability that Andrew will go to school early is 0.8. What is the probability that Andrew will be late for school?

Solution: Going to school early and being late are mutually exclusive as they cannot happen at the same time. It's either you are early or late.

Therefore, the probability that Andrew will be late for school is 1 − 0.8 = **0.2**

EXERCISE 13C

1) The probability that Sarah is late for work is 0.25. What is the probability that she is early?

2) A fair sided die is rolled, and the events are shown below.
P – a number less than 5 is rolled
Q – a 3 is rolled
R – a 4 is rolled
S – a prime number is rolled

State whether the pairs of events are mutually exclusive
a) P and Q   b) P and R   c) Q and R
d) P and S   e) Q and S

3) The table below shows the probability of each of the four colours on a spinner. What is the probability of getting a white?

Colour	Blue	White	Red	Green
Probability	0.15		0.2	0.35

4) The table shows the probabilities of winning prizes in a lucky dip. Copy and complete the table.

Colour	Blue	White	Red	Green
Probability	0.1	x	0.4	x

5) A student is chosen from a school. The probability the student wears glasses is $\frac{3}{15}$, the probability that the student has red hair is $\frac{1}{7}$, and the probability that the student is a boy is $\frac{4}{6}$. Find the probability that the student,
a) has no red hair   b) is a girl   c) do not wear glasses.

OUTCOMES AND TWO-WAY TABLES

It is easier to list **all the possible outcomes** when two events are happening at once. Events happening at the same time could be tossing a coin and rolling a die.

Example 1: Mushroom, sweetcorn, ham, pepperoni are toppings available in a pizza shop. Also, Coca Cola, Sprite and Fanta are drinks available.
a) How many combinations of toppings and drinks are there?
b) List all the possible combinations.
c) What is the probability of choosing Mushroom topping and Sprite?
d) What is the probability of choosing a pizza with a ham topping?

Solutions: a) There are 12 combinations or outcomes.
b) You may draw up a **two-way table** to show all the possible combinations.

	Mushroom(M)	Sweetcorn(S)	Ham(H)	Peperoni(P)
Coca Cola(C)	MC	SC	HC	PC
Sprite(S)	MS	SS	HS	PS
Fanta(F)	MF	SF	HF	PF

c) $\frac{1}{12}$, since there is only one combination of mushroom and sprite (MS).
d) $\frac{3}{12} = \frac{1}{4}$, since there are three combinations (HC, HS ad HF).

Example 2: a) List all the possible combinations when a coin and a six-sided die are tossed. b) What is the probability of getting a head and a 5?
Solution: You may also use a two-way table to list all the combinations.

a)

		Die					
		1	2	3	4	5	6
Coin	H	H1	H2	H3	H4	H5	H6
	T	T1	T2	T3	T4	T5	T6

b) $\frac{1}{12}$

Example 3: Two six-sided dice are rolled.
a) List all the possible total scores.
b) How many possible outcomes are there?
c) What is the probability of getting the following total scores i) 7 ii) 11
iii) a prime number?
Solutions: b) 36   c) i. $\frac{6}{36} = \frac{1}{6}$ ii. $\frac{2}{36} = \frac{1}{18}$
iii. $\frac{15}{36}$

b)

+	1	2	3	4	5	6
1	2	3	4	5	6	7
2	3	4	5	6	7	8
3	4	5	6	7	8	9
4	5	6	7	8	9	10
5	6	7	8	9	10	11
6	7	8	9	10	11	12

TREE DIAGRAMS

Drawing a **tree diagram** is another way of listing all possible outcomes and calculating the probability. It has wider reach as it can list outcomes of more than two events, unlike the two-way table which has its limitations.

Example 1: When a coin is tossed **twice**, the outcomes can be represented using a tree diagram.

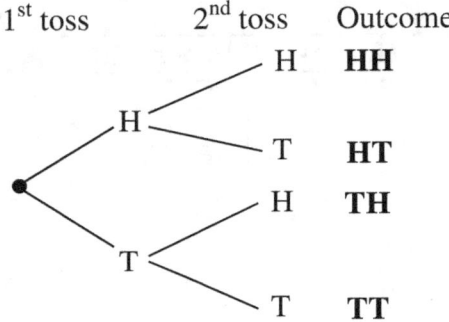

It can be seen that there are four possible combinations when two coins are tossed.

The probability of getting two heads is $\frac{1}{4}$

The probability of getting a head and a tail is $\frac{2}{4} = \frac{1}{2}$

The probability of getting a head and a tail **in that order** is $\frac{1}{4}$

EXERCISE 13D

1) A coin is tossed, and a six-sided die is rolled. Draw a two-way table to show all the possible outcomes.

2) Two six-sided dice are rolled and total scores recorded.

a) Show all the possible outcomes.
b) How many possible outcomes are there?
c) Find the probability of each of the following total scores.
i) 4   ii) 13   iii) more than 5
iv) a square number   v) a factor of 7

3) Two spinners are spun and total scores recorded.

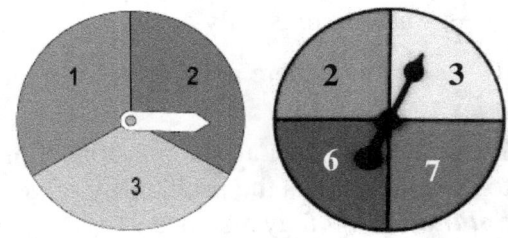

a) Show all the possible outcomes.
b) Find the probability of each of the following total scores.   i) 5   ii) 8   iii) 1
iv) a prime number

4) A box contains 5 cups: 2 red, 2 blue, and 1 yellow. A second box contains 3 cups: 1 red, 1 blue and 1 yellow. If a cup is taken at random from the box,

a) draw a table to show all the possible outcomes.
b) Find the probability of picking
i) 2 blue cups   ii) 2 cups of the same colour   iii) at least 1 red cup.

5) Two three-sided spinners are rolled.

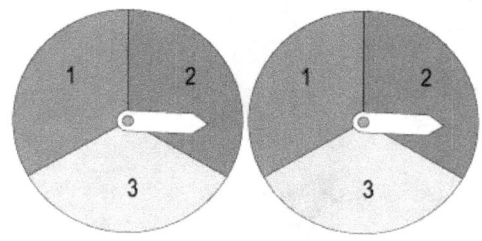

Show all the possible combinations using a **tree diagram**.

PROBABILITY FROM EXPERIMENTS

Probabilities from experiments or records of past events are also called **relative frequency**.

Experimental probability =

$$\frac{\text{Number of times the result happened}}{\text{Number of times the experiment was carried out}}$$

The more times we perform the experiments, the more accurate the estimate is. More trials are encouraged for more accurate probabilities.

Example 1: A coin is tossed 50 times and the results shown in the table.

Face	Head	Tail
Frequency	47	53

The estimated probability of getting a head is $\frac{47}{100}$ and for a tail is $\frac{53}{100}$.

Example 2: August 2000, it rained for 3 days. Estimate the probability that it rains on a day in August.

Answer: $\frac{3}{31}$ because there are 31 days in any January.

Example 3: 0.2 is the probability of a biased die landing on 4. How many times would you expect to roll a 4 if you rolled the die 150 times?

Solution: Multiply the number of times by the probability.
$150 \times 0.2 =$ **30 times**

EXERCISE 13E

1) Arsenal FC plays an average of 50 matches each season and loses 15 of them. Estimate the probability of losing a match.

2) A spinner with five sections is spun 80 times and results recorded as shown.

Colour	Blue	Red	White	Pink	Grey
Frequency	15	20	12	28	5

a) Find the relative frequency of
i) red   ii) white   iii) grey

b) Find the relative frequency of getting red or white.

c) Estimate the probability that the colours are not blue, red or white.

d) How could these estimates be made more accurate if the relative frequencies were used to estimate the probabilities?

3) The probability that a six-sided biased die lands on 3 is 0.45.
How many times would you expect the die to land on 3 when rolled?

a) 40 times   b) 60 times   c) 200 times?

4) 65% of students who buy fruits in shop **A** buy oranges. If 40 students buy fruits from shop **A** next week, how many of them would be expected to buy oranges?

5) A spinner with five sections is spun 200 times and results recorded as shown.

Colour	Blue	Red	White	Pink	Grey
Frequency	50	20	60	28	42

a) Find the relative frequencies of **all** the five colours

b) What is the theoretical probability of getting each colour assuming the spinner is fair?

c) Do you think the spinner is fair or biased? Explain.

Chapter 13 Review Section
Assessment

1) Steve rolls a six-sided die. What is the probability that he gets:
a) a five   b) a 7   c) a prime number   d) a multiple of 4? ....**4 marks**

2) Class 8D completed a survey on the mode of transportation to school, and the results are shown below. The pupils are selected at random.

Transportation	Number of pupils
Walk	15
Taxi	1
Cycle	5
Bus	4

What is the probability that they come to school:
a) by bus ……………………..**1 mark**
b) by walking ……………………..**1 mark**
c) by cycle ……………………..**1 mark**
d) not by cycle? ....**1 mark**

3) Four pilots play regular Scrabble game in a games club. The table below shows the outcomes.

Names	Lost	Won
Smith	6	10
Alfie	4	8
Deborah	5	9
Stella	3	2

a) Find the probability that i) Smith ii) Deborah wins a match.
    ………………………… **2 marks**
b) The umpire said that Smith is the best player. Is the umpire correct? Explain   ……………………..**2 marks**
c) Who is the worst player?   ………………………**2 marks**
d) Who do you think will win between Smith and Deborah?
    …………..……………**2 marks**

4)

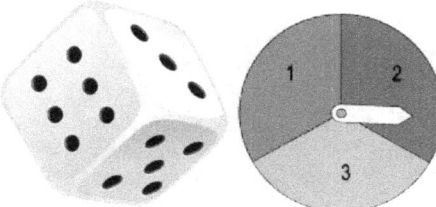

A six-sided die and a 3 sided spinner were spun together and total scores added.
a) List all the possible outcomes. ……….**3 marks**
b) Find the probability of each of the following total scores.
i) 4   ………………..**1 mark**
ii) 8   ………………..**1 mark**
iii) 15   ………………..**1 mark**
iv) a square number   ………………..**1 mark**

14 Reciprocal and Inverse

This section covers the following topics:

- Reciprocals
- Inverse
- Identity

LEARNING OBJECTIVES

By the end of this unit, you should be able to:

a) Find reciprocal of a number
b) Understand and use identities for addition and multiplication

KEYWORDS

- Reciprocal
- Multiplicative inverse
- Identity
- Chance

RECIPROCALS

When two numbers are multiplied, and the result (answer) is **1**, we say that each number is the **reciprocal or multiplicative inverse** of the other. 1 is the **identity** for multiplication.

Examples

1) $\frac{1}{7} \times 7 = 1$, we say that 7 is the reciprocal of $\frac{1}{7}$ and $\frac{1}{7}$ is also the reciprocal of 7.

2) $\frac{2}{3} \times \frac{3}{2} = 1$, we say that $\frac{3}{2}$ is the reciprocal or multiplicative inverse of $\frac{2}{3}$

3) Write down the reciprocals of
a) $\frac{5}{7}$ b) 9   c) $\frac{1}{8}$ d) 0.7   e) $4\frac{2}{5}$ f) -3

Answers

a) $\frac{5}{7}$: $1 \div \frac{5}{7} = \frac{7}{5}$

b) 9: $1 \div 9 = \frac{1}{9}$

c) $\frac{1}{8}$: $1 \div \frac{1}{8} = 8$

d) 0.7: $1 \div 0.7 = 1 \div \frac{7}{10} = \frac{10}{7}$

e) $4\frac{2}{5}$:
(First, change to improper fraction)
$\frac{(5 \times 4)+2}{5} = \frac{22}{5}$
Therefore, $1 \div \frac{22}{5} = \frac{5}{22}$

f) -3: $1 \div -3 = -\frac{1}{3}$

Therefore, swapping/interchanging the denominator and the numerator of a fraction gives the reciprocal of that fraction.

EXERCISE 14

1) Write down the reciprocal of the numbers below.

a) 2   f) $-\frac{5}{7}$ k) -5

b) 4   g) -11   l) $-\frac{1}{12}$

c) $\frac{1}{7}$ h) $\frac{9}{14}$ m) $\frac{2}{11}$

d) $\frac{5}{9}$ i) $\frac{4}{7}$ n) - 4

e) -2   j) 12   o) $\frac{3}{17}$

2) Write down the multiplicative inverse of the following:

a) - 0.9   f) $3\frac{5}{7}$

b) - 1.5   g) $-2\frac{1}{2}$

c) - 1.25   h) - 4.5

d) -100   i) $\frac{1}{w}$

e) $5\frac{1}{7}$ j) $\frac{n}{7}$

3) Andrew says "The reciprocal of 6 is greater than the reciprocal of $\frac{1}{4}$. Is Andrew correct? Explain fully.

4) a) Write $5\frac{2}{3}$ as an improper fraction.
b) What is the reciprocal of $5\frac{2}{3}$?
c) Find the reciprocal of $-5\frac{2}{3}$
d) Multiply your answers from part a and part b.

15 Graphs and Gradients

This section covers the following topics:

- Straight line graphs
- Gradient
- Intercept
- Quadratic Graphs
- Mid-point of a line segment

LEARNING OBJECTIVES

By the end of this unit, you should be able to:

a) Draw straight line graphs
b) Find gradients and y-intercepts
c) Calculate mid-point of a line segment
d) Find the equations of straight lines using $y = mx + c$
e) Understand parallel and perpendicular lines
f) Draw quadratic graphs

KEYWORDS

- Straight line
- Gradients
- Y-intercept
- Parallel and perpendicular lines
- Quadratic Graphs

15.1 STRAIGHT LINE GRAPHS

By using the knowledge of coordinates, you can find points and draw a graph for any relationship. Before plotting a graph, we need to find the coordinate points (x, y) to locate a point. The coordinate pairs are better found using a **table of values.** It is a table that contains the coordinate pairs needed for drawing straight line graphs.

Example 1: Draw the graph of $y = x + 1$. Use x values from 0 to 5.

Solution: Draw a table of values.

Table of values →

x	0	1	2	3	4	5
y = x + 1	1	②	3	4	5	⑥
Coordinates	(0,1)	(1,2)	(2,3)	(3,4)	(4,5)	(5,6)

The number 2 in circle, was obtained by substituting (replacing) the value of x (1) in the equation original equation $y = x + 1$. From $y = x + 1$, $y = 1 + 1 = 2$. Also, the number 6 in circle, was obtained by substituting the value of 5 in the equation $y = x + 1$. From $y = x + 1$, $y = 5 + 1 = 6$. Do same to all the y values.

Therefore, from the table of values, we proceed to draw the graph.

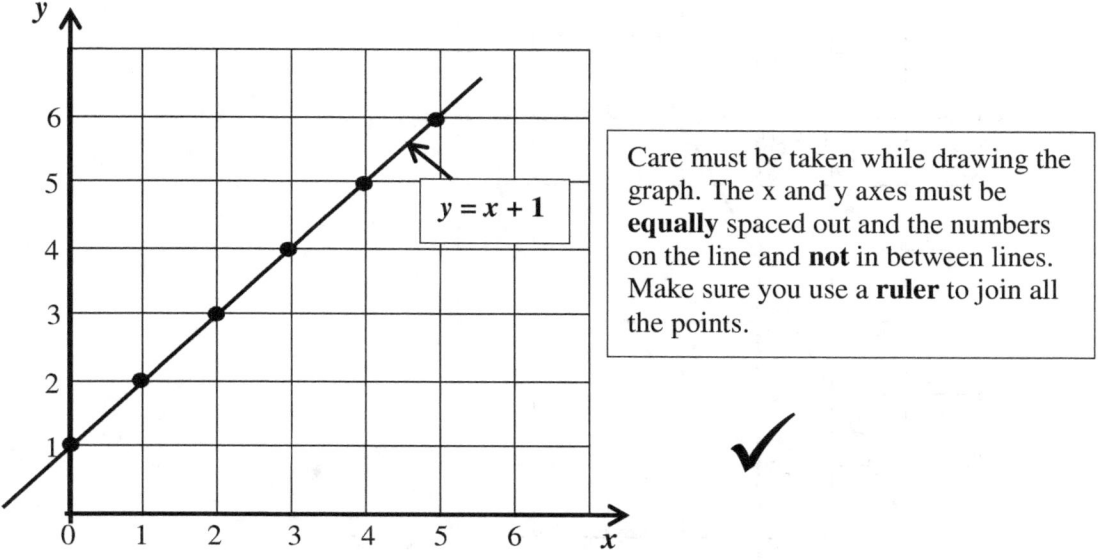

Care must be taken while drawing the graph. The x and y axes must be **equally** spaced out and the numbers on the line and **not** in between lines. Make sure you use a **ruler** to join all the points.

265

Example 2 Daw the graph of y = 2x - 2, x values from -2 to 3.

First, draw the table of values to get your coordinate pairs.

x	-2	-1	0	1	2	3
y = 2x − 2	ⓘ-6	-4	-2	0	ⓘ2	4
Coordinates	(-2, -6)	(-1, -4)	(0, -2)	(1, 0)	(2, 2)	(3, 4)

The number -6 in circle was obtained by substituting the value of x (-2) in the original equation, y = 2x - 2. y = (2 × -2) - 2 = -4 - 2 = - 6. Also, the value of 2 in circle was obtained by substituting the value of x (2) in the original equation y = 2x - 2. y = (2 × 2) - 2 = 4 - 2 = 2.

By using the coordinates from the table of values above, plot the graph. Remember: Since there are positive and negative numbers in the x and y coordinates, there must be four quadrants.

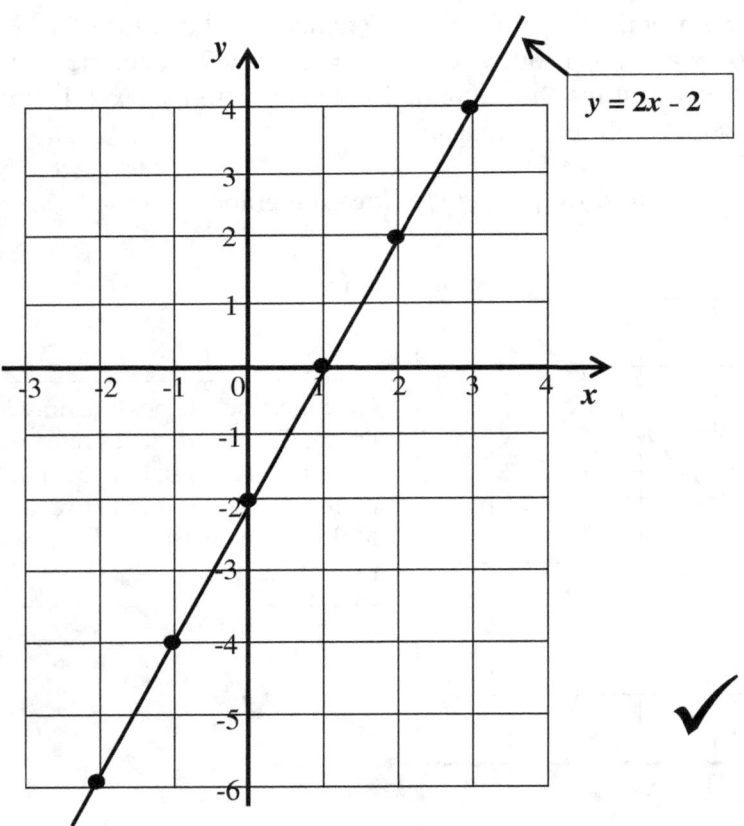

EXERCISE 15 A

1a) Copy and complete the table of values below for equation y = 3x + 1.

x	1	2	3	4	5
y = 3x + 1			10		
Coordinates			(3,10)		

b) Draw the graph of y = 3x + 1.
c) What is the coordinate of the point where the line touches the y-axes?

2a) Copy and complete the coordinates below using equation y = x + 3.

(-1, ?), (0, ?), (2, ?), (6, ?)

b) Draw a set of axes with x-values from -1 to 7 and y values from 1 to 10.

c) Plot the points with coordinates found in question 2a.

d) What is coordinate of the point where the line crosses the y-axis?

3) Draw the graph of the following equations. Use x values from -3 to 3.
a) y = x + 4
b) y = 2x
c) y = 5 – x
d) y = 3x - 3
e) $y = \frac{1}{2}x + 1$
f) $y = \frac{1}{2}x - 5$

4) A physicist uses the relationship C = 3d to convert the diameter (d) of a circle to its circumference (c).
a) Copy and complete the table below.

d	0	5	13	20
C				

b) Using d on the horizontal axis and c on the vertical, plot the points from 4a.

c) Copy the following table and use the graph to fill in the missing entries.

Diameter	Circumference
4	
16	
90	
	3

5) Two points P and Q are shown.

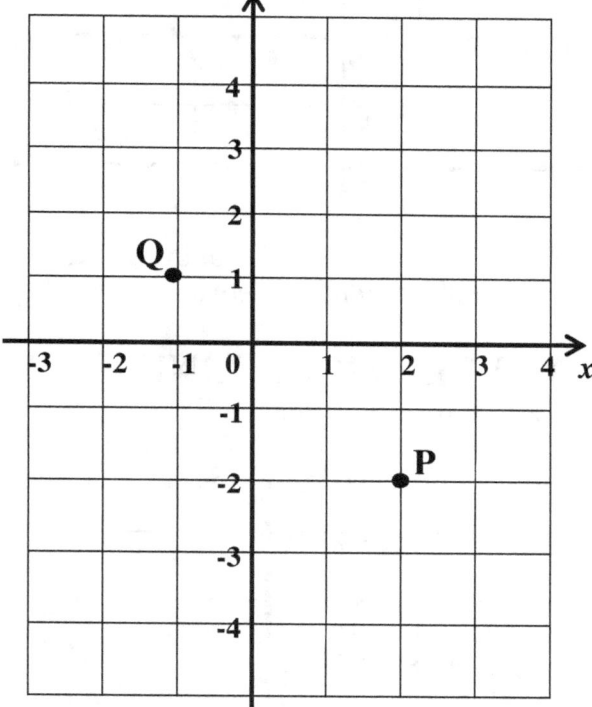

a) Write down the coordinates of P and Q.
b) Copy the graph above and join line PQ with a ruler.
c) Draw the graph of y = 3x – 4 on the same graph in 5b above.
d) Write down the coordinates of the point where line PQ cuts the graph of y = 3x - 4.

15.2 HORIZONTAL LINES

Horizontal lines have equations of the type y = 2, y = - 2 and so on as shown below. It starts with y because it touches the y axes.

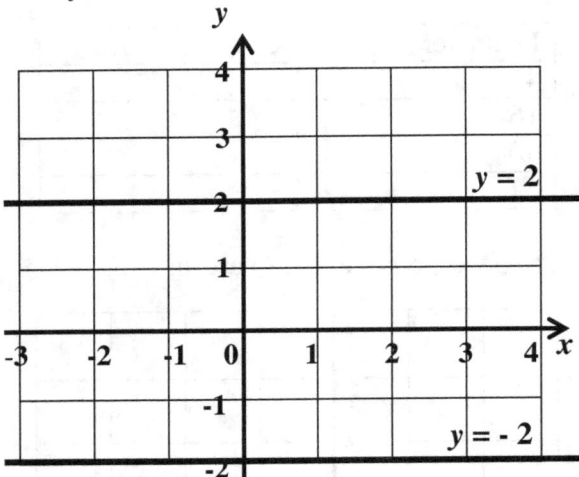

The horizontal line on top of the **x axes** and parallel to it has the equation **y = 0**.

15.3 VERTICAL LINES

Vertical lines have equations of the type $x = 1$, $x = -2$ and so on. It starts with x because it touches the x axes.

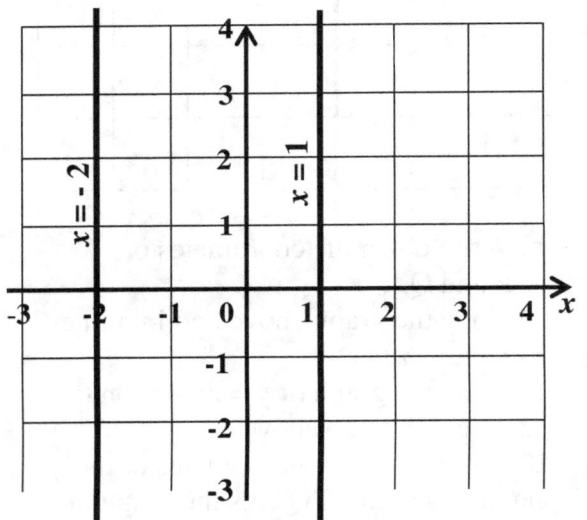

The vertical line on top of the **y axes** and parallel to it has the equation **x = 0**.

EXERCISE 15B

1) Write down the equation of each line in the diagram below.

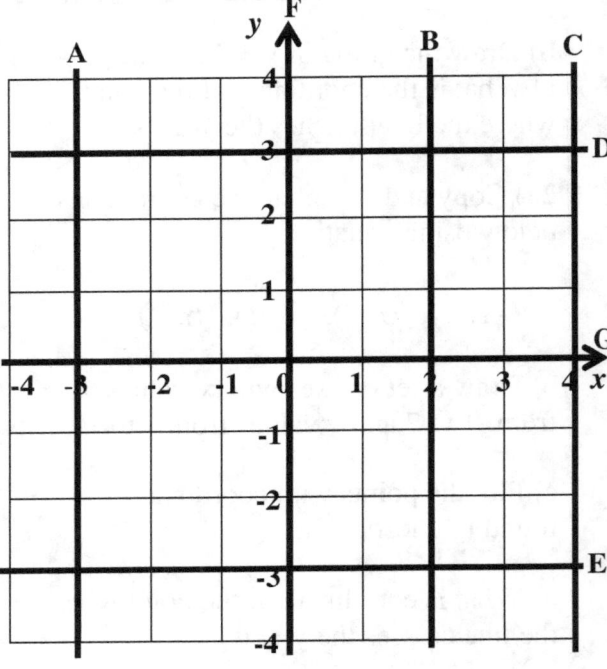

2a) Draw a rectangle with coordinates (1, 4), (-3, 4), (-3, -2) and (1, -2)

2b) Write down the equation of the lines that form the sides of the rectangle.

3a) Draw the lines y = 3, x = 3 and y = x on the same graph.

3b) Write down the coordinates of the triangle formed.

3c) Work out the area of the triangle formed if the lengths are in cm.

4) The table below shows the conversion rate from Pounds Sterling (£) to Euro (€).

£	0	5	10	15	20
€	0	10	20	30	40

a) Draw the conversion graph.

b) Use the graph to convert
i) £6   ii) £14   iii) £17 into Euros.

c) Using the graph, convert these amounts into Pounds Sterling:
i) €5   ii) €22   iii) €37

15.4 EQUATIONS OF STRAIGHT LINES

Equations of straight lines can be obtained by finding the **gradient** or **slope** of the line and the **y – intercept**.

The intercept of a graph is the value of **y** at the point where the graph crosses the y-axes.

Example 1

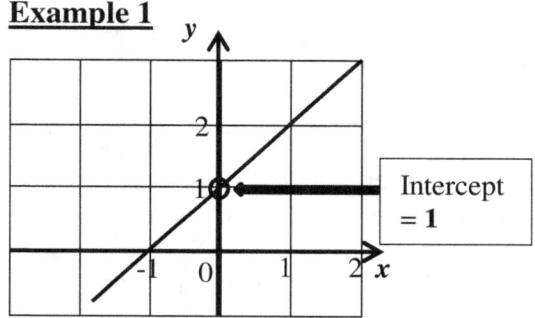

Intercept = 1

Example 2

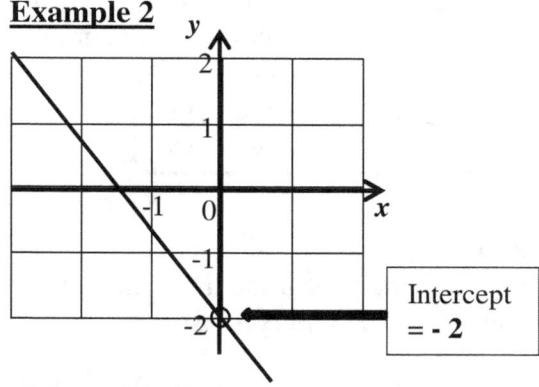

Intercept = - 2

Example 3

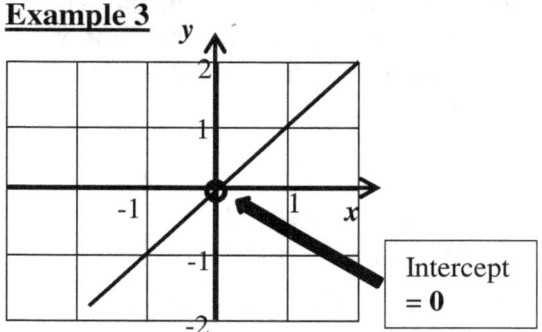

Intercept = 0

GRADIENT/SLOPE

The **gradient** of a line describes how steep the line is. The gradient can be positive, negative or zero and the direction of the line helps in determining the type of gradient.

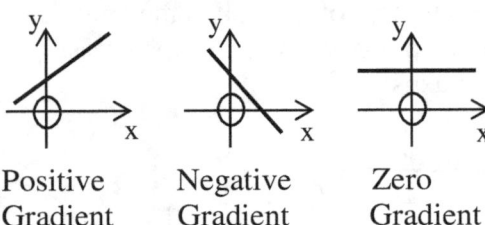

Positive Gradient | Negative Gradient | Zero Gradient

In other words, the gradient of a graph is the mathematical way of measuring its **steepness** or rate of change.

Note: If a line is vertical, the gradient **cannot** be specified.

The gradient can be measured between any two convenient points on the graph.

$$\text{Gradient} = \frac{\text{Change in y}}{\text{Change in x}}$$

Example 1:
Find the gradient of the line joining the points A (1, 1) and B (4, 3).

Solution: You may work this out with drawing a graph. Identify the x and y coordinates and apply the formula above.
For points A (1, 1) and B (4, 3)
          x, y   x1, y2

Change in y = 3 – 1 = 2
Change in x = 4 – 1 = 3

Therefore, gradient of AB = $\frac{2}{3}$

To use the graph to find the gradient of line AB, join the point with a ruler. Draw a perpendicular line from B to meet the horizontal line at C.

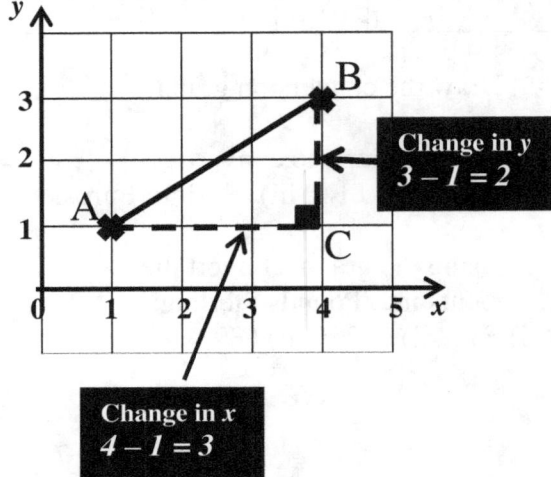

Gradient of AB = $\frac{2}{3}$ and it is **positive**.

Example 2: Find the gradient of the line joining the points P and Q.

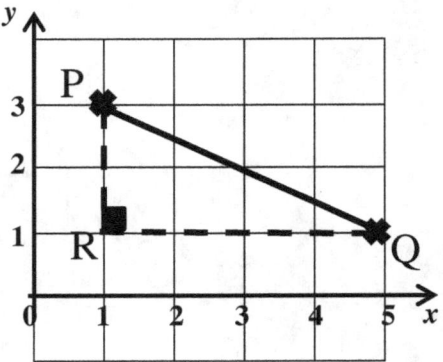

Gradient = $\frac{3-1}{1-5} = \frac{2}{-4} = -\frac{1}{2}$

The gradient is negative as the line slopes downwards.

270

Example 3: Without drawing the graph, work out the gradient of the line joining points A (-3, 1) and B (-1, 5).

Gradient = $\dfrac{Change\ in\ y}{Change\ in\ x} = \dfrac{5-1}{-1-(-3)}$

$= \dfrac{4}{2} = \mathbf{2}$

Example 3: Find the gradient of the line joining the points (0, 6) and (2, 1).

Gradient = $\dfrac{Change\ in\ y}{Change\ in\ x} = \dfrac{1-6}{2-0}$

$= \dfrac{-5}{2} = \mathbf{-2.5}$

The negative gradient indicates that the graph slopes downwards.

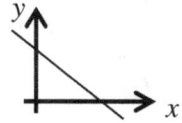

Example 4: Find the gradient of the line

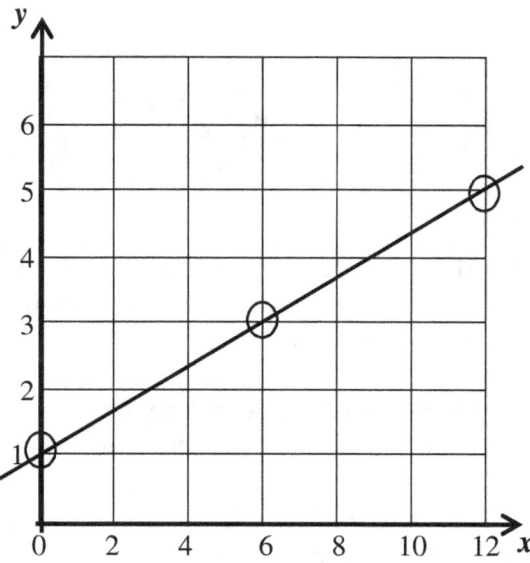

Three **convenient** points are circled.

No other point(s) is convenient as we must know the exact coordinates (integers).
Now, choose **any** of the two convenient points and draw a right angle triangle.

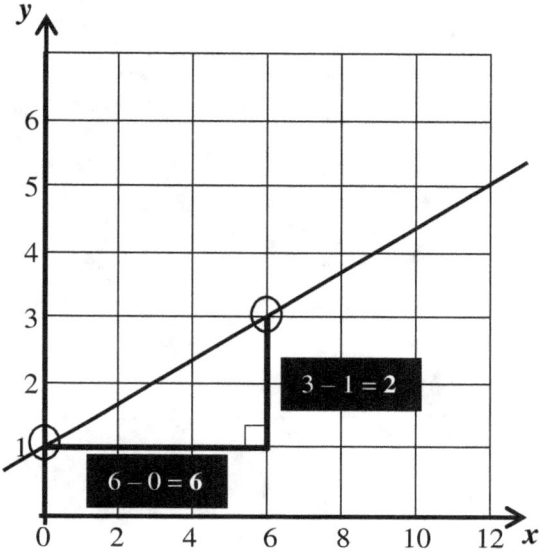

Gradient = $\dfrac{Change\ in\ y}{Change\ in\ x} = \dfrac{2}{6} = \dfrac{1}{3}$

EXERCISE 15 C

1) Calculate the gradients of the lines joining each of these pairs of coordinates.

a) (1,1) and (2,2)   i) (2,3) and (3,7)
b) (4,4) and (7,7)   j) (-1,4) and (-2,5)
c) (2,1) and (3,7)   k) (1,4) and (-1,2)
d) (1,2) and (4,8)   l) (-2,3) and (6,1)
e) (3,1) and (5,4)   m) (-1,8) and (-2,4)
f) (3,0) and (4,3)   n) (-3,5) and (-1,1)
g) (0,4) and (2,-6)   o) (6,11) and (-2,-5)
h) (-7, 10) and (0,0) p) (-5, -3) and (1,9)

2) Find the gradient of each line.

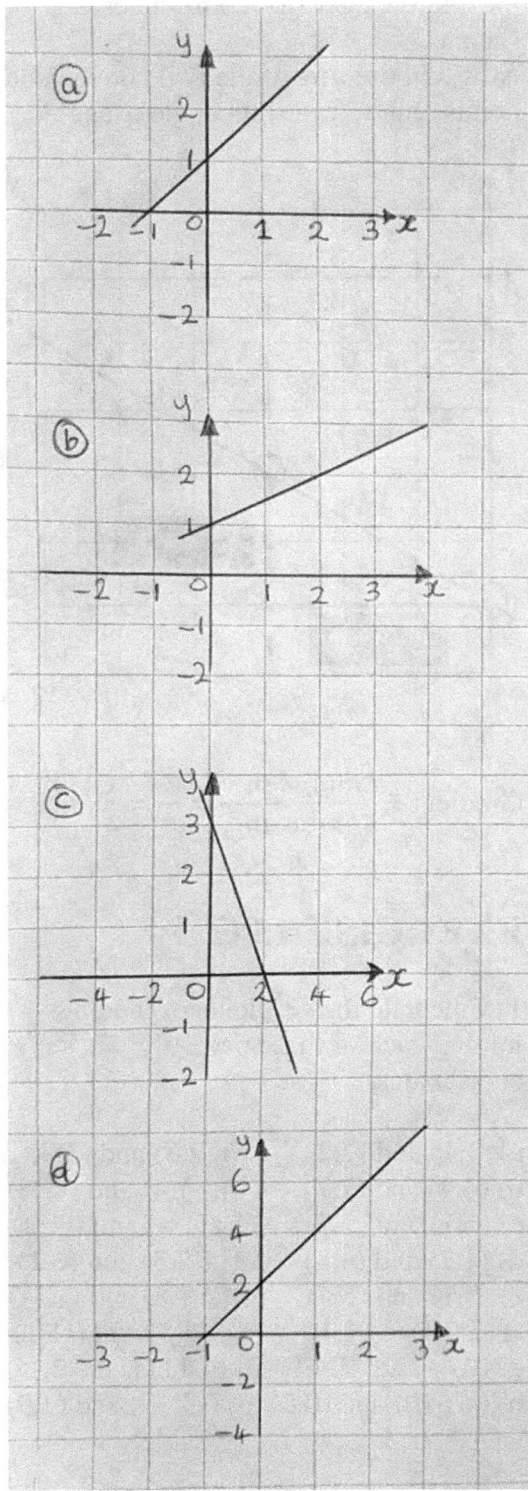

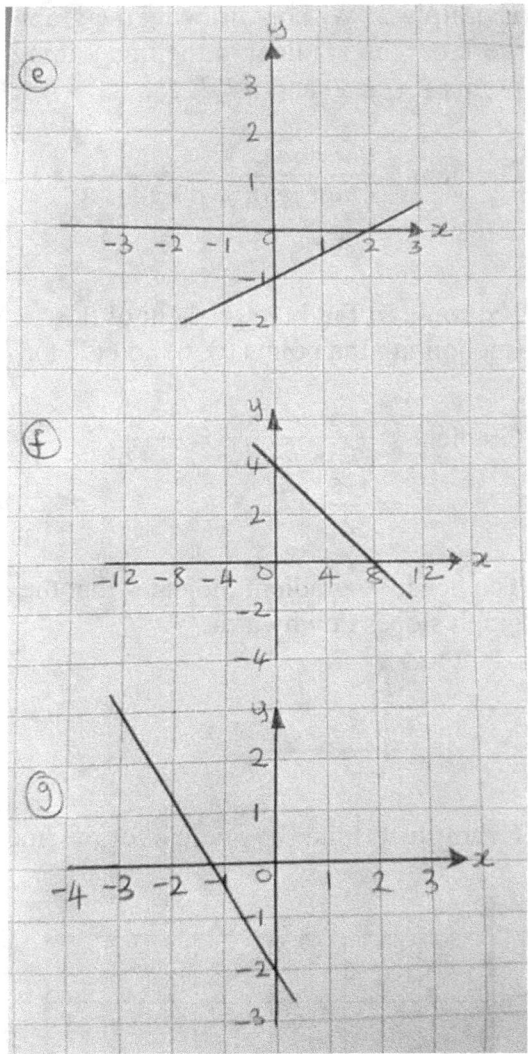

3) A quadrilateral ABCD has coordinates A (2, 9) B (1, 5) C) (4, 1) and D (10, 6). Find the gradient of each side of the quadrilateral.

4) For each pair of coordinate values, work out the slope of the line that joins them.
a) (b, b) and (0, 0)
b) (-p, 4p) and (2p, - 2p)
c) (0, 0) and (7q, 7q)
d) (g, 3g) and (-4g, - 4g)

15.5 MIDPOINT OF A LINE SEGMENT

The middle of a line segment is its **midpoint.** Remember that a **line segment** has a beginning and an end.

Example 1: Find the coordinates of the midpoint of line segment joining P(2, 5) and Q(7, 10).

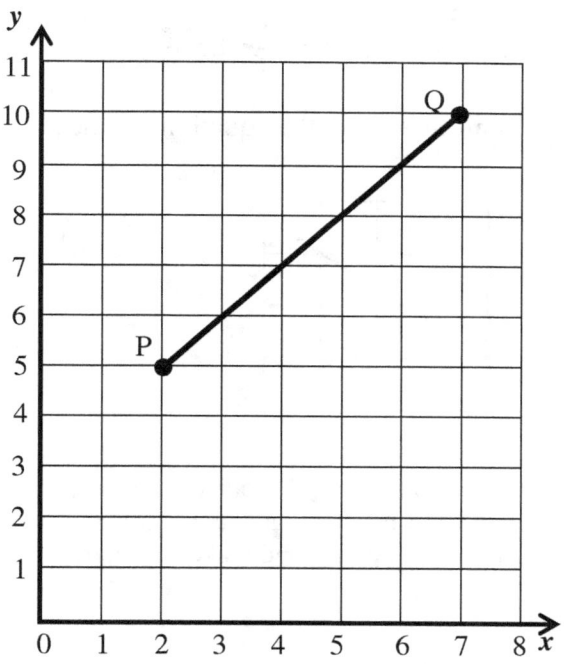

Mid-point (x) = $\frac{2+7}{2} = \frac{9}{2} = 4.5$

Mid-point (y) = $\frac{5+10}{2} = \frac{15}{2} = 7.5$

Therefore, the coordinates of the midpoint of PQ = **(4.5, 7.5)**

EXERCISE 15D

1) Calculate the coordinates of the midpoint of the line segment joining the pairs of points below.

a) (1, 6) and (7, 6) d) (0, 5) and (4, 1)
b) (1, 6) and (1, 2) e) (-2, -5) and (1, 6)
c) (0, 2) and (7, 6) f) (-3, -4) and (5, 0)

2) Work out the coordinates of the midpoints of the given line segments.

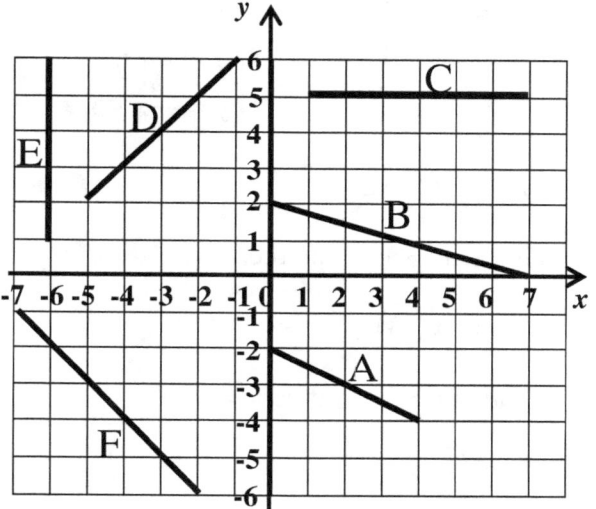

3) The coordinates of the midpoint of line AB are (-2, 2.5). If A is at (0, 0), work out the coordinates of B.

4) The coordinates of the vertex of a triangle are A(1,2), B (1, 6) and C(5, 2).
a) Work out the coordinates of the midpoint of AB, BC and AC.

b) Work out the area of the triangle if the length of each side of the square is 1cm.

15.6 Y = MX + C

The equation of a straight line is usually written in the form **y = m*x* + c,** where **m** is the gradient of the line and **c** the intercept on the y-axis.

$$y = mx + c$$

Gradient (m) — Intercept on the y-axis (c)
or slope

Example 1: Write down the gradient and intercept of the following:
a) y = 3x + 5   d) y = 3 − 2x
b) y = 2x − 3   e) y = 7 + $\frac{1}{2}$x
c) y = − 4x   f) y = − $\frac{3}{4}$x − 5

Solutions:

a) y = 3x (+5)

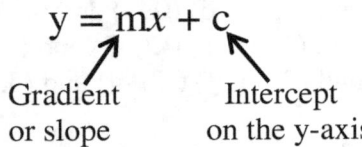

Gradient = **3**   Intercept = **5**

b) y = 2x (−3)

Gradient = **2**   Intercept = **−3**

c) y = (−4)x

Gradient = **− 4**   Intercept = **0**

d) y = 3(−2)x

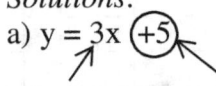

Intercept = **3**   Gradient = **−2**

e) y = 7 (+$\frac{1}{2}$)x

Intercept = **7**   Gradient = **$\frac{1}{2}$**

f) y = (−$\frac{3}{4}$)x (−5)

Gradient = **− $\frac{3}{4}$**   Intercept = **−5**

Example 2: Write down the equation of the graph given that the gradient is 2 and intercept is -3.

Solution: Remember the form of a straight line graph: y = mx + c. Substitute the values of m and c. The equation is **y = 2x − 3**

Example 3: Write down the equation of the graph with a gradient of -$\frac{1}{4}$ and 0 intercepts.

The equation is **y = -$\frac{1}{4}$x** or **-$\frac{x}{4}$**

Example 4: Find the equation of line A.

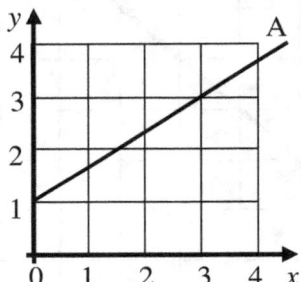

We need to find the gradient and the y-intercept. For the gradient, choose two convenient points and make a triangle.

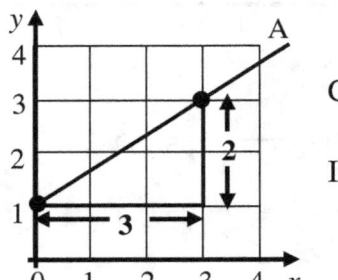

Gradient = $\frac{2}{3}$

Intercept = +1

Therefore, the equation of line A using
y = mx + c
$y = \frac{2}{3}x + 1$

Example 5: Find the equation of the straight line which passes through points (0, 2) and (6, 8).

Solution: You may decide to draw a graph and then find the gradient by drawing a triangle as shown earlier.

However, you may work out the gradient without drawing a graph as follows:

Gradient (m) = $\dfrac{\text{Change in } y}{\text{Change in } x}$

$m = \dfrac{8-2}{6-0} = \dfrac{6}{6} = 1$

Since one of the coordinate pairs is (0, 2), it means that the line must pass through y axis at 2. Therefore, c = 2

Using the $y = mx + c$, $y = 1x + 2$.
The equation of the line is **$y = x + 2$**

Example 6: Sketch the line $y = 2x + 1$.
Solution: We know that the graph must pass through +1 on the y axis (intercept). Since the gradient is positive, the line will slope upwards from left to right.

Also, the gradient is 2 from the equation which means that the steepness of the line must be 1 across and 2 upwards.

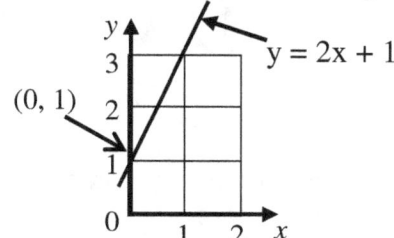

EXERCISE 15 E

1) Write down the equations of the straight lines with
a) gradient of 4 and through point (0,4)
b) gradient of -5 and through point (0,5)
c) gradient of 4 and y – intercept of -6
d) gradient of $\dfrac{2}{3}$ and y-intercept of 4
e) gradient of -4 and y-intercept of -5
f) gradient of $-\dfrac{5}{6}$ and y-intercept of -4

2) Find the equation of the line which passes through
a) (0,4) and (3,2)   b) (5, 2) and (0, -2)
c) (0, -1) at a gradient of 3

3) A shop has a conversion table below.

Pounds (£)	5	10	15	20	25
Naira (₦)	8	21	34	47	60

a) Draw a graph for the information.
b) Find the equation of the line connecting £ and ₦.

4) Find the equation of each line.

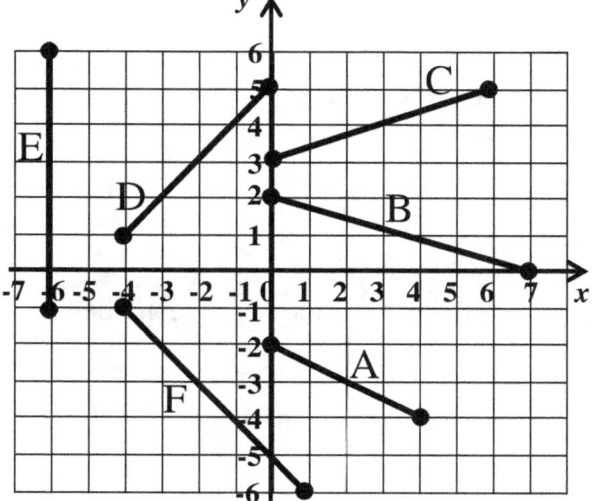

15.7 PARALLEL LINES

Parallel lines can never meet. They have the same distance apart and are always seen with arrows.

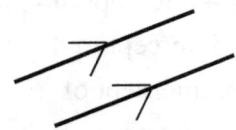

Lines with the **same gradients** are parallel.

Example 1: These lines are parallel because they have the same gradient.

y = **3**x   y = **3**x – 1   y = 9 + **3**x
   ↑   ↑   ↑
gradient   gradient   gradient

All have 3 as their gradient(s).

Example 2: Is the line 2y – 6x = 8 parallel to the line y = 3x – 1?

Solution: First put 2y – 6x = 8 in the form y = mx + c by making y the subject of the formula.

2y – 6x = 8
 (+6x)   (+ 6x)
    2y = 6x + 8
  (÷2)   (÷2)
     y = 3x + 4

The gradient is 3 likewise the gradient for y = 3x – 1.

Therefore, the two lines are parallel.

15.8 PERPENDICULAR LINES

Two straight lines are perpendicular (at right angles) if the **product** of their gradients is **-1**.

$$m \times m_1 = -1$$

If a line has gradient, **m**, then a line perpendicular to it must have a gradient of $-\frac{1}{m}$.

The two lines below are perpendicular. Hence, the product of their gradients 1 × -1 = -1.

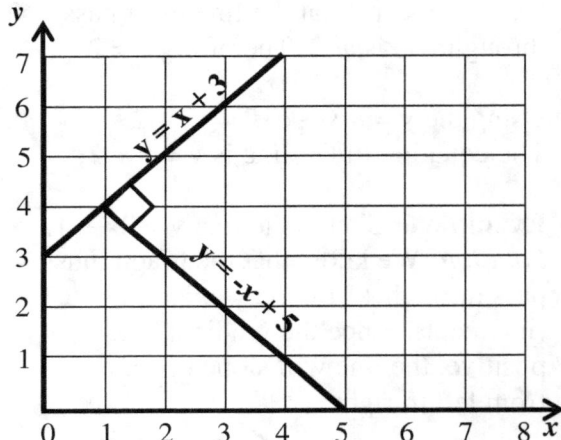

Example 3: Show that the two lines y = 4x + 5 and y = $-\frac{1}{4}$x + 3 are perpendicular.

y = 4x + 5   y = $-\frac{1}{4}$x + 3
    ↑   ↑
Gradient = 4   Gradient = $-\frac{1}{4}$

4 × ($-\frac{1}{4}$) = - 1

Since the product of the two gradients is **-1**, the two lines are perpendicular.

Example 4: A line joins (1, 2) and (5, -6). Find the gradient of a line perpendicular to this line.

Solution: Find the gradient of the line with points (1, 2) and (5, -6).
Gradient (m) = $\frac{-6-2}{5-1} = \frac{-8}{4} = -2$

Therefore $-2 \times m_1 = -1$
$$m_1 = \frac{-1}{-2} = \frac{1}{2} \checkmark$$
....Where m_1 is the gradient of the line perpendicular to the original line
Check: $-2 \times \frac{1}{2} =$ **-1**

EXERCISE 15 F

1) Look at the pairs of equations below:
P) $y = 2x - 2$ and $y = -\frac{1}{2}x + 5$
Q) $y = 4x + 3$ and $y = 5 + 4x$
R) $y = 5x - \frac{1}{2}$ and $y = -\frac{1}{5}x + \frac{1}{2}$
S) $y = 4 + 9x$ and $y = 9x - 3$
T) $y = 8x - 7$ and $y = 7x - 7$

State which of the lines are
a) parallel to each other
b) perpendicular to each other
c) neither perpendicular nor parallel to each other.

2) Draw the line $y = 2x + 1$. State its **gradient** and draw a line perpendicular to the line. What is the gradient of the line perpendicular to $y = 2x + 1$?

3) Line $y = 5x + c$ passes through the point (3, 20). What is the value of c?

4) A straight line parallel to $y = x + 1$ passes through the point (0,4) What is the equation of the line?

15.9 QUADRATIC GRAPH

Quadratic functions will always have an x^2 **term** with no other higher powers. The graph of the quadratic function is a curve known as a **parabola**.

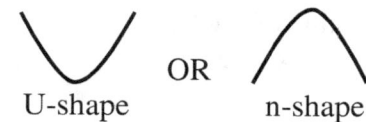

U-shape   OR   n-shape

Drawing a quadratic graph is like drawing a linear graph by substituting the values of *x* to find the corresponding *y* values. The difference is that the graph of a quadratic function will always be a U or n-shape while that of a linear function will always be a straight line.

Example 1: Draw the graph of $y = x^2 + 1$, x values from -2 to 2

Solution: First draw a table of values.

x	-2	-1	0	1	2
x^2	4	1	0	1	5
$y = x^2 + 1$	5	2	1	2	5

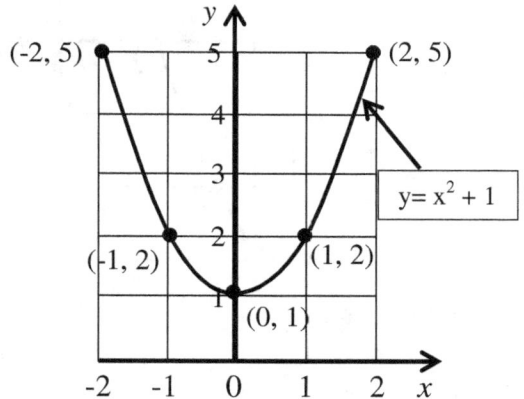

Note that the points are joined with a smooth curve.

Example 2
a) Draw the graph of $y = x^2 - 3$, x values from -3 to 3.
b) Use your answer from part a to estimate the value of y when
i) x = 0.5   ii) x = -2.5

x	-3	-2	-1	0	1	2	3
x^2	9	4	1	0	1	4	9
$y = x^2 - 3$	6	1	-2	-3	-2	1	6

a)
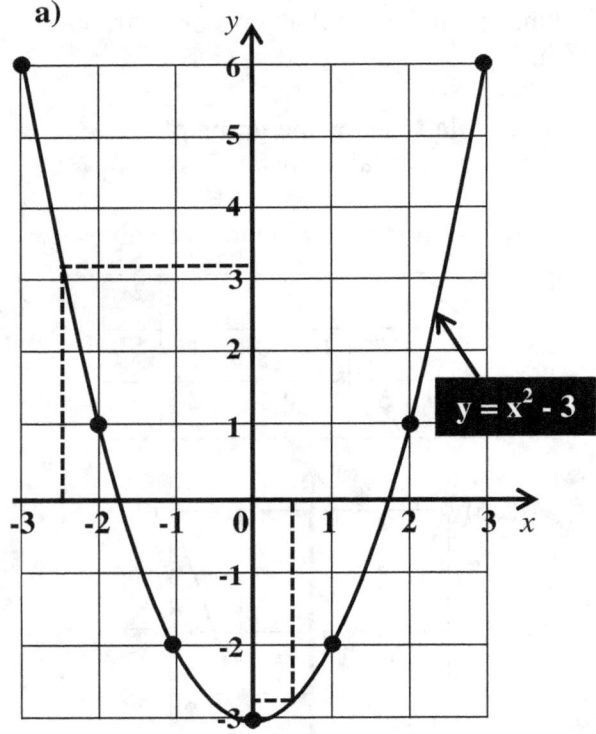

From the graph,

b) i) when x = 0.5, y = -2.75
   ii) when x = -2.5, y = 3.25

EXERCISE 15 G

1a) Copy and complete the tables below to find the value of $y = x^2 + 4$.

x	-3	-2	-1	0	1	2	3
x^2	9					4	
$y = x^2 + 4$			5				

1b) Draw the graph of $y = x^2 + 4$.

2) For x values from -3 to 3, plot the graph of
a) $y = x^2 - 5$ c) $x^2 + x$
b) $y = x^2 - 1$ d) $5 + x^2$

3a) Draw the graph of $y = x^2 - 4$, x values from -5 to 5.
b) Use your answer from part **a** to estimate the value of y when
i) x = 1.5   ii) x = -2.5

EXTENSION WORK

4) Draw the graph of the following equations, x values from -3 to 3.

a) $y = x^2 + 3x + 1$
b) $y = x^2 + 2x - 1$
c) $y = -x^2 - x - 2$

5a) Draw the graph of $y = x^2 + 1$, x values from -3, to 3.

b) Draw the graph of y = 5 on the same axes as in part **a** above.

c) Use your graph to solve the equation $x^2 + 1 = 5$

Chapters 14 & 15 review Sections
Assessment

1) a) Write $3\frac{2}{3}$ as an improper fraction.   b) What is the reciprocal of $3\frac{2}{3}$?
   c) Find the reciprocal of $-3\frac{2}{3}$ d) Multiply your answers from part a and b. ...**4 marks**

2a) Copy and complete the table of values below for equation $y = 5x - 1$.

x	1	2	3	4	5
y = 5x - 1			14		
Coordinates			(3,14)		

.... **3marks**

b) What is the coordinate of the point where the line touches the y-axes? **1mark**

3) A quadrilateral ABCD has coordinates A (2, 5) B (5, 9) C) (9, 5) and D (5, 1).

Find a) the gradient of line CD   ………...**2 marks**

   b) the equation of line CD   ………...**2 marks**

4) Calculate the coordinates of the mid-point of the line segment joining the pairs of points. a) (1, 6) and (1, 2) b) (-2, -5) and (1, 6)   …………. **4 marks**

5) Find the equation of the line which passes through
a) (0,7) and (3,2)   b) (0, -1) at a gradient of 4   …………. **4 marks**

6) Show that the two lines $y = 4x + 5$ and $y = -\frac{1}{4}x + 3$ are perpendicular......… **3 marks**

16 Diagrams

This section covers the following topics:

- Scatter graphs
- Line of best fit
- Stem-and-leaf diagram

LEARNING OBJECTIVES

By the end of this unit, you should be able to:

a) Identify types of correlations
b) Draw a scatter graph/diagram
c) Use line of best fit
d) Draw and interpret a stem-and-leaf diagram

KEYWORDS

- Scatter diagram/graph
- Correlation
- Positive
- Negative
- Line of best fit
- Stem-and-leaf diagram

CORRELATIONS

Scatter graphs are used to investigate if there is a **correlation** (link) or connection between sets of data. In other words, two variables are correlated if they are related to each other in any way.

TYPES OF CORRELATION

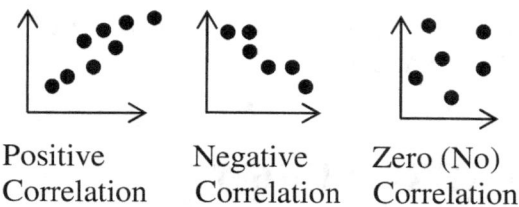

Positive Correlation   Negative Correlation   Zero (No) Correlation

POSITIVE CORRELATION

We think of a **positive** correlation as a relationship between two variables where if one variable increases, the other one increases. Also in a positive correlation, if one variable decreases, the other also decreases.

If the weight and height of a group of children are recorded, the graph would show a positive correlation and looks like this:

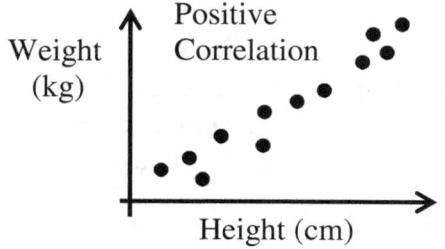

NEGATIVE CORRELATION

A **negative** correlation is when one quantity increases and the other decreases.

If the GCSE grades are recorded against absenteeism of a class, the graph will show a negative correlation.

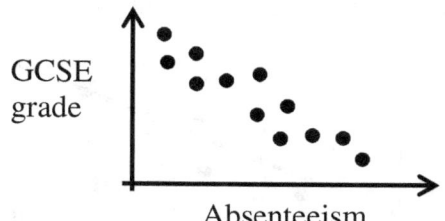

NO (ZERO) CORRELATION

When there is no link or relationship between two variables, the graph will show a zero or no correlation.

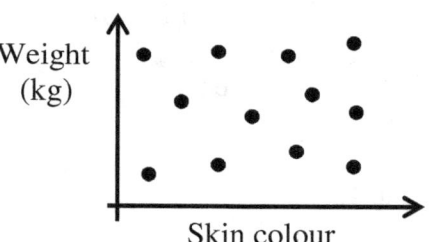

A positive or negative correlation can also be **weak** or **strong** depending on the closeness of the dots in a scatter diagram. The closer the dots are, the stronger the correlation. The further away the dots are, the weaker the correlation.

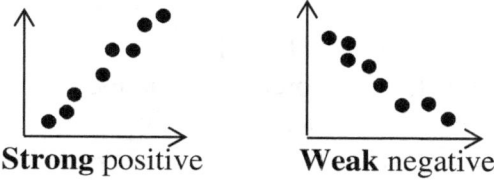

Strong positive   **Weak** negative

LINE OF BEST FIT

A **line of best fit** can only be drawn when a scatter graph shows either positive or negative correlation. It is a line **straight line** that passes close to most of the points. As a guide, there should be approximately the same number of points on each side of the line.

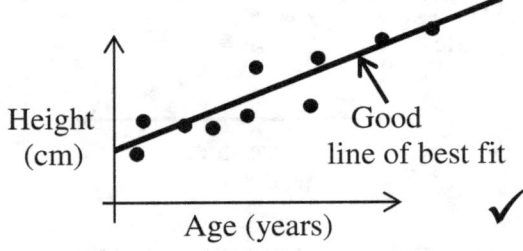

Good line of best fit ✓

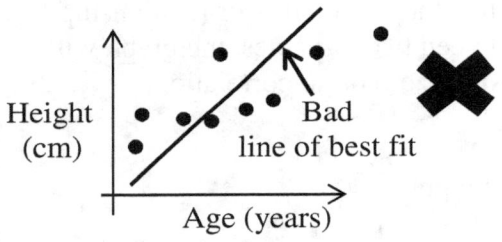

Bad line of best fit ✗

Example 1:

a) Plot the points from the table below on a scatter diagram.

Maths test	5	30	17	7	25	25
Chemistry test	7	25	15	10	23	20

b) Draw a line of best fit
c) If a student scored 20 in maths, estimate the student's score in Chemistry.

a)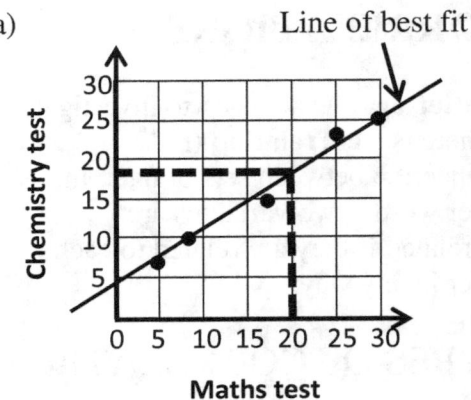

b) See diagram above for line of best fit.
c) 18. See diagram

EXERCISE 16 A

1a) Describe the correlations shown on the graphs below.

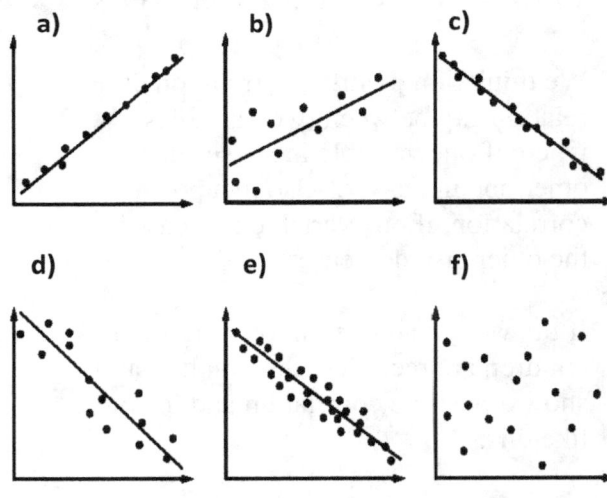

2a) Plot the points on a scatter graph.
b) Describe the correlation, if any, between the values of x and y.

x	3	10	4	20	15	12
y	4	12	4	18	16	13

282

3a) Plot the points on the scatter graph for each table and state the type of correlation.
b) Draw a line of best fit for i and ii.
c) Estimate the value of q for i and ii when p = 15.

i)

P	5	12	18	21	23	5	8
q	22	18	14	10	9	24	21

ii)

P	5	19	18	8	15	24	11
q	8	18	20	6	17	24	13

iii)

P	3	16	18	4	21	4	7
q	14	8	6	13	5	14	12

4) The graph below shows how many hours on Friday and Saturday that some students spent watching TV and doing homework.

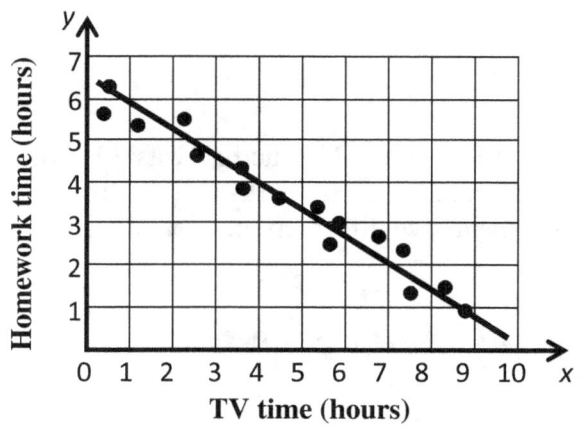

a) Describe the correlation between time spent doing homework and time spent watching TV.
b) If Isabel did $5\frac{1}{2}$ hours of homework, predict how many hours she spent watching TV.

STEM-AND-LEAF DIAGRAM

This is a way of arranging data using a key to explain what the stem and leaf represents.

Example 1: The **ordered** stem and leaf diagram below shows the number of students on each of the 15 boat cruises.

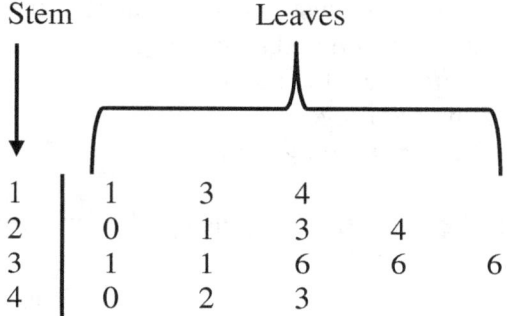

Key: 1|2 represents 12 students

The above stem-and-leaf diagram is "ordered" because it is written in order. Therefore, the numbers from the stem-and-leaf diagram represents

11, 13, 14, 20, 21, 23, 24, 31, 31, 36, 36, 36, 40, 42, 43

The **mode** is the number that occurs the most and it is **36**.

Range is highest value take away lowest value: 43 − 11 = **32**

Median is the middle value which is **31**
It can also be worked out by
$\frac{(15 + 1)}{2}$ th value = $\frac{16}{2}$ = 8

Therefore, the 8th number is **31**.

EXERCISE 16 B

1a) Draw an ordered stem and leaf diagram to represent the information below.
5   2   12   12   15   17   9   22   30   25   27   29

b) Work out   **i)** the mode   **ii)** the median   **iii)** the range

2) Prices for breakfast meals are recorded below, in Pounds (£).
1.50   2.25   3.00   1.75   4.05   1.80   3.33   2.35   3.70   1.90   3.00

a) How many customers bought breakfast?
b) Draw an ordered stem and leaf diagram for the information above.
c) Write down the mode.
d) Find the range
e) Work out the median.

3) The number of bicycles parked in a field each day at 8 am is recorded in the stem and leaf diagram below.

3	4	3	4	4	
2	6	2	1		
1	1	3	4	2	

a) Draw an ordered stem and leaf diagram.
b) How many bicycles were parked at 8 am?
c) Work out the median.
d) Find the range.
e) Work out the mode

Key: 2|6 represents 26 bicycles.

EXTENSION WORK

4) Andrew recorded the scores in a maths test of 13 students. The median was 30 marks. The range was 35 marks. However, five students scored over 38 marks and the lowest mark was 15. Draw a stem and leaf diagram for the above information. Create/design the entries to finish the diagram.

5) The number of customers using a gymnasium for a week was recoded for men and women in a **back-to-back** stem and leaf diagram.

Men	Stem	Women
6 5 4 2 1	1	2 4 5 6
6 5 3	2	1 4 4 4 4
5 5 4 1	3	2 3 5

Key: 1|2 = 21 customers

Key: 1|2 = 12 customers

a) Calculate the mode for the men.
b) Work out the mode for the women.
c) Calculate the mean for the men.
d) Find the range for the women.
e) Wok out the median for the women.

17 Averages and Range

This section covers the following topics:

- Mean
- Median
- Mode
- Range

LEARNING OBJECTIVES

By the end of this unit, you should be able to:

a) Calculate the mean of a set of numbers
b) Find the median of a set of numbers
c) Identify the mode from a given set of data
d) Write down mode from a bar chart
e) Find the range from a bar chart
f) Work out the mean from a frequency table
g) Find the mode from a frequency table
h) Find the median from a frequency table
i) Find the averages and range from grouped data

KEYWORDS

- Averages
- Mean
- Median
- Mode
- Range
- Grouped data

17.1 AVERAGES

An **average** is a central number that is representative of all the numbers in a set. The **mean**, **median** and **mode** are the three most commonly used averages.

There is also a concept known as the *range,* which **is not** an average, but the difference between the highest and lowest numbers.

Example 1:
Joe completed seven mathematics homework and his marks out of 20 recorded as shown below.

9   11   7   8   15   14   6

A set of values like the one above is called a **distribution**.

MEAN
The mean of a probability distribution is obtained by adding all the numbers in the distribution and then divide by the number of items.

The mean for Joe's marks would be
$$= \frac{9 + 11 + 7 + 8 + 15 + 14 + 6}{7}$$
$$= \frac{70}{7} = 10 \checkmark$$

Example 2: Calculate the mean of 6, 7, 3, 4, 1, 9.
$$\text{Mean} = \frac{6 + 7 + 3 + 4 + 1 + 9}{6} = \frac{30}{6} = 5 \checkmark$$

Example 3:
Calculate the mean of 4, 7, 4, 2, 7.
$$\text{Mean} = \frac{4 + 7 + 4 + 2 + 7}{5} = \frac{24}{5} = 4.8 \checkmark$$

EXERCISE 17 A

1) Calculate the mean of each distribution.
a) 3, 7, 2
b) 10, 5, 12, 6, 12
c) 2, 4, 6, 6, 8, 10
d) 7, 9, 2, 4, 2, 1, 3
e) 5, 8, 4, 9, 10, 12, 23, 35

2) In a sale, Eric bought: three books at ₦250 each, two books at ₦450 each, six books at ₦125 each and four books at ₦300 each.
a) How much did Eric pay for all the books he bought?
b) What was the mean cost of the books?

3)

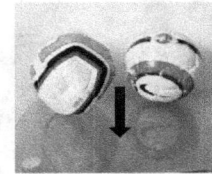 

410 g each   450 g

a) Calculate the overall mass of the three balls.
b) Calculate the mean mass of the three footballs.

4)

Two baby pandas weigh 225 g and 350g respectively. What is the average weight of the two pandas?

5) Work out the mean of the following distributions. Leave some answers as *mixed numbers* in their simplest form.

a) 7, 5, 9, 9, 5, 6, 4, 3
b) 10, 7, 12, 7, 24, 20
c) 9, 7, 4, 3
d) 12, 3, 4, 5, 2, 5, 4
e) 7 m, 2 m, 10 m, 3 m, 8m
f) 3.75, 0.5, 0.8, 0.95

6) Five students took a geography examination and scored the following marks: 12%, 33%, 50%, 30%, and 60%. Calculate the mean percentage mark.

7) The mean of three numbers is 9.
a) What is the sum of the three numbers?
b) If two of the numbers are 15 and 4, what is the third number?

8) Ifeoma covered a number from the set of numbers below and said: "The mean of the numbers is 6.34."

3.5   4   9.2   8

What is the value of the covered number?

9) To one decimal place, work out the mean of these numbers.
a) 8, 11, 3, 4, 3, 2
b) 10, 3.3, 0.7, 6.5, 9.5, 3, 6, 7
c) 6, 5, 7, 4, 3, 2, 1, 6, 7, 8, 9
d) $\frac{1}{2}, 2\frac{1}{4}, 5\frac{1}{5}$

10) Enugu Rangers football team's goal average was 2.3 after 30 matches. How many goals did Enugu Rangers football team score?

MEDIAN

The **median** is the middle value in the list of numbers when arranged in **order of size**. The order could be ascending or descending.

However, if there are two middle numbers, find the average of the two numbers (add them and divide by 2). That would give the median of the numbers.

Example 1: Ebuka sometimes cycles to school. He recorded the number of times he cycled to school in the last seven weeks, as shown below.

1, 2, 5, 0, 3, 2, 4
To find the median, first, arrange in order of size.

It becomes: 0, 1, 2, ②, 3, 4, 5

There are seven numbers in the list, the 4th number ($\frac{7+1}{2}$) is the median. So, the 4th number which is also the middle value is 2; therefore the median number of times Ebuka cycled to school is **2**. ✓

Notice that you would still get the same answer if you had arranged from highest to lowest. 5, 4, 3, ②, 2, 1, 0

Example 2:
Find the median of 5, 4, 7, 2, 3, 1

First arrange in order: 1, 2, ③, ④, 5, 7

Notice that 3 and 4 are the middle numbers. (3 + 4) ÷ 2 = 7 ÷ 2 = 3.5
Therefore, **3.5** is the median. ✓

EXERCISE 17 B

1) Write down the median value of each set of numbers.

a) 3, 6, 5, 2, 9, 8, 7
b) 9, 7, 1, 0, 3, 2
c) 12, 8, 4, 3, 1, 7, 6, 4, 2
d) 8, 1, 6, 4
e) 10, 20, 80, 40, 50
f) 4, 3, 2, 8, 7, 6, 5, 4

2) Two students, Chuba and Tochukwu, had five maths tests over a term. The tests were marked out of 30, and the results are shown below:

Chuba: 19, 23, 17, 16, 20
Tochukwu: 6, 21, 27, 3, 20

a) Find the median mark for Chuba.
b) Find the median mark for Tochukwu.
c) Comment on the marks of both students.

3) The morning temperatures (°C) in Lagos for a week in January were:

27, 19, 30, 16, 21, 15, 20
Find the median temperature in Lagos.

4) The show sizes of eight sailors are:
12, $10\frac{1}{2}$, 10, 11, 8, 10, 9, 7
Find the median shoe size for the data.

5) The height of some tables in metres was recorded as 0.9, 1, 1.2, 0.8, 0.85
Work out the median of the heights.

6) Find the median value of
a) 8, 9, 12.5, 4.5, 3, 2.5, 7, 10.5, 4, 3
b) 60, 70, 45, 35, 75

MODE

The most occurring number or value in a distribution is called the **mode** or **modal value**.

You may also have two modes which are explained in Example 2 below. The distribution with two modes is sometimes said to be **bimodal**. Other distributions with more than two modes are said to be **multimodal**. The mode can also represent qualitative data.

Example 1:
Find the mode of the set of numbers.
5, 3, 4, 5, 6, 5, 5, 7, 2

It is always advisable to arrange the distribution in order of size (Not a must). It helps in identifying the mode or modes quicker.

In order of size, the above distribution is 2, 3, 4, 5, 5, 5, 5, 6, 7

It is very clear from the numbers above that the most occurring number is 5, as it appeared four times and no other number or numbers appeared four times. Therefore, **5** is the mode. ✓

Example 2:
Find the mode of the numbers below.
6, 9, 10, 15, 6, 3, 10, 7

In order of size: 3, **6, 6**, 7, 9, **10, 10**, 15
Two numbers appeared twice.
Therefore, the modes are **6** and **10**. ✓

Example 3: Work out the mode for 1, 4, 3, 5. There is **no mode**. ✓

EXERCISE 17 C

1) From the list of numbers below, work out the mode.

a) 2, 5, 4, 1, 2, 1, 1, 7,
b) 3, 5, 8, 9, 3,
c) 4, 2, 12, 7, 12, 7, 5
d) 1, 9, 8, 15, 9, 3, 9, 4, 9
e) 20, 30, 40, 45, 58

2) Kolade asked six people what size of shoes they wear, and he recorded their sizes. 7, $8\frac{1}{2}$, 9, 8, $10\frac{1}{2}$, 8. What is the mode of their shoe sizes?

3) In a school survey, six students said their favourite colours. The results are given below.
   Blue Red Blue Yellow Pink Blue
Work out the mode of their colours.

4) A pyramid was designed with different colours as shown. What is the mode of the colours?

5) Write down the mode for each set of data below.

a) 13, 13, 43, 34, 53, 13, 50, 34
b) 7, -8, 6, -3, -8, 2, 3, 8
c) 3.4, 3.5, 1.2, 3.5, 6.7, 2.8
d) $\frac{2}{5}, \frac{1}{5}, \frac{2}{5}, \frac{3}{5}, \frac{3}{6}, \frac{2}{5}, \frac{2}{4}, \frac{2}{5}, \frac{2}{4}$

6) Write down the modal type for each set of data below.
a) Blue, white, yellow, black, blue, red,
b) π, £, √, ₦, π, Θ, π, Θ
c) Dog, cat, rabbit, cat, mouse, dog

RANGE

The **range** is not an average, but the difference between the highest and lowest values in the **distribution**.

Example 1: Work out the range of the set of numbers: 4, 7, 2, 9, 13, 4, 2

Solution:
You may decide to first arrange the numbers in order of size before looking for the highest and lowest numbers.
In order of size: 2, 2, 4, 4, 7, 9, 13

Highest number = 13,
Lowest number = 2
Therefore, the range is 13 - 2 = **11** ✓

EXERCISE 17 D

1) Work out the range of the set of numbers below.

a) 6, 11, 8, 2, 10
b) 4, 5, 3, 12, 15
c) 19, 23, 35, 14, 8, 5
d) 90, 40, 30, 85, 20, 10, 60

2) The number of goals scored in seven football matches in one state is shown.
   2 1 3 1 2 2 1
Work out the range.

3) In a small class of five pupils, their heights are 140, 156, 134, 170, and 165. Work out the range of their heights in centimetres.

4) In exercise 20 C, work out the range of the data in question 5.

17.2 AVERAGES AND RANGE FROM FREQUENCY TABLES AND DIAGRAMS

AVERAGES AND RANGE FROM A FREQUENCY TABLE

Example 1: The frequency table below shows the marks obtained by a class in a history test.

Marks	Frequency
5	1
8	2
12	5
15	3

The **modal mark** is the mark with the highest frequency. The highest frequency is 5, therefore the modal mark is **12**.

The **range** is
(Highest mark - the lowest mark) 15 – 5 = **10**

For the **median mark**, first, add up the frequencies. 1 + 2 + 5 + 3 = 11.
The median is $\left(\frac{n+1}{2}\right) th$ value. (11 + 1) ÷ 2 = 6.

It means that the median lies on the **6th value.** Go back to the frequency and start from the top, keep adding the numbers until you get to 6 or more and read off the corresponding *Mark's* value.

1 + 2 = 3……..But 3 is not up to 6, so consider adding the next number which is 5.
3 + 5 = 8 this is more than 6 but still within limits. Read across to give 12. Therefore, the **median mark** is **12.**

To calculate the mean, extend the column as shown below.

Mark	Frequency	Mark × Frequency
5	1	5 × 1 = 5
8	2	8 × 2 = 16
12	5	12 × 5 = 60
15	3	15 × 3 = 45
Total	11	126

Mean = $\frac{\text{total of (mark} \times \text{frequency)}}{\text{total number of frquency}}$ = $\frac{126}{11}$ = **11.45** to 2 decimal places.

Note: A common mistake is to divide by 4. Always divide by the total of all the frequencies and in the above example, 11.

EXERCISE 17 E

1) The table below shows the weights of the apples in a bag.

Weight (g)	Frequency
100	5
110	7
115	10

Work out the
a) modal weight   b) median weight
c) range   d) mean weight.

2) A six-sided dice is rolled in an experiment and the results shown in the frequency table below.

Score	Frequency
1	10
2	20
3	6
4	25
5	12
6	15

Work out the following:
a) the modal score   b) the range
c) the median score d) the mean score.

3) The table below shows the number of goals scored in each football match by Enyimba FC for a season in Nigeria.

Goals	Frequency
0	8
1	19
2	11
3	2

a) Work out the mean number of goal scored per match.
b) How many matches did Enyimba FC play in the season?

4) In a primary school, pupils are awarded merits for 90% attendance or more. The table shows the number of merits awarded to pupils in two different year groups.

Number of merits	Year 3 frequency	Year 4 frequency
0	5	3
1	15	27
2	20	12
3	14	8
4	7	6

a) How many students are in Year 3?
b) How many students are in Year 4?
c) How many merits were awarded in total to pupils in Year 4?
d) Work out the mean number of merits given per pupil for pupils in year 3?
e) How many more pupils received three merits in Year 3 than in year 4?

5) The table below shows the number of off days taken by some employees at a company in August 2017.

Number of off days	0	1	2	3	4	5
Frequency	5	4	1	0	3	2

a) Calculate the modal number of off days.
b) Calculate the mean number of off days.
c) Calculate the range
d) Work out the median number of off days.

AVERAGES AND RANGE FROM DIAGRAMS

Example 1: Some adults were asked about their favourite hair colour, and the bar chart shows the results.

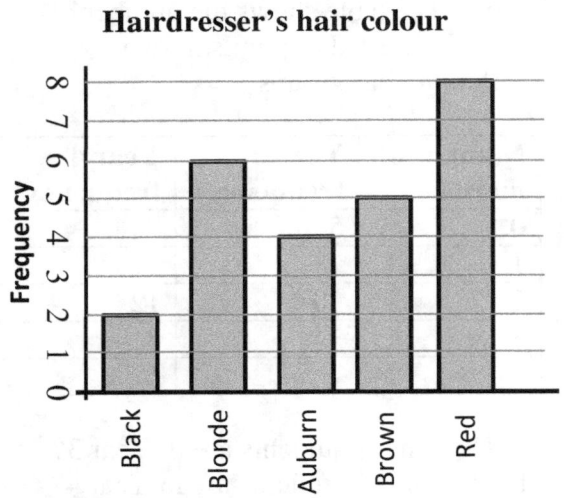

a) What is the modal hair colour?
The longest bar is red. Therefore, **red** is the modal colour.

b) How many adults took part in the survey?
2+6+4+5+8 = 25
25 adults took part in the survey.

c) How many adults have blonde hair?
6 adults

2) Study the bar chart shown below.

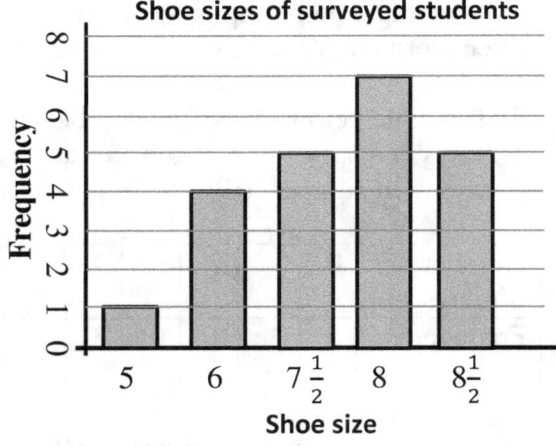

a) How many students were surveyed?

1 + 4 + 5 + 7 + 5 = **22 students**

b) What is the modal shoe size? The longest bar has shoe size of 8. Therefore, **8** is the modal shoe size.

c) Complete the frequency table below.

Shoe size	5	6	$7\frac{1}{2}$	8	$8\frac{1}{2}$
Frequency	1	4	5	7	5

d) Work out the median shoe size.
1 + 4 + 5 + 7 + 5 = 22…….(22 + 1) ÷ 2 =11.5. The median lies on the 11.5th value which is 8 (Refer to section 20.2 example 1). Therefore, the median shoe size is **8**.

e) How many more students wear size eight shoes than size 6?
7 – 4 = **3 more students**

EXERCISE 17 F

1) Michael stood in front of his house and counted the number of children in each of the cars that passed for 20 minutes. He drew a bar chart to represent his findings as shown below.

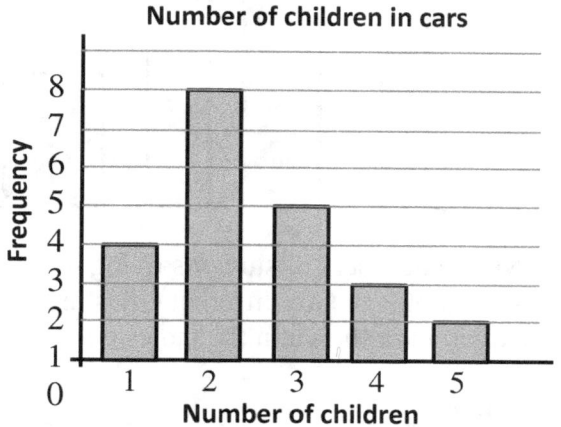

a) How many cars passed Michael's house in total?

b) Find the modal number of children in the car.

c) Work out the range for the data.

d) Calculate the median number of children in the car.

2) The pie chart shows types of pets owned by some students.

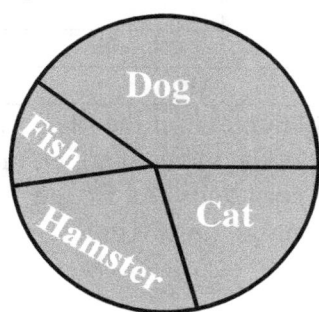

Write the mode of this data.

3) The pie chart below shows information about mock grades of some SSS3 students in a school.

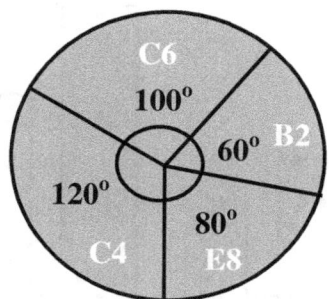

a) What fraction of the students achieved grade E8? Simplify your answer to its lowest form.
b) What grade was the mode?
c) If 12 students achieved grade B2,
i) how many students took part in the survey?
ii) how many students achieved grades C6 and C4?

4) The pictogram represents the number of apples sold each day by Tunji.

Key: ⊕ Represent 8 apples

Monday	⊕ ⊕
Tuesday	⊕ ⊕ ⊕ ᶜ
Wednesday	⊕ ⊕ ᵅ
Thursday	⊕ ⊕ ⊕ ⊕
Friday	⊕ ⊕ ⊕ ⊕ ⊕ ᶜ

a) Work out the range.
b) How many apples were sold on Tuesday?
c) On which day were the most apples sold?
d) How many apples were sold altogether?

5) The bar chart represents the number of horror books owned by four boys.

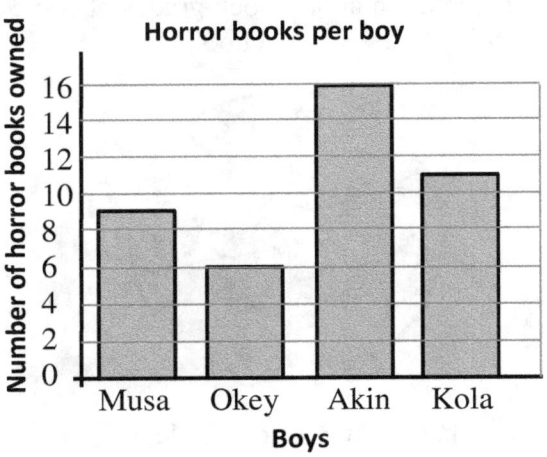

a) Who owned the most horror books?
b) Work out the range of the number of horror books owned by the boys.
c) Work out the mean number of horror books.
d) Musa and Kola owned how many horror books?

6) An advert in the Sun Newspaper reads:

Profession	Wages per month(₦)
Carpenter	40 000
Electrician	65 000
Handy person	35 000
Cleaner	45 000
Plumber	55 000

a) What is the median wage?
b) Work out the mode of these wages.
c) Work out the mean wage.

The advert further says "Carpenter needed, average salary more than ₦53 000 per month."

d) Show why the advert is deceptive.

7) Name three things that are wrong with this bar chart.

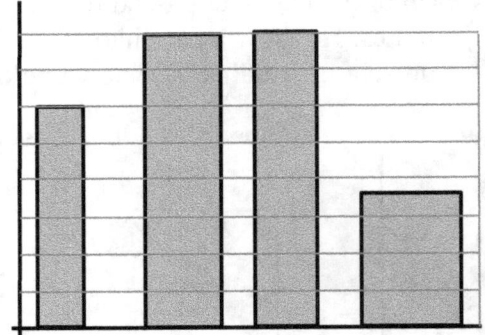

8) The numbers of students living in 50 households in **two** different cities in Nigeria are shown in the tables below.

Lagos

No. of Students	Frequency
0	7
1	15
2	20
3	5
4	2
5	1

Enugu

No. of students	Frequency
0	12
1	16
2	14
3	6
4	2
5	0

a) Draw bar charts for the two cities.
b) Work out the mean, median, mode and range for Lagos and Enugu.
c) Comment on the two cities.

17.3 AVERAGES FROM GROUPED DATA

When data is **grouped**, we can only estimate the mean using the midpoint of each group. For the median, we can only say which group it lies in and not the exact numerical value. You will get to know the **modal group** instead of mode since the data is in a group. For grouped data, class intervals are equal.

Example 1: The physics test results for 20 people are shown below.

5	10	12	4	10	6	18	20	22	25
7	19	5	22	14	18	3	9	19	20

a) Draw a grouped frequency table to represent the information.
b) Work out the modal group.
c) Which group contains the median?
d) Find an estimate for the mean.
e) Why is the answer to part d above only an estimate?

a)

Score	Tally	Frequency
0 – 5	////	4
6 – 10	ЖТ	5
11 – 15	//	2
16 – 20	ЖТ /	6
21 - 25	///	3

Score	Frequency (F)	Midpoint (M)	F × M
0 - 5	4	2.5	10
6 - 10	5	8	40
11-15	2	13	26
16-20	6	18	108
21-25	3	23	69
	20		**253**

Mean = $\dfrac{253}{20}$ = **12.65**

b) The modal group is the group with the highest frequency. It is **16 - 20**

c) There are 20 values (add up all the frequencies). Therefore, the median lies halfway between the 10th and 11th values. Starting from the top of the frequency, start adding until you get to 10.5 or more. Then, read of the group across. This means that the median is in the group of **11 – 15.**

e) The mean is only an **estimate** since we used the midpoints instead of the actual values.

d) To work out the mean, the **midpoint** of each interval must be used. Then multiply the midpoint and the frequency and add them together. For 0 - 5, midpoint = **2.5.**

EXERCISE 17G

1) The table below shows information about the weights of some oranges.

Weight (g)	Frequency
0 – 5	2
6 – 10	4
11 – 15	7
16 - 20	10
21 - 25	12

a) Work out the modal class of the weight of the oranges.
b) Which group contains the median?
c) Find an estimate for the mean.

2) The weights of some pupils were recorded and the results shown below.

Weight (kg)	Frequency
$40 \leq w < 45$	2
$45 \leq w < 50$	4
$50 \leq w < 55$	7
$55 \leq w < 60$	10
$60 \leq w < 65$	12

a) Estimate the median weight.
b) What is the modal class?
c) Find an estimate for the mean.
d) Why is the answer to part *c* only an estimate?

3) The lengths of a certain type of grass are given below, in millimetre.

| 30 | 31 | 36 | 39 | 40 | 43 | 47 | 52 | 55 | 60 | 61 | 65 |
| 70 | 71 | 73 | 75 | 78 | 79 | 81 | 83 | 85 | 87 | 88 | 89 |

a) Work out the mean length.
b) In class intervals of 30 – 39, calculate the mean length.
c) What are your findings from parts **a** and **b**?

4) The table shows the heights of some sailors.

Height (m)	2.5 – 2.6	2.7 – 2.8	2.9 – 3.0	3.1 – 3.2
Frequency	3	5	2	1

a) Find the modal class.
b) Estimate the median.

18 Pythagoras' Theorem

This section covers the following topics:

- Pythagoras' theorem
- Length of a line segment

LEARNING OBJECTIVES

By the end of this unit, you should be able to:

a) Calculate the missing lengths in a right-angled triangle
b) Solve problems using Pythagoras' theorem
c) Work out the length of a line segment

KEYWORDS

- Pythagoras' theorem
- Hypotenuse
- Right-angled triangle
- Square

PYTHAGORAS' THEOREM

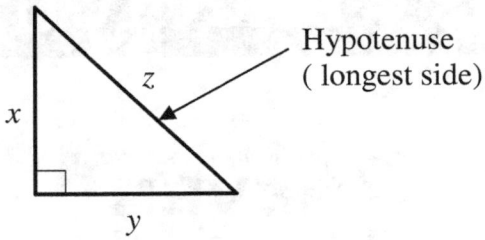

x and *y* are the shorter sides.

In a right-angled triangle, the sum of the squares on the other two sides is equal to the square of the hypotenuse. This is Pythagoras' theorem.

$$z^2 = x^2 + y^2$$

Note:
* Pythagoras' theorem can only be applied to a right-angled triangle.
* Identify the longest side before applying Pythagoras' theorem.

Example 1: Identify the longest sides (hypotenuse) of these right-angled triangles.

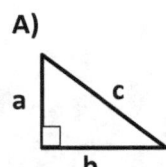 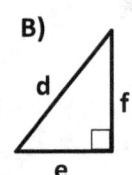 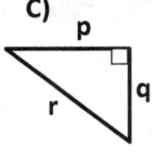

Solutions: A = c,   B = d,   C = r

Example 2: Calculate the missing length.

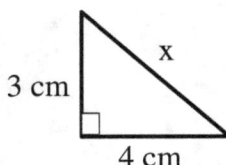

$x^2 = 3^2 + 4^2$
$x^2 = 9 + 16$
$x^2 = 25$
$x = \sqrt{25}$
x = 5 cm

Example 3: Calculate the missing length.

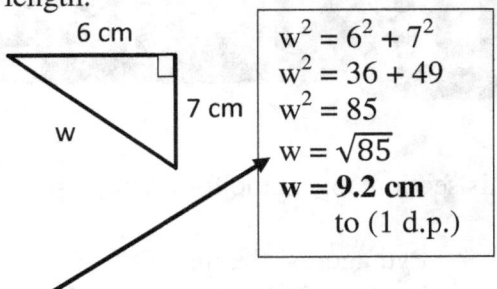

$w^2 = 6^2 + 7^2$
$w^2 = 36 + 49$
$w^2 = 85$
$w = \sqrt{85}$
w = 9.2 cm
   to (1 d.p.)

$\sqrt{85}$ **cm** is the **exact** answer. Working out the decimal part is only an approximation. Always check the question for the format.

EXERCISE 18A

1) Calculate the side marked **w**. All lengths are in cm and give your answers correct to two decimal places where applicable.

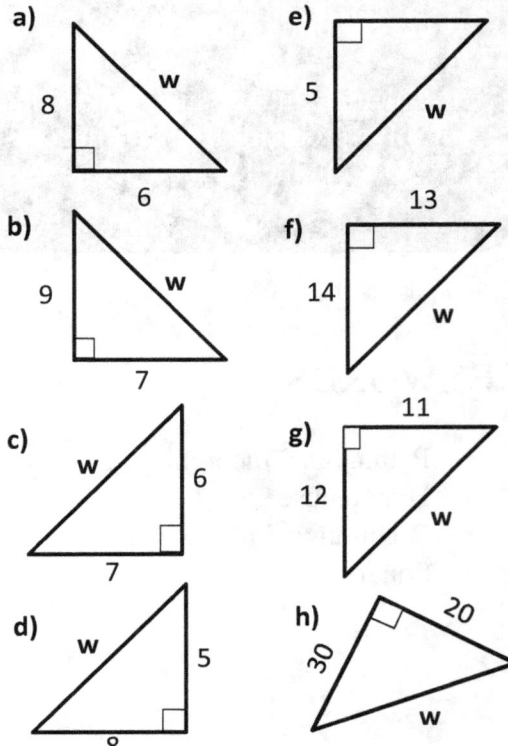

298

WHEN FINDING A SHORTER SIDE

Example 4: Calculate the missing length.

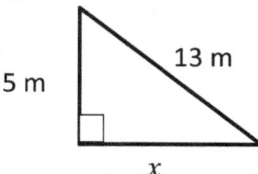

It is obvious we are not looking for the longest side. We still apply Pythagoras' theorem. Identify the hypotenuse, which is 13 m.

	Note: When looking for any of the shorter sides, **take away**.
$x^2 + 5^2 = 13^2$	$x^2 = 13^2 - 5^2$
$x^2 + 25 = 169$	$x^2 = 169 - 25 = 144$
$x^2 = 169 - 25$	$x = \sqrt{144}$
$x^2 = 144$	**x = 12 m**
$x \sqrt{144}$	
x = 12 m	

EXERCISE 18 B

Calculate the side marked x. All lengths are in **m** and give your answers correct to two decimal places where applicable.

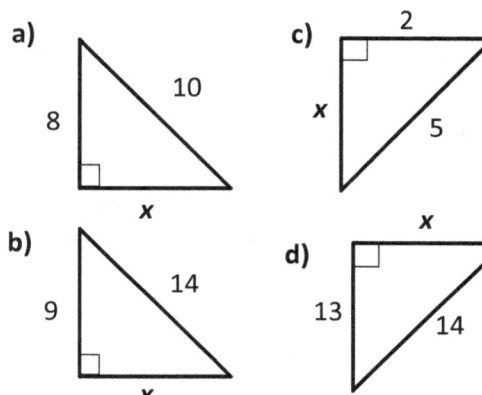

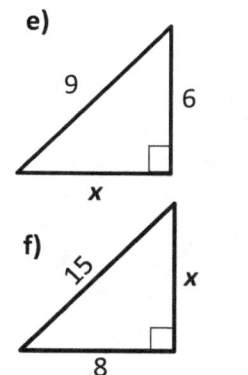

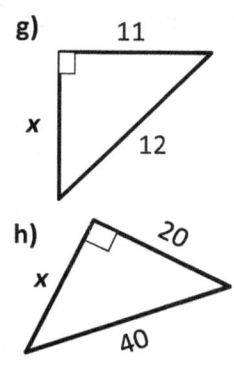

MIXED PROBLEMS

Example 5: Calculate the diagonal of the rectangle below.

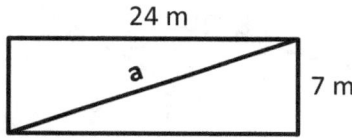

Half of the rectangle is a right-angled triangle. Therefore, we can apply Pythagoras' theorem.

$a^2 = 7^2 + 24^2$
$a^2 = 49 + 576$
$a^2 = 625$
$a = \sqrt{625} = $ **25 m**

Example 6: A cone is shown below. Work out the radius to 1 decimal place.

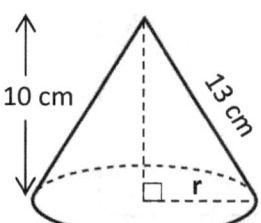

	$r^2 = 13^2 - 10^2$
	$r^2 = 169 - 100$
	$r^2 = 69$
	$r = \sqrt{69}$
	r = 8.3 cm

EXERCISE 18 C

1) The front of a school field looks like this. Work out the missing side, to one decimal place.

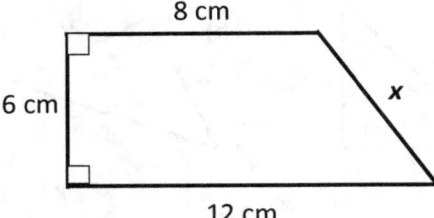

2) Calculate the length of the side marked *m* in the diagram below.

a)

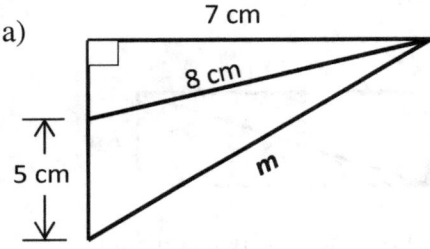

3) A square EFGH is drawn inside a bigger square PQRS, as shown below. Work out the length of the bigger square.

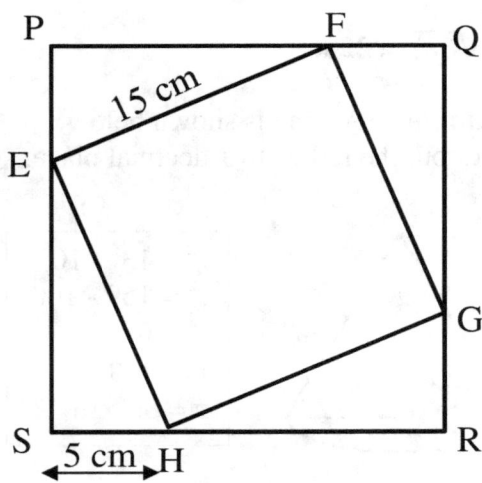

4) Work out the missing length *w*, to 2 decimal places.

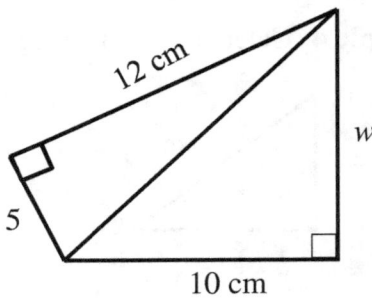

5) The length of a ladder is 15 m and it rests against a vertical wall. The foot of the ladder is 5.6 m from the wall. How far up the wall does the ladder reach?

6) Find the length of the sides marked *y* in the diagrams below.

a)

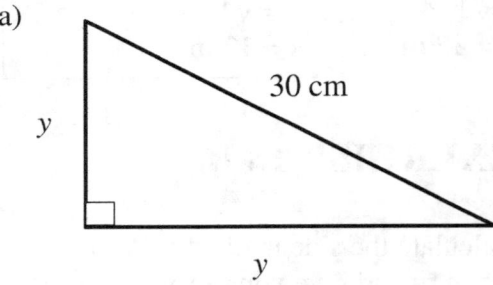

b)

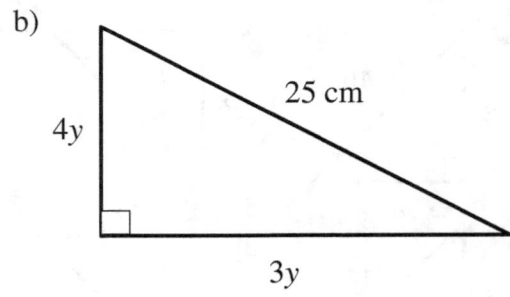

LENGTH OF A LINE SEGMENT

A line that has a beginning and an end is called a **line segment**.

Example 1: Calculate the length PQ.

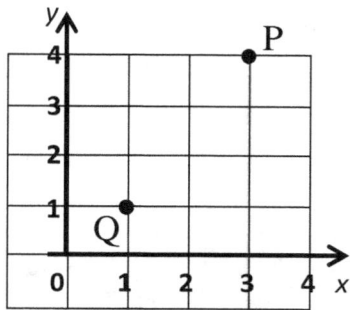

Solution: Join the points P and Q with a straight line and draw a triangle with a right angle at R as shown below.

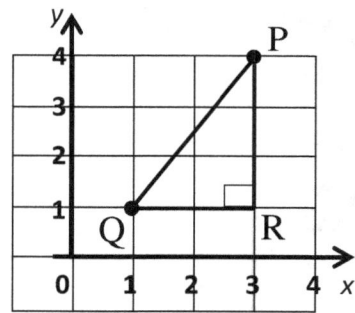

PR = 4 − 1 = 3
QR = 3 − 1 = 2

Using Pythagoras' theorem,
$PQ^2 = PR^2 + QR^2$
$PQ^2 = 3^2 + 2^2$
$PQ^2 = 9 + 4 = 13$
$PQ = \sqrt{13}$ units (Exact length) ✓

PQ = **3.61** to 2 decimal places

EXERCISE 18D

1) Work out the length of the line segments shown to two decimal places where applicable.

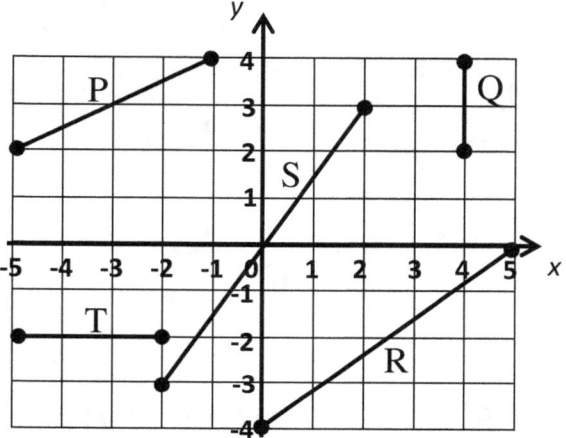

2) Work out the **exact** lengths of these line segments. Do not draw a sketch/graph.

a) A (1, 1) and B (5, 4)
b) P (-5, 1) and Q (-1, 4)
c) D (0, -2) and E (-5, 0)
d) S (3, 4) and T (0, 7)
e) L (-3, -2) and M (3, 3)

19 BEARINGS

This section covers the following topics:

- Bearings

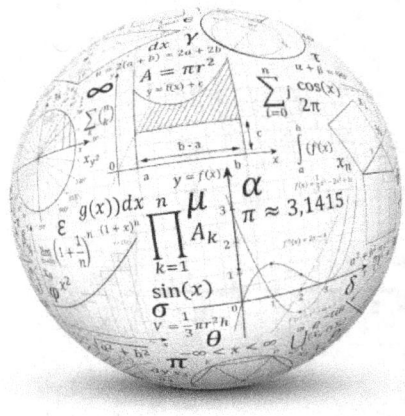

LEARNING OBJECTIVES

By the end of this unit, you should be able to:
a) Understand bearings
b) Measure bearings
c) Calculate bearings of return journeys

KEYWORDS

- Bearings
- North
- Clockwise direction
- Angle

BEARINGS

Bearing are angles measured **clockwise** from the **North**. Bearings are always written with three digits. 7° is 007° and 35° is 035°.

COMPASS BEARINGS

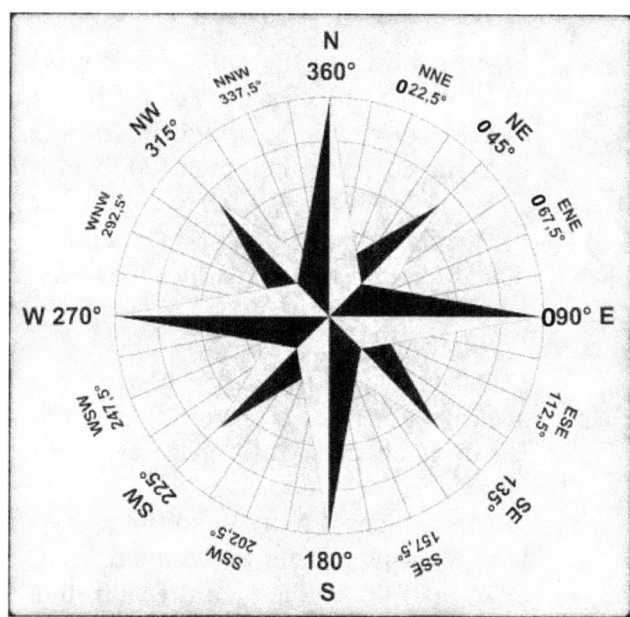

Points of a compass are useful in understanding bearing and angles around it. Bearings are used by sailors and other navigational purposes.

Example 1: Measure the bearing of B from A.

Solution: "From A" is a key word. Start from **A** and draw a North line.

Place a protractor from the north and measure the angle.

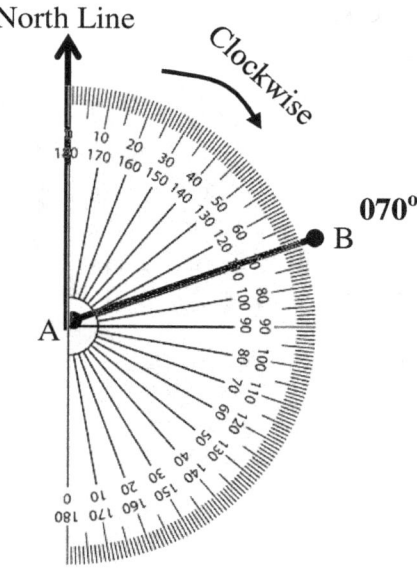

The bearing of B from A is **070°**

Example 2: Work out he bearing of
i) P from Q   ii) Q from P

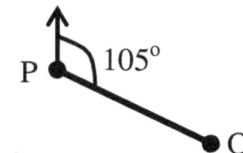

Solution: i) P from Q.
Start from Q and draw a North line.

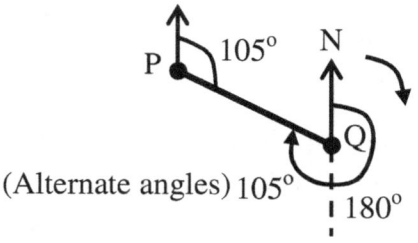

(Alternate angles) 105°

Therefore, the bearing of P from Q is
180° + 105° = **285°**

ii) Bearing of Q from P is **105°** as we go clockwise from the North to the line.

EXERCISE 19A

1) Measure the bearing of
i) B from A   ii) A from B

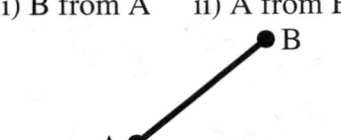

2) Work out the bearing of
i) P from Q   ii) Q from P

3) Below is a map of four towns.

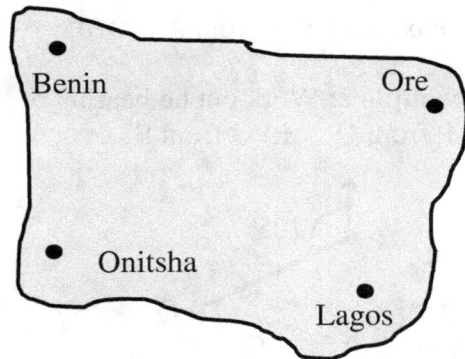

Find the bearings of
a) Benin from Ore
b) Benin from Lagos
c) Onitsha from Ore

4) Work out the bearing of
i) P from Q   ii) Q from P

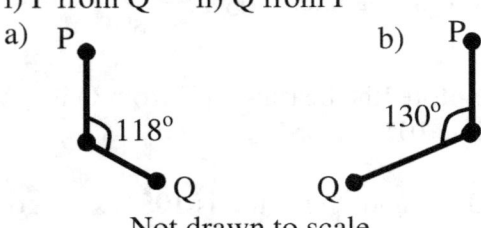

Not drawn to scale

5)

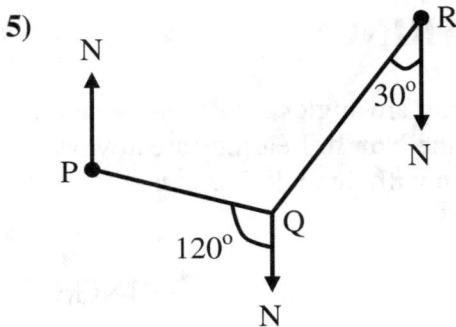

a) Write down the bearing of Q from P.
b) Write down the bearing of P from Q.
c) Write down the bearing of R from Q.
d) Write down the bearing of Q from R

6) Birmingham is 200 km on a bearing of 055° from London. Milton Keynes is 80 km on a bearing of 300° from London.
By using a scale of 1 cm = 40 km, draw accurate positions of London, Birmingham and Milton Keynes.

7) Mrs Ginger is a pilot and flies 400 km from Manchester to Scotland, on a bearing of 060°. She then flies a further 650 km to Wales on a bearing of 150°.

a) Calculate the direct return **distance** from Wales to Manchester to two decimal places.

b) By using a scale of 1 cm to 50 km, draw an accurate diagram for Mrs Ginger's flight.

c) **Measure** the return distance from Wales to Manchester from your diagram in part b.

d) Compare the result from part c to that of part a. What do you notice?

20 Simultaneous Equations

This section covers the following topic:

- Simultaneous Linear Equations

LEARNING OBJECTIVES

By the end of this unit, you should be able to:
- Solve a pair of simultaneous equations

KEYWORDS

- Solve
- Simultaneous equation

SOLVING SIMULTANEOUS EQUATIONS

These are equations with two unknowns. They also have two solutions which are true at the same time.

Example 1: Solve the pair of simultaneous equations
$x + 2y = 15$
$x + y = 8$

First label the equations

$x + 2y = 15$ ....(1)
$x + y\ \ = 8$ ....(2)
$\ \ \ \ \ y = 7$

Think of what letter to eliminate first. Since the x's have the same coefficients, eliminate x first by subtracting.

Next, find the value of x by replacing (substituting) y = 7 into equation 1 or 2. Put x y = 7 into equation 2. $\quad x + 7 = 8$ implying that $x = 1$.
Therefore, the solution to the simultaneous equations is **x = 1** and **y = 7**

Check: put x = 1 and y = 7 into (1) or (2). Using equation (2), $1 + 7 = 8$

Example 2: Solve the pair of simultaneous equations
$\quad\quad 4c + 3d = 49$ ....(1)
$\quad\quad c - 2d = 4$ .... (2)

Solution: It is easier to eliminate **c**. Therefore, multiply equation (2) by 4.

$\quad\ \ 4c + 3d = 49$
$-\ 4c - 8d = 16$
$\quad\quad\ \ 11d = 33$
$\quad\quad\quad\ \ \mathbf{d = 3}$

To get the value of c, substitute the value of d = 3 into equation (1).
$\quad\quad 4c + 3(3) = 49$
$\quad\quad 4c + 9 = 49$
$\quad\quad 4c = 49 - 9$
$\quad\quad 4c = 40 \quad\quad$ Therefore, **c = 10**

The solutions to the simultaneous equations is c = 10 and d = 3.

Check: Replace c = 10 and d = 3 in equation (1)
$\quad\quad 4(10) + 3(3) = 49$
$\quad\quad\ \ 40 + 9\ \ = 49$

Since Left Hand Side is equal to Right Hand Side, the solutions are correct.

Example 3: The difference between the ages of Paul and Sasha is 12 years. The sum of their ages is 62 years.
    a) Form an equation to find their ages.
    b) Solve the equation to find the ages of Paul and Sasha.
    c) What is the product of Paula and Sasha's ages?

Solution:
a) Let Paul's age = x
   Let Sasha's age = y
Difference between their ages is x − y = 12
Sum of their ages is x + y = 62
The equations are **x − y = 12** and **x + y = 62**

b) Solve simultaneously
      x − y = 12 ......(1)
    − x + y = 62 ......(2)
        −2y = − 50
         y = 25

Replace y = 25 in equation (2)
    x + 25 = 62
    x = 62 - 25
    x = 37

Paul is 37 years and **Sasha, 25 years**.

c) The product of their ages is **925 years**.

EXERCISE 20

Solve the simultaneous equations for Questions 1 to 6.

1) x + y = 7
   x − y = 3

2) x + y = 13
   x − y = 3

3) x − y = 15
   x + y = 25

4) 2x + y = 13
   3x − y = 22

5) 4x − 2y = 10
   3x + y = 10

6) 2x − 3y = 4
   5x − 2y = 21

7) Try and solve the simultaneous equations. Why do you think you were not able to solve the equations?
    y − 2x = 4
    y − 2x = 5.

8) The difference between the ages of Tom (t) and Harry (h) is 37 years. The sum of their ages is 67 years.
a) Form an equation to find their ages.
b) Solve the equation to find the ages of Tom and Harry.
c) What is the product of Tom and Harry's ages?

ANSWERS

EXERCISE 1A

1)
a) 2 b) 3 c) 5 d) 9
e) 4 f) 3 g) 5 h) 5

2)
a) 4 b) 8 c) 6 d) 0
e) 0 f) 8 g) 0 h) 9

3)
a) 70   b) 800
c) 3   d) 2
e) 0.4   f) 70
g) 40000   h) 0.08

4)
a) 4000   b) 9
c) 5.7   d) 3.2
e) 8.91   f) 12.6

5a)
D Africana   1000 m
Hi-Impact   900 m
T.W.Bunch   1000 m
Klubdelag   800 m
Fun Factory 1000 m
Omu Resort   500 m

b) 1230 m
c) 200 m
6) 65
7
a) 40
b) 55
c) 0.9
d) 90
e) 6790
f) 60
g) 800
h) 0.88
i) 543
j) 1.1
k) 400
l) 4.9

EXERCISE 1B

1
a) 70
b) 200
c) 400
d) ₦45
e) 300

2) ₦50 000 to ₦52 000

3) ₦45 000
4a) 30   b) 4
c) 80   d) 20
5a) 2   b) 40
c) 250   d) 3
6a) 12   b) 0.2
c) 105
7a) 8   b) 81
c) 14   d) 70
e) 50   f) 10
g) Chukwudi should round 95 pence to £1. Then multiply £1 by 8 to give £8. Since he rounded to £1, he must expect more than £2... (From £10 – £8).

EXERCISE 1C

1a) 104   b) 126   c) 176
d) 427   e) 380   f) 168
g) 160   h) 228   i) 84
j) 490   k) 77   l) 603
m) 408   n) 741   o) 7068

2a) 300   b) 240   c) 60
d) 120   e) 180   f) 150
g) 125   h) 210   i) 210
j) 80   k) 180   l) 560
m) 60   n) 72   o) 220

3) 390   4) 3318   5) 75
6) 2280
7a) 240   b) 3819
c) 49725 d) 9432
e) 10680 f) 6732

8) 1820
9) ₦306 000
10)

×	3	6	7	9	10
2	6	12	14	18	20
8	24	48	56	72	80
7	21	42	49	63	70
11	33	66	77	99	110

11) 1998 m^2 12) 200
13) ₦109 200
14a) 7 b) 10 c) 3015
15a) ₦2 700 b) ₦6 750
c) ₦63 450   d) 67 500
16) 300 eggs
17a) 49152 b) 128
c) 32, 128, 4096
d) 16 or 4096
18) ₦15 000
19) 1645 seats

20)

×	4	7	8	9
2	8	14	16	18
5	20	35	40	45
7	28	49	56	63
8	32	56	64	72

EXERCISE ID

1a) 20   b) 3   c) 4
d) 20   e) 12   f) 13
g) 10   h) 34   i) 129
j) 16   k) 75   l) 400
m) 700   n) 74   o) 2222

2) ₦2 400   3) 24 kg
4) 120   5) ₦650
6) 749   7) ₦987
8a) 6 coaches
b) 9 coaches
9) Students own answers that add up to ₦9 250 e.g. ₦4 000, ₦3 000 and ₦2 250.

EXERCISE 1E

1a) 30   b) 430   c) 770
d) 3190 e) 4000
f) 5190 g) 760
h) 4310 i) 7800
j) 2340 k) 9800

l) 1340   m) 5550
n) 11110   o) 789650

2a) 3 000   b) 4 200
c) 600   d) 7 500
e) 31 900   f) 5 000
g) 41 900   h) 91 700
i) 700   j) 4 700
k) 928 800   l) 111 200
m) 97 600   n) 5 900
o) 8 669 700

3a) 3 000   b) 5 000
c) 77 000   d) 318 000
e) 96 000   f) 7 000
g) 431 000   h) 96 000
i) 419 000   j) 60 000
k) 8 776 000   l) 92 000
m) 1112000   n) 59 000
o) 486 000

4a) 3   b) 7   c) 4.6
d) 9.6   e) 1   f) 46.5
g) 543.21   h) 67.8
i) 128   j) 40   k) 831
l) 5700   m) 11   n) 5
o) 50

5a) 9   b) 86   c) 32.450
d) 100   e) 654   f) 4590
g) 7000   h) 234   i) 870
j) 31.8   k) 0.83   l) 8.9
m) 0.452   n) 98.74
o) 40.8

FUNCTIONAL MATHS
1) £8.97   2) £9.49
3a) £11.98   b) £8.02
4) 240   5a) 800 g
5b) £2.04   6) £45

CHAPTER 1 REVIEW SECTION

1) 35, 57, 76, 89, 789
2) 123, 245, 621, 705, 947
3a) 40   b) 50   c) 80
d) 140   e) 280   f) 310
g) 460   h) 790   i) 840
j) 1460   k) 4570
l) 5560
4a) 9000 m   b) 7000 m
c) 6000 m   d) 8200 m

5) No, Mbakwe is wrong. The digit, 4 is not up to 5.

3490 would round to 3000 to 1 significant figure and not 4000.

6a) 50   b) 87
c) 432   d) 733

7a) 50 paintings
b) ₦81 750
c) ₦81 750

8a) 991 tins
b) £51.45

9a) 2300   b) 768000
c) 900890   d) 540
e) 6867   f) 23576

10) ₦551 200

11)

×	2	5	6	7	9
3	6	15	18	21	27
2	4	10	12	14	18
7	14	35	42	49	63

12)

11	6	13
12	10	8
7	14	9

13) ₦12 154 214
14) ₦1 219 104
15a) ₦31 050
b) ₦11 610
c) ₦62 660

16a) £41.75
b) £275.50   c) £30
d) £42   e) £25.10
17) 289 students
18) 49
19a) Port Harcourt ➡ Warri ➡ Benin route is the quickest. 198 + 97 = 295 km.

It takes longer to take Port Harcourt ➡ Aba ➡ Owerri ➡ Benin route. 66 + 69 + 227 = 362 km

b) 361 km   c) 657 km
d) Yes, Achike is correct. 69 + 227 = 296 km through Owerri.

Through PH would take longer: 66 + 198 + 97 = 361 km
20a) 100   b) 600   c) 500
d) 400   e) 300   f) 2000
21a) 120   b) 580   c) 530
d) 350   e) 330   f) 1600
22a) 1.3   b) 23.5   c) 1.0
d) 897.3   e) 3.1   e) 98.6
23a) 70   b) 1400   c) 5
24a) 45   b) 24
c) 358   d) 658

EXERCISE 2A

1a) 1, 2, 3, 4
b) 2, 4, 6, 8
c) 5, 10, 15, 20
d) 7, 14, 21, 28
e) 9, 18, 27, 36
f) 10, 20, 30, 40
g) 13, 26, 39, 52
h) 17, 34, 51, 68
i) 20, 40, 60, 80

2a) 2, 12, 16, 22, 30, 36, 54, 70
b) 12, 30, 36, 54
c) 12, 16, 36
d) 5, 30, 70
e) 7, 49, 70

3) Any number that fits the requirements, e.g. 9, 15, ….
4) None
5) Examples: 10, 20…
6) 49 and so on
7) 8 and 20
8a) 33 and 55
b) 27 and 63
c) 16
9) YES, because 6 x 3 = 18
10) NO, Sanusi is incorrect. The first multiple of a number is the number itself. Therefore, the first five multiples of 4 are: 4, 8, 12, 16, and 20

11a) 7, 14, 21, 28, 35
b) 3, 6, 9, 12, 15, 18, 21, 24, 27, 30
c) e.g. 21
12) 12 and 30
13) 18, 36, 63

EXERCISE 2B

1a) 4   b) 6   c) 35
d) 70   e) 24   f) 30
g) 60   h) 20
2a) 12   b) 210   c) 240
d) 56   e) 315   f) 84
g) 90   h) 300

3) 360 minutes

4a) $2 \times 2 \times 3 \times 3 \times 3$
b) $2 \times 2 \times 3 \times 3 \times 5 \times 5$
c) $2 \times 2 \times 2 \times 3 \times 3 \times 3 \times 5 \times 5$
d) $5 \times 5 \times 7 \times 7 \times 7 \times 11 \times 11$
f) $2 \times 2 \times 3 \times 3 \times 3 \times 3 \times 4 \times 4 \times 5 \times 5 \times 5$

5) Jude is wrong. LCM of 3 and 6 is 6 and not ($3 \times 6 = 18$)
6) 560
7) 40
8) 4 cups

EXERCISE 2C
1)
a) 1, 3
b) 1,2,4
c) 1,2,7,14
d) 1,2,3,4,6,8,12,24
e) 1,2,3,5,6,10,15,30
f) 1,5,7,35
g) 1,2,3,4,6,9,12,18,36
h) 1,2,3,4,6,8,12,16,24,48
i) 1,2,4,7,8,14,28,56
j) 1,2,3,4,6,7,12,14,21,28,42,84
k) 1,2,4,5,10,20,25,50,100
l) 1,2,3,4,5,6,8,10,12,15,20,24,30,40,60,120

2)
a) 3, 6
b) 3, 7
c) 3, 6, 9
d) 3, 9
e) 3, 5, 6
f) 3, 6, 9

3) 1 or 5
4) Could be 2, 6 or 18
5) 5, 10 and 25
6) 1 and 5
7) 4 and 18
8) Yes, Anthony is correct. $7 \times 8 = 56$
9) 72

EXERCISE 2D

1)
a) 3, 7, 19
b) 2, 5, 13
c) 31, 41
d) 37, 59, 97
e) 11, 17

2)
a) 2
b) 2 and 3
c) 17 and 2
d) 2 and 5

3) 2 or 3
4) 2 or 3
5a)

1	2	3	4	5	6	7
8	9	10	11	12	13	14
15	16	17	18	19	20	21
22	23	24	25	26	27	28
29	30	31	32	33	34	35
36	37	38	39	40	41	42
43	44	45	46	47	48	49

5b)

1	2	3	4	5	6	7
8	9	10	11	12	13	14
15	16	17	18	19	20	21
22	23	24	25	26	27	28
29	30	31	32	33	34	35
36	37	38	39	40	41	42
43	44	45	46	47	48	49

5c) 71
5d) Yes, 71 is a prime number because it has only two factor, 1 and 71.

EXERCISE 2E

1a) $2 \times 3 \times 5$
b) 5×11
c) $2 \times 5 \times 7$
d) $2 \times 2 \times 2 \times 2 \times 3 \times 3 = 2^4 \times 3^2$
e) $2 \times 2 \times 2 \times 2 \times 3 \times 3 \times 5 = 2^4 \times 3^2 \times 5$
f) $2 \times 2 \times 5 \times 47 = 2^2 \times 5 \times 47$
2a) 2, 5 and 7
b) $2 \times 2 \times 5 \times 7 = 2^2 \times 5 \times 7$
3a) $2 \times 3 \times 5^2$
b) $2^2 \times 5^3$
c) $2 \times 3 \times 5 \times 17$
4a) 2×3
b) 3×7
c) 11×13
5a) 3, 3, 5
b) 3, 7, 2, 5
6a) $2^3 \times 3^3$
b) $2^2 \times 5 \times 7^4$
c) $3^4 \times 5^2$
7) NO, he is incorrect. For example;

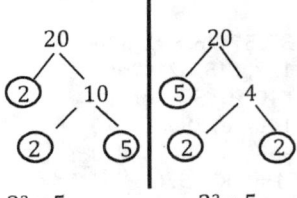

$2^2 \times 5$ $2^2 \times 5$

8a) $2^2 \times 3 \times 5^2 \times 11$
b) $2^2 \times 5 \times 7 \times 11$

EXERCISE 2F

1a)
6: |1| |2| 3 6
10: |1| |2| 5 10

HCF = 2

b) 12: |1 2 3 4 6|
   18: |1 2 3 6| 9 18

HCF = 6

c) 5: |1| |5|
   25: |1| |5| 25
HCF = 5

d) 27: |1| 3 9 27
   49: |1| 7 49
HCF = 1

e) 30: |1, 2| 3, 5, 6, 10, 15, 30
   76: |1, 2| 4, 19, 38, 76
HCF = 2

f) 36: 1,2,3,4,6,9,12,18,36
48: 1,2,3,4,6,8,12,16,24,48
Common factors are 1,2,3,4,6 and 12
HCF = 12

2a) 20 b) 30 c) 140

3a) $3 \times 5 \times 5 \times 5 \times 7 \times 7$
    $= 3 \times 5^3 \times 7^2$
3b) $2 \times 2 \times 3 \times 3 \times 3$
    $= 2^2 \times 3^3$
3c) $5 \times 7 \times 13$
4a) 6 b) 5
5a) 8 b) 9 c) 30 d) 5
6a) 1,2,4 b) 1,2,4
c) 1,3,9 d) 1,2,4
7) 35 cm by 35 cm
8) No, Kenechukwu is incorrect. The HCF of 25 and 50 is 25.

EXERCISE 2G

1a) YES, last digit is even
b) YES, last digit is even
c) NO, last digit (7) is not 0 or even
d) NO, last digit (9) is not 0 or even
e) YES, last digit is 0
f) NO, last digit (5) is not 0 or even

2a) YES, sum of digits is divisible by 3. 1 + 2 = 3 and 3 ÷ 3 = 1
2b) YES, same reason as above
2c) NO, 3 + 4 = 7 and 7 is not divisible by 3
2d) NO, 4 + 6 = 10 and 10 is not divisible by 3
2e) YES, 1 + 8 + 0 = 9 and 9 is divisible by 3
2f) YES, 6 + 5 + 4 = 15 and 15 is divisible by 3

3) YES, last digit is 5.
$5 \times 2 = 10$. $28794 - 10 = 28784$ and $28784 \div 7 = 4112$

4a) Yes, b) Yes, c) No
d) No   e) Yes, f) Yes
5) NO, the last digit is not 0 or 5
6) No, the last three digits (451) is not divisible by 8.
7) Yes, last digit is 0 and even
8a) Yes b) No

EXERCISE 2H

1a) 1 or 64
b) 1 or 125
c) 13
d) 3
2) 55
3a) 81 b) 2.25 c) 169
d) 121
4a) 15 b) 17 c) 21
d) 2.5
5) NO..because two identical numbers cannot multiply to give a negative number.
6) 109   7) 85
8a) 125 b) -64 c) 1/8
d) 8/125
9a) 4 b) 8 c) 343
d) 1.728
10a) 121 b) 7 c) 25 d) 1022
11a) 64 b) 22 c) 98 d) 65
e) 2 f) 1.6 g) 0.44 h) 7
i) 14.96 j) 5
k) a … 2, b)… -14 c).. - 30

EXERCISE 2I

1a) 5^3 b) 6^5 c) 13^7
d) $4^3 \times 7^2$
2a) 1 b) 8 c) 289 d) 1000
e) 1 f) 125000 g) 484
h) 27000
3a) 40 b) 40 c) 255 d) 16
e) 77 f) 16 g) -63 h) 81
4a) 2^8 b) 4^7 c) 10^9 d) 17^{11}
e) 3^1 f) 9^9 g) e^8 h) y^{16}
i) w^5 j) $6d^{10}$ k) $105h^8$ l) y^4
5a) 4^3 b) 3^7 c) n^2 d) c^6
e) y^6 f) $3w^5$ g) 5^4 h) $3w^8$
i) $3d$ j) 5^{-6}
6a) ¼ b) 1/3 c) 1/25
d) 1/2197
7a) 5^{-2} b) 5^{-1} c) 10^2 d) 5^5
e) n f) 8^5 g) 11^6 h) 15^{-12}
8a) false b) true c) true
d) false e) true f) false
9a) $12x^6$ b) $5a^4$ c) $16x^2y^2$
d) $25x^2$ e) $8p^2y$

EXERCISE 2J

1) 2×10^2 2) 5×10^2
3) 4×10^3 4) 1.2×10^5
5) 6.5×10^2 6) 2.345×10^3
7) 3.037×10^3 8) 1×10^4
9) 3.4×10^5 10) 3.9×10^4
11) 4×10^1 12) 5.4×10^1
13) 9.89×10^2
14) 1.345×10^3 15) 3×10^3
16) 6×10^4 17) 1×10^9
18) 7.893×10^6
19) 9.05×10^3
20) 9.999×10^3 21) 2.1×10^8
22) 1.68×10^{-24}
23) 2×10^5 m/s 24) b
25) 4×10^{12}

EXERCISE 2K

1) 2×10^{-2} 2) 3×10^{-3}
3) 5×10^{-5} 4) 1.23×10^{-4}
5) 8.09×10^{-2} 6) 7×10^{-6}
7) 6.4×10^{-4} 8) 4.44×10^{-5}
9) 1.2×10^{-2} 10) 4.5×10^{-2}
11) 3.23×10^{-3} 12) 9×10^{-3}
13) 8.4×10^{-5}
14) 6.128×10^{-2}
15a) 0.000018 b) 0.0067
c) 0.000205 d) 0.073
e) 0.89 f) 0.00502

EXERCISE 2L
1a) 2×10^3 b) 2.8×10^3
c) 3.5×10^2 d) 6×10^1
e) 2.5×10^4 f) 1.6×10^3
g) 2.86×10^8 h) 2.25×10^{10}
i) 8×10^3 j) 2.76×10^{-11}
2) 6×10^{17} b) 1.5×10^{-5}
c) 2.00003×10^{11}
d) 1.99997×10^{11}
3a) 8.6×10^4, 2.3×10^5, 1.5×10^8
3b) 2.5×10^{-6}, 3.7×10^{-3}, 6.5×10^{-3}, 6.9×10^{-3}
3c) 6.1×10^{-9}, 7.1×10^{-9}, 3.4×10^{-6}, 4.5×10^{-6}
4a) 2×10^3 b) 5×10^2
c) 4×10^{-2} d) 2×10^{-9}
e) 3×10^1 f) 2×10^{-12}
g) 4.035×10^{-4}
h) 8.778×10^{-3}
5a) 2.78×10^6 km
5b) 2.43×10^{13} km^2
6) 6.05×10^{-8} m^2
7a) 5.9736×10^{24} kg
7b) 1.1036×10^{24} kg
7c) 460.2 million km^2
7d) 1.989×10^{30} kg
8a) 60000000000 cm^3
8b) 6×10^{10} cm^3
8c) 1.2×10^7 cm^2

CHAPTER 2 REVIEW SECTION

1a) 3,6,9,12,15
b) 4,8,12,16,20
c) 14,28,42,56,70
2a) 1, 13
b) 1,2,3,4,6,9,12,18,36
c) 1,2,4,8,13,26,52,104
3a) 7   e) None
 b) 7,9
 c) 6, 72
 d) 2, 7, 23
4) $2 \times 5 \times 5 \times 5$
   $= 2 \times 5^3$
5a) 9   b) 100
6a) 14 b) 350 c) 5460
7a) LCM = 132
    HCF = 66
b) LCM = 1680
    HCF = 140
8a) $2 \times 3^3 \times 5$ b) 7
9) Azubuike is wrong. LCM of 2 and 4 is **4**
10) 9

11) 165 seconds
12a) $2 \times 3 \times 3 \times 3$
   b) 2×3^3
   c) 18
13) 60
14) Henry is not correct. 1 is not a prime number since it has only one factor. Also, 2 is the only even prime number with factors 1 and 2
15) 37, 41, 43, 47, 53
16a)

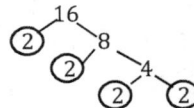

16b)

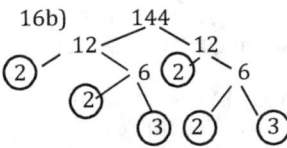

17) Yes.
$1+8+9+3+5+7 = 33$
$33 \div 3 = 11$ (Whole number)

18) $2 \times 3 \times 5 \times 5 \times 5$ which implies 2, 3 and 5
19a) 47 b) 54
c) 16 or 32
d) 12 or 16
e) 16
20)

x	1	4	9	16	25	144
x^2	1	16	81	256	625	20736
$\sqrt{x}$	1	2	3	4	5	12
-	-	-2	-3	-4	-5	-12
$\sqrt{x}$	1					

21a) $(-3)^2$ b) None, they are the same. c) 4^2 d) 2^3 e) 0.3^2
22a) 7^{10} b) 3^8 c) 12^{-4} d) 4^{-7}
e) a^{-16}
23a) 3.45×10^2
b) 2.8×10^4
c) 8.98×10^9 d) 1.2×10^{-3}
e) 9.8×10^{-6} f) 4.05×10^{-1}
g) 8.732×10^3 h) 6×10^9
i) 9.15×10^{-5}
24a) 6000 b) 0.42
c) 0.00007 d) 0.0000066
e) 0.00094 f) 0.014
25a) 1.52×10^9
b) 1.505×10^{-5} c) 2×10^{-2}
d) 5×10^{-4} e) 9.6×10^8
f) 5.67×10^4

EXERCISE 3A

1) ¼ 2) 2/4 3) 7/12
4) 4/6 5) 1/5 6) 5/6
7)

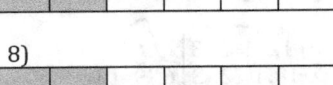

8)

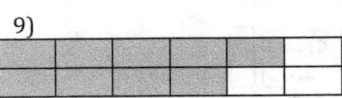

9)

10) $1\frac{1}{2}$ 11) $2\frac{2}{6}$ or $2\frac{1}{3}$
12) $3\frac{2}{11}$ 13) $3\frac{2}{40}$ or $3\frac{1}{20}$
14) $\frac{8}{5}$ 15) $\frac{7}{3}$ 16) $\frac{32}{5}$
17) $\frac{68}{7}$ 18) 12 19) 14
20) 6 21) 14 22) 21
23) 5 24) 35 25) 5
26) 1 27) 2 and 4
28) 3, 4, 5

EXERCISE 3B

1a) $\frac{1}{2}$, b) $\frac{2}{3}$ c) $\frac{1}{3}$
d) $\frac{2}{11}$ e) $\frac{1}{4}$ f) $\frac{2}{9}$
g) $\frac{5}{24}$ h) $\frac{5}{8}$ i) $\frac{1}{3}$
j) $\frac{3}{10}$ k) $\frac{4}{15}$ l) $\frac{11}{20}$
m) $\frac{1}{5}$ n) $\frac{2}{5}$ o) 1

2a) $\frac{12}{15}$ and $\frac{10}{15}$
b) $\frac{5}{30}$ and $\frac{24}{30}$
c) $\frac{6}{8}$ and $\frac{4}{8}$
3a) $1\frac{3}{5}$ b) $4\frac{1}{7}$
c) $18\frac{1}{3}$ d) $1\frac{5}{6}$
e) $1\frac{4}{5}$ f) $4\frac{1}{4}$
g) $10\frac{12}{13}$ h) $13\frac{5}{15}$

4a) $\frac{4}{9}$ b) $\frac{3}{7}$ c) $\frac{5}{6}$
d) 0 e) $\frac{7}{12}$ f) $\frac{14}{45}$
g) $\frac{13}{15}$ h) 0 i) $\frac{26}{45}$
j) $\frac{34}{40} = \frac{17}{20}$ k) $\frac{9}{14}$
l) $\frac{47}{70}$ m) $1\frac{1}{12}$

n) $1\frac{1}{10}$ o) $1\frac{3}{20}$
5a) $\frac{2}{3}$ b) $\frac{8}{9}$ c) $\frac{1}{3}$
d) $\frac{2}{17}$ e) $\frac{7}{13}$ f) $\frac{4}{19}$

6a) $\frac{5}{6}, \frac{2}{3}, \frac{1}{2}$
b) $\frac{7}{8}, \frac{1}{2}, \frac{1}{4}$
c) $3\frac{2}{3}, 2\frac{2}{3}, 2\frac{1}{3}$

7a) ¼ b) ¼ c) ¼
d) ¼
8) $\frac{7}{15}$
9a) $1\frac{1}{19}$ m b) $\frac{12}{19}$ m
10) $\frac{5}{45}$. In its lowest form, $\frac{5}{45}$ = $\frac{1}{9}$ but the rest are $\frac{4}{15}$ to their lowest terms.
11a) 3 b) $2\frac{1}{2}$ c) 1
d) 1 e) $2\frac{1}{2}$

EXERCISE 3C

1) $3\frac{1}{3}$ 2) $4\frac{1}{15}$ 3) 8
4) 3 5) $6\frac{1}{2}$ 6) 5
7) $1\frac{37}{105}$ 8) $3\frac{13}{21}$ 9) 0
10) $4\frac{13}{15}$ 11) $5\frac{5}{12}$

EXERCISE 3D

1a) $\frac{2}{15}$ b) $\frac{4}{35}$ c) $\frac{18}{28} = \frac{9}{14}$
d) $\frac{10}{54} = \frac{5}{27}$ e) $\frac{9}{25}$ f) $\frac{7}{32}$
g) $\frac{40}{63}$ h) $\frac{3}{96} = \frac{1}{32}$ i) $\frac{12}{63}$
j) $\frac{12}{105}$ k) $\frac{24}{40} = \frac{3}{5}$
l) $\frac{4}{400} = \frac{1}{100}$

2) 12 m 3) $\frac{12}{55}$ m² 4) $\frac{1}{6}$
5) P: $\frac{3}{21}$ m² Q: $\frac{15}{63}$ m²
   R: $\frac{5}{63}$ m² S: $\frac{25}{189}$ m²

6a) $\frac{1}{27}$ b) $\frac{6}{60} = \frac{1}{10}$
c) $\frac{36}{143}$
7a) Nnaemeka is wrong. Correct answer should be $\frac{5}{18}$

b) Nnaemeka added the numerators and denominators instead of multiplying them.

EXERCISE 3E

1)
a) 1 b) 4 c) 9 d) 21
e) 8 f) 10 g) 50 h) 15
i) 40 j) 55 k) 350 l) 1

2a) ₦40 b) ₦50 c) $50
d) 6 kg e) 40 f) ₦1200
g) £15.50 h) 1020 kg

3a) £40
b) Arinze is correct. $\frac{4}{5}$ of £40 = £32 and £32 is more than £30

EXERCISE 3F

1a) $\frac{7}{4}$ b) $\frac{28}{5}$ c) $\frac{103}{9}$
d) $\frac{52}{3}$
2a) $1\frac{14}{25}$ b) $3\frac{31}{35}$ c) $4\frac{14}{25}$
d) $4\frac{24}{50}$ e) $5\frac{1}{4}$ f) $17\frac{1}{7}$
g) $110\frac{1}{4}$ h) $6\frac{19}{25}$
3a) $1\frac{10}{21}$ b) $6\frac{3}{25}$ c) $17\frac{2}{3}$
d) $60\frac{3}{4}$
4) $3\frac{1}{10}$
5a) $27\frac{2}{5}$ kg b) 29 kg

EXERCISE 3G

1a) 2 b) $1\frac{1}{5}$ c) $1\frac{1}{6}$
d) $4\frac{2}{7}$ e) 14 f) $1\frac{2}{3}$
g) 21 h) $7\frac{21}{45}$ i) $11\frac{1}{9}$
j) $1\frac{97}{153}$ k) $2\frac{8}{81}$ l) $3\frac{9}{77}$
2a) 15 b) 21 c) 27
d) 45
3a) 60 b) 80 c) 110
d) 200
4) $1\frac{1}{2}$ m 5) $12\frac{3}{5}$ m
6) 9 7) $2\frac{26}{235}$

CHAPTER 3 REVIEW SECTION

1a) $\frac{2}{3}$ b) $\frac{1}{3}$ c) $\frac{1}{3}$
2) $\frac{3}{12} = \frac{1}{4}$

3) $\frac{1}{3}$
4a) $\frac{10}{40}$ b) $\frac{15}{40}$ c) $\frac{32}{40}$
d) $\frac{36}{40}$
5a) $\frac{15}{7}$ b) $\frac{13}{8}$ c) $\frac{37}{3}$
d) $\frac{229}{11}$
6a) $\frac{5}{7}$ b) $1\frac{5}{21}$ c) $1\frac{17}{42}$
d) $\frac{65}{77}$
7a) $\frac{1}{2}$ b) $\frac{1}{3}$ c) $\frac{3}{5}$
d) $\frac{6}{7}$
8a) 12 b) 60 c) 72 d) 12
9a) $2\frac{1}{3}$ b) $5\frac{3}{4}$ c) $8\frac{3}{7}$
d) $10\frac{4}{20} = 10\frac{1}{5}$
10a) $\frac{8}{9}$ b) $\frac{5}{9}$ c) $6\frac{3}{5}$
d) $3\frac{1}{3}$ e) $1\frac{5}{9}$ f) $\frac{5}{6}$
g) $5\frac{29}{54}$ h) $1\frac{44}{81}$
11a) 12 b) 27 kg c) ₦300
d) 140
12a) Q = $\frac{6}{77}$ m²   R = $\frac{8}{77}$ m²
   S = $\frac{36}{77}$ m²
12b) 1 m² 12c) Square Same lengths for the four sides. 1 metre each
13a) 2 b) 3 c) $9\frac{9}{35}$
14a) $19\frac{5}{6}$ kg b) $38\frac{1}{2}$ kg
c) $5\frac{1}{2}$ kg d) ₦115 500

EXERCISE 4A

1a) 0.6 b) 0.8 c) 0.3 d) 0.1
e) 0.28 f) 0.125 g) 0.25
h) 0.625 i) 0.06 j) 0.76
k) 0.75 l) 0.8 m) 0.35
n) 0.48 o) 0.6 p) 0.5 q) 0.1
r) 0.875 s) 0.02 t) 0.8

2a) $\frac{1}{50}$ b) $\frac{3}{100}$ c) $\frac{1}{10}$ d) $\frac{3}{20}$
e) $\frac{9}{50}$ f) $\frac{1}{5}$ g) $\frac{1}{4}$ h) $\frac{3}{10}$ i) $\frac{7}{20}$
j) $\frac{23}{50}$ k) $\frac{1}{2}$ l) $\frac{11}{20}$ m) $\frac{3}{5}$ n) $\frac{7}{10}$
o) $\frac{3}{4}$ p) $\frac{79}{100}$ q) $\frac{4}{5}$ r) $\frac{81}{100}$ s) $\frac{9}{10}$
t) $\frac{49}{50}$

EXERCISE 4B

1a) 25% b) 20% c) 40%
d) 30% e) 80% f) 15%
g) 60% h) $37\frac{1}{2}$% i) 35%

j) 36% k) 32% l) 60%
m) 24% n) 64% o) 65%
p) 10% q) 50% r) 50%
s) 25% t) 5%

2a)
i) $\frac{7}{10}$ ii) 70% iii) $\frac{3}{10}$ iv) 30%
2b) i) $\frac{3}{25}$ ii) 12% iii) $\frac{22}{25}$
iv) 88%

2c)
i) $\frac{1}{2}$ ii) 50% iii) $\frac{1}{2}$ iv) 50%

EXERCISE 4C

1) $\frac{5}{7}$ 2) $\frac{4}{20} = \frac{1}{5}$ 3) $\frac{5}{200} = \frac{1}{40}$
4) $\frac{30}{300} = \frac{1}{10}$
5a) $\frac{3}{13}$ b) $\frac{1}{4}$ c) $\frac{1}{30}$ d) $\frac{17}{50}$
e) $\frac{5}{48}$ f) $\frac{2}{125}$ g) $\frac{1}{150}$

EXERCISE 4D

1a) 40% b) 25% c) 20%
d) $33\frac{1}{3}$%
2a) 10% b) 50% c) 25%
d) 100%
3) $1\frac{2}{3}$% 4) 15% 5) 30%
6a) 62.5% b) 37.5%
7a) 20% b) 50% c) No, Okonkwo is wrong. Uduak scored 70%

CHAPTER 4 REVIEW SECTION

1) 27%
2a)
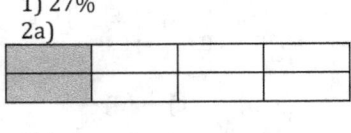

2b)
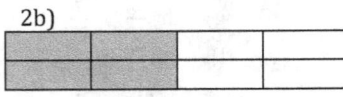

2c)
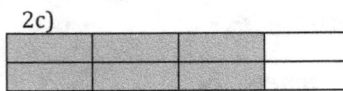

3) $\frac{1}{50}$
4a) 28.8% b) $\frac{1}{4} \times 100 = 25\%$
c) 50%

5a) 50% b) 60% c) 70%
d) 15%
6a) Maths – 68%
Economics – 5%
English - 60%   Physics – 35%
6b) 5%, 35%, 60%, 68%
7)

Fractions	%	Decimals
$\frac{2}{5}$	40%	0.4
$\frac{7}{20}$	35%	0.35
$\frac{3}{5}$	60%	0.6
$\frac{6}{25}$	24%	0.24

8) 0.65, $\frac{3}{5}$, 51%, $\frac{1}{2}$
9) $\frac{8}{25} \times 100 = 32\%$
$\frac{8}{25}$ is bigger
10a) 75% b) 25%

EXERCISE 5A

1a) 9,5,4,-3,-8,-10
b) 40,10,9,-2,-7,-15
c) 42,17,-10,-15,-20
d) 5,3,-1,-5,-7
e) 9,8,4,-4,-8,-9
f) 30,14,10,-12,-16,-18
2a) A : 9°C, B : 13°C,
C : -9°C, D : -2°C
3a) – 12°C b) 14°C
c) 4°C d) 9°C
4a) > b) > c) < d) =
5a) 8°C, b) 16°C c) 3°C
d) 5°C e) 3°C f) 3°C
g) 13°C h) 6°C
6a) -1°C,-3°C,-7°C,-10°C, -15°C
b) 5°C c) 5°C d) -15°C
e) 20°C
7a) Fallen by 6°C
b) Increased (risen) 15°C
c) Fallen by 10°C
d) Increase by 23°C

8a) 5 b) 1 c) -2 d) -4
e) -9 f) 4

EXERCISE 5B

1a) -10, -12 b) -2, -5
c) -5, -10 d) -4, -10
e) -1, 1 f) -9, -11
2a) -9,-4,-3, 0, 5

b) -50,-30, 20, 40, 55
c) -7,-3,-2,-1, 0, 1, 4

3)

Temp °C	Change °C	New Temp °C
-3	+4	1
-3	-3	-6
-7	-3	-10
+6	+6	12
-9	-9	-18

4a) 2°C b) 9°C c) 28°C
5a) Kano b) Essex c) Essex, Manchester, Lagos, Onitsha, Abuja, Kano   d) Essex: 1°C Onitsha: 31°C

EXERCISE 5C

1a) 11 b) 0 c) -1 d) -4 e) -4
f) 3 g) 5 h) -4 i) 3 j) -2
k) -5 l) 7 m) 3 n) 2 o) 2
p) 0 q) 0 r) 1
2a) 2 b) 2 c) -3 d) -1 e) 0
f) 0 g) -7 h) -3 i) -1 j) -7
k) -8 l) 4 m) -5 n) 7 o) -6
p) -6 q) 0 r) 3
3a) -6 b) -16 c) -10 d) -6
e) -3 f) -16 g) -13 h) -13
i) -11 j) -17 k) -11 l) -10
m) -15 n) -9 o) -5 p) -12
q) -10 r) -23
4a) -11 b) -15 c) -7 d) -12
e) -22 f) -10 g) -15 h) -15
i) -10 j) -5 k) -14 l) -16
m) -8 n) -11 o) -9 p) -6
q) -5 r) 15
5a) -2 b) -1 c) -2 d) -2
e) -2 f) -6 g) 7 h) -4 i) 7
j) 1 k) -130
6a) 13 b) 5 c) 4 d) 12
e) 19 f) 10 g) 10 h) -5

7)

×	-4	-3	-2	-1	0	1	2
+4	0	1	2	3	4	5	6
-5	-9	-8	-7	-6	-5	-4	-3
+3	-1	0	1	2	3	4	5
-2	-6	-5	-4	-3	-2	-1	0
0	-4	-3	-2	-1	0	1	2
+8	4	5	6	7	8	9	10
-10	-14	-13	-12	-11	-10	-9	-8
-7	-11	-10	-9	-8	-7	-6	-5
-6	-10	-9	-8	-7	-6	-5	-4

8a) -14 b) -4 c) -4 d) -1   9) ₦22 259 overdrawn

EXERCISE 5D

1a) 6 b) 20 c) 7 d) 30 e) 21 f) 25 g) 120 h) 45
i) 14 j) 72 k) 8 l) 56
2a) -6 b) -70 c) -15 d) -36 e) -36 f) -8 g) -30
h) -18 i) -63 j) -6 k) -16 l) -33
3a) 2 b) 3 c) 3 d) 3 e) 10 f) 4 g) -3 h) -10
i) -4 j) -4 k) -5 l) -5
4)

×	8	-2	-5	7	-3
-3	-24	6	15	-21	9
4	32	-8	-20	28	-12
5	40	-10	-25	35	-15
-10	-80	20	50	-70	30
-2	-16	4	10	-14	6
-7	-56	14	35	-49	21

5)
-3 × 7 ⟶ -21
-6 × 6 ⟶ -36
-6 × -6 ⟶ 36
-7 × -3 ⟶ 21
-10 ÷ -2 ⟶ 5
10 ÷ -2 ⟶ -5

6a) 1 b) 6 c) 2 d) -10 e) -32
f) -27 g) 56 h) -9

CHAPTER 5 REVIEW SECTION

1) 2 2) -5 3a) -3 3b) -3 3c) -3 3d) -14
4) -9°C 5a) 5 5b) 2 5c) 9 5d) 11
6a) -5, -3, -2, -1, 7, 10   b) -7, -5, -4, 0, 3, 5, 8
c) -17, -5, -2, 1, 2, 4, 5

7)

×	-4	+3	-8	2
-2	8	-6	16	-4
-7	28	-21	56	-14
+8	-32	24	-64	16
+5	-20	15	-40	10

EXERCISE 6A

1a) 3 b) 0.4 c) 6 d) 0.8 e) 1.6 f) 4 g) ₦8 h) ₦30
i) 6 j) 0.8 k) 12 l) 1.6 m) 16 n) 18 o) 16 p) 50
2a) 1 b) 7 c) 60 d) 2 e) 1.9 f) 0.24 g) 1220
h) 28 kg

EXERCISE 6B

1a) 1.3 b) 1.05 c) 16 d) 24 e) 6.93 f) 7.7 g) 5.2
h) ₦21 i) £33 j) 675 k) 1.2 l) 28 m) 24.4 kg
n) 62.3 o) ₦460 p) 0.36
2a) £1.89 b) 1.792 c) 5.6 d) 80.5 e) 0.75 f) 9
g) 2.38 kg h) ₦325

EXERCISE 6C

1a) 22 b) 88 c) ₦220 d) ₦374 e) 924 f) 3300
2a) 23 b) 92 c) ₦230 d) ₦391 e) 966 f) 3450
3a) 32 kg b) 73.6 litres c) £98.40 d) $146.42
e) 3200 g f) 196 cm   4) ₦2500 5) £3150
6) b ⟶ 50 to 70
7) For £15, sale price = £9.75
For £32, sale price = £20.80
For £45, sale price = £29.25
8) 1.94 kg 9) 33.3% 10) ₦5610 11) ₦5220
12) 82.8 kg 13) £772.20

EXERCISE 6D

1a) £21 b) £320 c) ₦2400 d) ₦2500 e) ₦900
F) $412.50 g) $490 h) ₦9000
2a) ₦125 b) ₦5125
3a) £1750 b) £5250
4) Bank A: £1500
5a) ₦840 000 b) ₦4 340 000 c) ₦120 555.56

EXERCISE 6E

1a) £16 b) £96   2) £90   3a) £420 b) £14 400
c) £1080   4a) £70 b) £420

PROBLEM SOLVING

1a) £193 b) i) £393 ii) £2 2a) £3400 b) £6000
c) £9400 3) £40 4a) £2100 b) £375 c) £600
d) £2475

EXERCISE 6F
1a) £2970.52 b) £3059.64 c) £3443.65
2) £779.14 3a) £7334.57 b) £20834.57
4) £5324.46 5a) 4 years
5b) 215 456 800.80 6a) Niamh
6b) Niamh will pay £141.55 more.

EXERCISE 6G
1) £710.78 2) 246.57 cm³ 3) £500
4) A £64   B £80
5a) £483.50 b) £96.70

CHAPTER 6 REVIEW SECTION

1a) 13 b) 0.34 c) 180 d) ₦1260
2) ₦3600 3) 67.5kg 4a) 25% 4b) 4%
4c) 4.2% 5) 10% 6) ₦1068.48
7a) ₦24 500 7b) ₦350 000 7c) ₦31 500
8a) £26 b) £156 9) £114
10a) £360 b) £18 000 c) £1320
11a) £7334.57 b) £20834.57
c) £29386.67

EXERCISE 7A

1a) 22 b) 11 c) 47 d) 8 e) 4 f) 5
g) 18 h) 22 i) 44 j) 9 k) 39 l) 3
m) 18 n) 22 o) 14 p) 10 q) 15
r) 11 s) -2 t) 300
2a) 5 + (6÷2) = 8   b) 21 ÷ (3+4) = 3
 c) 2 × (7-4) = 6   d) 8 – (4÷4) = 7
e) 9 + (3+3) × 2 = 21
f) (40-8) × 7 = 224
3a) 12w + 10 b) 3 + 8y c) 47c
d) 5 + 3x e) 5n – 1 f) 7w – 2
g) 18 h) 12m + 10 i) 47p – 3
j) 4w + 5 k) 51x + 2 l) 10x – 2
m) 27p – 9 n) 24 – 2n o) 35k – 21
p) 10 q) 44b r) $24g^2 - 3$
s) -2x t) 28v

EXERCISE 7B

1a) 20 b) 30 c) 13 d) 7 e) 10 f) 5
g) 100 h) 60.25 i) ¼ j) 18 k) 130
l) 5
2a) 21 b) 33 c) 37 d) 49 e) 0 f) 70
g) 42 h) -6 i) 21 j) 9 k) 49 l) 20
3a) i) 22 cm ii) 10 m 3b) i) 18 cm
ii) 9m c) i) 36 cm ii) 19.5 m
d) i) 11 cm ii) 6.6 m
4a) 18 b) 116 c) 10 d) 140 e) 3
f) -24 g) 2 h) 10
5a) 25 b) 3 c) 31 d) 36 e) 2 f) -10
6a) 10 b) 270 c) 63 d) 169
7a) 10 b) 4.25 c) 1/8 d) -5.75

EXERCISE 7C

1a) -9 b) -6 c) -2 d) 13 e) -10 f) 24 g) -103 h) -9 i) -27
j) 0 k) 1 l) -58
2a) -2 b) -11 c) 28 d) 1 e) -1 f) 8 g) 95 h) -21 i) -48 j) 4
k) 34 l) 19
3a) -18 b) -18 c) -30 d) 6 e) -6 f) -24 g) -13 h) -24 i) 9
j) 2 k) 1 l) 34

EXERCISE 7D

1a) -5 b) 0 2) 2 3a) 50 b) -30 4) 19 5) 4 6a) 201 b) 170
7a) 81.225 b) 6
8) $4y^3 + 20 = 128$, $10y - 5 = 25$, $3y^3 = 81$, $5 - 2y = -1$, $2(y - 1) = 4$
$6y - 3y = 9$
9a) 5/6 b) 5 10a) 1 b) -7.6

EXERCISE 7E

1a) c = 2 b) c = 13 c) c = 15 d) x = 1 e) x = 11 f) x = 91
g) x = 18 h) w = 10 i) w = 22 j) w = -9 k) u = -13 l) x = -16
m) u = -7 n) r = 53 o) g = -5 p) w = 23 q) y = -16 r) r = 64
s) y = 30 t) n = 22 u) e = -7 v) x = 24 w) c = 10 x) x = 6.8
y) x = 7.8 z) u = 344

EXERCISE 7F

1a) x = 2 b) x = 3 c) x = 6 d) x = 7 e) x = 11 f) w = 3 g) w = 9
h) c = -2 i) c = -2 j) w = -4 k) x = 8 l) x = -10 m) x = 7 n) n = 3
o) n = 42 p) x = 30 q) w = -18 r) n = 3 s) x = 7 t) n = -9
u) n = -2 v) m = 3 w) w = 4.5 x) z = 1/3 y) n = 10 z) y = -94

EXERCISE 7G

1a) c = 4 b) c = 2 c) x = 4 d) w = 2 e) x = 1 f) x = -1 g) x = 3
h) w = 1 i) y = 6 j) x = 2 k) u = -1 l) x = -1 m) y = 2 n) y = 10
o) n = 6 p) v = 3 q) x = 2 r) x = 3 s) c = 5 t) w = 2 u) x = 8
v) n = 7 w) w = 8 x) x = 3 y) c = 4 z) x = -1

EXERCISE 7H

1a) n = 15 b) n = 40 c) n = 10 d) n = 5 e) n = -9 f) n = 300
g) n = -63 h) n = 5 i) n = 14 j) n = 120 k) n = 3 l) n = 78
m) n = 32 n) n = 4 o) n = 2 p) n = 4/5 q) n = -4 r) n = -18
s) n = 5 t) n = 4 u) n = 5 v) n = 1 w) n = 1 x) n = 4 y) n = 28
z) n = -8.

EXTENSION QUESTIONS
1) 9 cm 2) 36 cm² 3a) d = -15 b) w = -10 c) n = 1

EXERCISE 7I
1) 3 2) 1 3) 5 4) 12 5) 3 6) 1 7) -1/8 8) $8\frac{1}{3}$ 9) 1 10) 5
11) $1\frac{21}{44}$ 12) 2 13) 2 14) 1 15) 10 16) -4 17) 3 18) -1
19) 3/7 20) $14\frac{8}{21}$

EXERCISE 7J
1) 4 2) 3 3) $-1\frac{1}{6}$ 4) 2 5) 3 6) -2 7) $-1\frac{7}{9}$ 8) $3\frac{2}{7}$ 9) $-3\frac{1}{2}$ 10) 3
11) $2\frac{3}{4}$ 12) 7 13) 3/8 14) ½ 15) -5 16) 6 17) 2 18) 1/3
19) $10\frac{2}{3}$ 20) $2\frac{3}{4}$ 21) 0 22) -1/2 23) -4/7 24) $-1\frac{2}{5}$

EXERCISE 7K

1) 2/3 2) 4 3) 3 4) 2 5) ½ 6) -2 7) 9
8) -1 9) -2/5 10) 18 11) 2 12) 16
13) 10 14) -2 15) -7

EXERCISE 7L

1) $2x + 10 = 44$, $x = 17$ 2) $7(x + 3) = 70$, $x = 7$
3) $\frac{3x + 6}{3} = 2x$, $x = 2$
4) $\frac{x - 6}{4} = x - 4$, $x = \frac{10}{3}$

5) $2(3x + 2) = 2x + 12$, $x = 2$
6) 21, 22, 23
7a) $t + 10$ years b) $t + 35$ years
c) $t - 5$ years d) 25 year
8a) $4x + 6 = 22$ b) 4
c) Length = 9 cm, width = 2 cm d) 18 cm²
9) £15 000 10) $\frac{x + 38}{2} = 10x$, $x = 2$
11a) $7(x + 6) - 40 = 16$, $7x + 2 = 16$
11b) $x = 2$ 12) $2(x - 9) = \frac{x}{5}$, $x = 10$
13a) $3w + 30 = 180°$, $w = 50°$
13b) $4w + 40 = 360°$, $w = 80°$
13c) $4w + 80 = 180°$, $w = 25°$
14a) $4x + 6 = 450$ b) 111, 112, 113, 114
15a) $3x + 28$ b) $3x + 28 = 49$ c) $x = 7$
15d) 10 cm, 12 cm and 27 cm

EXERCISE 7M

1) $x > -1$ 2) $x \le 4$ 3) $x \ge 0$ 4) $x \ge -1$
5) $-1 \le x < 5$ 6) $x < -1$ and $x \ge 2$
7) $x \le 0$ and $x > 3$

EXERCISE 7N

1a) $x < 5$ b) $x < 3$ c) $x > 1$ d) $x \le -2$
e) $y \ge 1$
2a)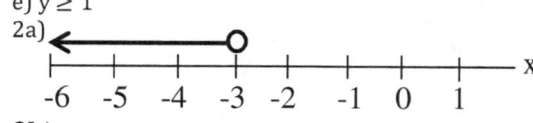
2b)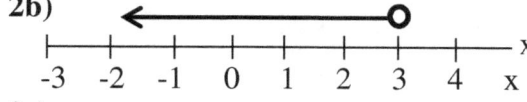
2c)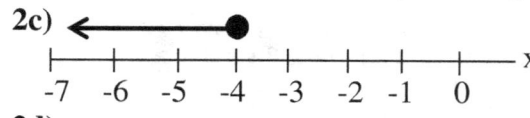
2d)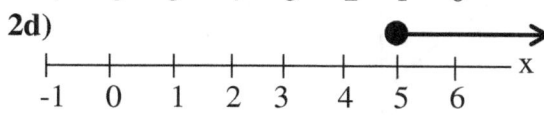
2e)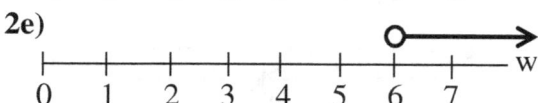

2f)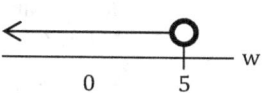

3a) True b) False c) True d) True e) True f) True
g) False h) False
4a) 2 b) -1 c) 5 d) -9   5a) 1 b) 1 c) -1 d) -4

EXERCISE 7O

1) $x < 5$ 2) $x < 5$ 3) $x \le -6$ 4) $x \ge 12$ 5) $x > 1$
6) $x > 1/3$ 7) $x < -2$ 8) $x \ge -2$ 9) $x < -13$ 10) $x > 2\frac{1}{2}$
11) $n \le 4$ 12) $w > ½$
13) $n \ge -2$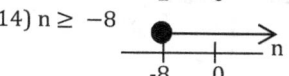
14) $n \ge -8$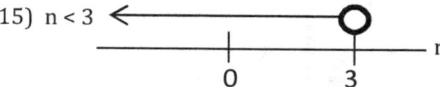

15) $n < 3$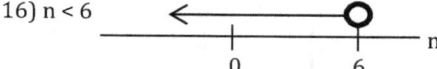

16) $n < 6$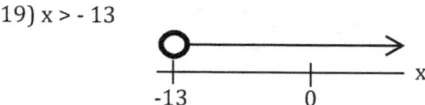

17) $1 < x < 2$

18) $x > 1/3$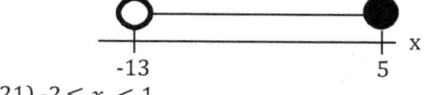

19) $x > -13$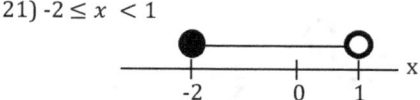

20) $-13 < x \le 5$

21) $-2 \le x < 1$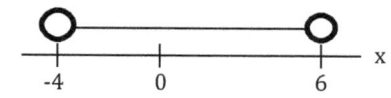

22) $-4 < x < 6$

23) -1, 0   24) 6, 7, 8, 9, 10   25) Not possible
26) x < 5 and x ≥ −1 .... -1, 0, 1, 2, 3, 4
27) x > 2 and x < 5 ....   3, 4
28) x > -3 and x < 6 ....-2, -1, 0, 1, 2, 3, 4, 5

EXERCISE 7P

1) x > 3

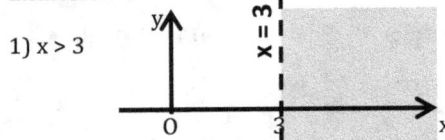

2) x ≤ 5

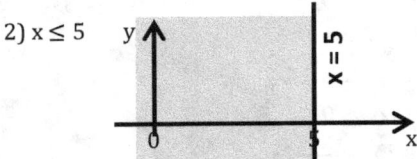

3) x > -1

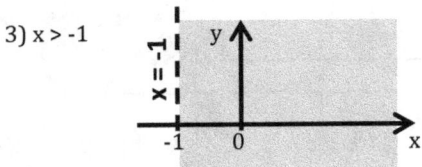

4) y > 5

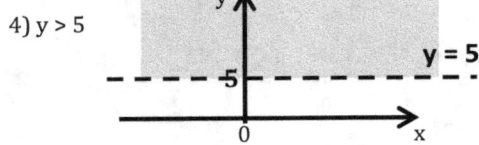

5) y > x + 2

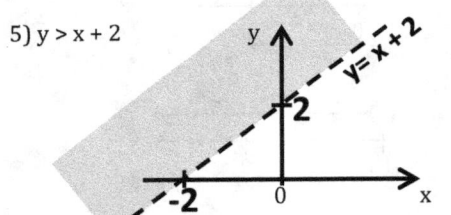

6) x + y ≤ 3

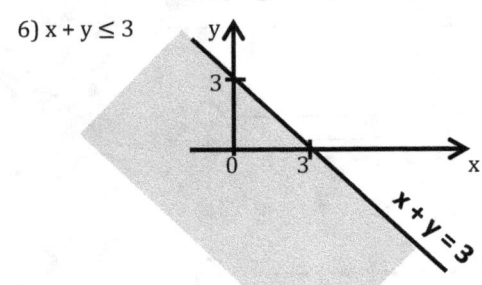

7) x + y ≥ −2

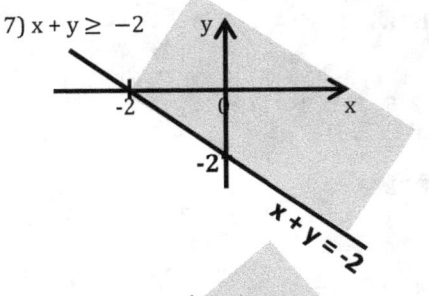

8) y > 3 − 2x

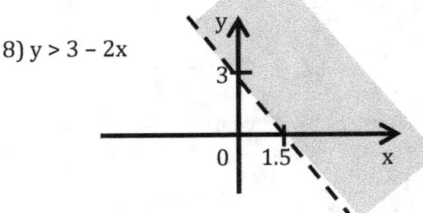

9) x > 1,   y ≤ 3,   y ≥ x − 3

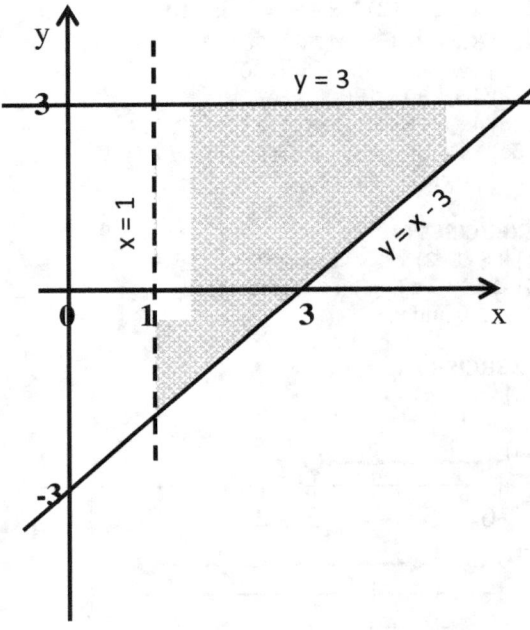

10) x ≥ 2   11) x < 3   12) y ≤ 2 and x < 3
13) x + y ≥ 3   14a) s + c ≤ 8, s ≥ 3, c ≥ 1
14b)

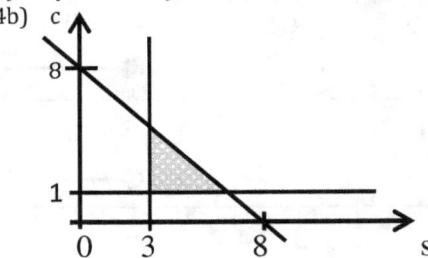

CHATPER 7 REVIEW SECTION

1a) 16 b) -3 c) 7
2a) c − d + e − f b) 13w + 6n
3a) f + 4b b) 7a − 3d + 2e c) $8w^2$ d) $3a^2b$
4a) 12 b) 44 c) 23
5a) 3 b) 7 c) 4 d) 9
6a) n = 2 b) n = 11 c) n = 27 d) x = 4 e) x = -9
f) x = -1 g) y = 14 h) x = 10
7a) n = 15 b) x = 70 c) x = 7 d) x = -45
e) x = -60 f) x = 18 g) y = 48 h) x = 3
8a) 3 b) -4 c) 3 d) -3 e) 2/3 f) -1/7 g) -2
h) 4/9
9) 10/3 = $3\frac{1}{3}$ 10a) 8x + 14 = 30   10b) x = 2
10c) Length = 10 cm, width = 5 cm   10d) 50 cm^2
11a) < b) > c) > d) > e) <
12a)   12b)
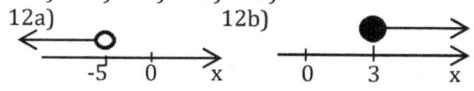
c)   12d)
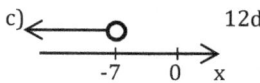   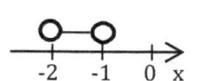
12e)
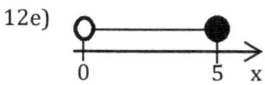

13a) x > $1\frac{1}{6}$ b) x ≤ −10 c) w ≥ −7
d) a < -25 e) -8 < x < 2
14a) 4, 5, 6, 7, 8 b) -4, -3, -2, -1, 0, 1, 2
c) 3, 4, 5, 6   d) -2, -1, 0

15)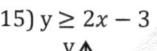
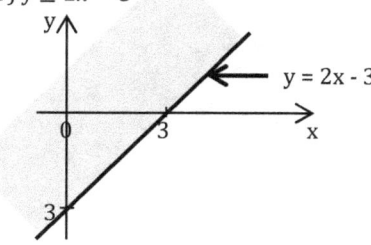

16a) x ≥ 3 b) x + y < 3 c) y < x or x > y
d) y ≥ 2

EXERCISE 8A

1a) 30° b) 10° c) 145° d) 44° e) 60°
f) 100° g) 50° h) 141° i) a= 27°,
b = 153°, c = 153° j) x = 30° k) 125°
l) w = 90°, y = 47° m) 215°
n) n = 90°, x = 40°

EXERCISE 8B

1a) a = 100° 2) b = 40°
3) c = 42°, d = 42°
4) e = 70°, f = 70°, g = 110°
5) h = 75°, i = 30°
6) j = 60°, k = 60°, l = 60°
7) m = 32°, n = 50°, o = 82°, p = 98°
8) q = 60°, r = 60°, s = 60°, t = 120°, u = 120°
9) v = 33°
10a) scalene………….three different angles
b) Right-angled……….have a 90° angle
c) Isosceles……two equal lengths
d) Equilateral…….all sides the same
e) Isosceles……two base angles are equal
f) Equilateral…….three equal sides
11a) 25° b) 25° c) 59° d) 56° e) 84°
12) NO, Obiora is incorrect. 54 + 85 + 42 = 181° but angles in a triangle add up to 180°.
13a) 94° b) 30° c) 81° d) 10° e) 40° f) 60° g) 90°
h) 81° i) 39° j) 55°
14a) 9w = 180 b) w = 20° c) 100° d) 80
15a) i) 55° ii) 70° iii) 18°
15b) i) Isosceles triangle because two base angles are the same (55° and 55°).
ii) Scalene triangle because all three angles are different.

EXERCISE 8C

1a) Rhombus b) Trapezium c) Rectangle d) Square e) kite
f) Parallelogram g) Trapezium

2) 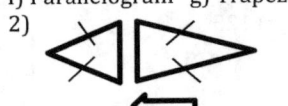   3)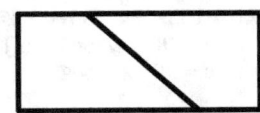

4) A square has all the corners at 90° while the corners of a rhombus are not 90°.
5) No. A quadrilateral has four sides but shape B has three sides.
6a) Trapezium b) Rhombus c) Kite d) Square e) Rectangle
f) Parallelogram
7) No, Kunle is incorrect. A trapezium has only a pair of parallel sides. The shape drawn is a parallelogram.

EXERCISE 8D

1a) p = 141° b) w = 65°, x = 95° c) a = 93°, b = 75°, c = 92°
d) f = 90° e) d = 43°   2) 140°   3a) 3n = 156, n = 52°
3b) 8x = 280, x = 35° 3c) 6x = 210, x = 35° 3d) 10c = 360, c = 36°

EXERCISE 8E

1) NO. A polygon has straight sides but a circle is curved.
2) Any four quadrilaterals, for example; Pentagon – 540°, Hexagon – 720°, Heptagon – 900°, Octagon – 1080°
3) Dodecagon – 1800°
For questions 4 - 7): Students to construct circles accurately, using the radii given. They must use a pencil, ruler and a pair of compasses for the construction. Teachers/parents to check the accuracy of the circles drawn.
8) No, Okoro is incorrect. A circle with a diameter of 20 cm must have a radius of 10 cm. *2 × radius = diameter*

EXERCISE 8F

1a) a = 47°........ Alternate angles are equal
b) b = 107°...... alternate angles are equal
c) c = 38° .....Angle on a straight line where 142° lies is 38° (180 – 142). Therefore,
Angle C = 38° ....alternate angles. Other Mathematical reasons are also encouraged.
d) d = 102°... corresponding angles are equal
e) e = 113° ....alternate angles
   f = 70° ....corresponding angles
f) g = 96° ....corresponding angles
g) h = 30° ....alternate angles
   i = 47° ....angles on a straight line
   j = 133° ....corresponding angles
h) k = 110°
 Angle next to K is 70° ....Corresponding angles.
Angle on a straight line = 180 – 70 = 110°
i) l = 40° ....angles on a straight line

n = 14° ....corresponding angles
m = 166° ... from (180 – 14) = 166°, then corresponding angles are equal.
m = 40° ......alternate angle to l
j) p = 58° ...co-interior angles add up to 180°

CHAPTER 8 REVIEW SECTION

1a) x = 85°   b) a = 109°, b = 71°, c = 71°
d = 109° c) e = 63°, f = 117°, g = 63°, h = 54°
i = 63°
2a) j = 92°, k = 110°, l = 70°   b) m = 95°
c) n = 48°, o = 89°, p = 48°
3a) 3 b) CD c) AB d) Regular hexagon
e) 6 f) obtuse

Secondary Mathematics 11 – 13 Years
Arinze Oranye

EXERCISE 9A

1) Teacher's supervision
2a) 9.4 cm b) 7.1 cm c) 8.1 cm d) 5.1 cm
3a) 22 cm b) 44 cm c) 110 km d) 132 cm
e) 176 m f) 6.3 km
4a) 25 cm b) 90 cm c) 3.6 m d) 94 cm
e) 35.7 cm
5) 131.9 cm 6) 502.4 cm 7) 2.2 cm
8) 159 rev.

EXERCISE 9B

1a) 21 cm^2 b) 24 cm^2 c) 22 cm^2 d) 4 cm^2
2) A = 3cm^2, B = 6 cm^2, C = 2 cm^2, D = 3.5 cm^2
3) P = 2 cm^2, Q = 11.5 cm^2
4) Could be 4 cm and 2 cm or 8 cm and 1 cm.
5) 6 m 6) 127 cm^2 7a) 22 m^2 b) ₦37 300
8)

Area	Length of side
	1 cm
	3 m
64 cm^2	
18.49 cm^2	
	9 cm

EXERCISE 9C

1a) 15 cm^2 b) 6 cm^2 c) 24 cm^2 d) 14 cm^2
e) 17.5 cm^2 f) 18 cm^2
2) 63.5m^2 3) 6 cm
4) For example; 14 cm and 4 cm or any two lengths that will multiply to give 56.
5a) 18 cm^2 b) 12 cm^2 c) 14 cm^2
d) 240 cm^2 e) 40 cm^2
6a) 18 cm^2 b) 20 cm^2 c) 28 cm^2
d) 75 cm^2 e) 36 cm^2 f) 175 cm^2
7a) 5 m b) 5 cm c) 5 m d) 2 m
8a) 3.52 cm^2 b) 171 cm^2 c) 30 cm^2
d) 153 cm^2 e) 67.5 cm^2 f) 6.3 cm^2
g) 19.22 cm^2

EXERCISE 9D

1a) 80 cm^2 b) 45 cm^2 c) 19.5 cm^2 d) 45.5 cm^2 e) 21.2 cm^2
f) 17.5 cm^2 g) 48 cm^2 h) 117 cm^2
2a) 6 m b) 5 m c) 7 m d) 20 m

EXERCISE 9E

1a) 28 cm^2 b) 36 cm^2 c) 126.5 cm^2 d) 148 cm^2 e) 24 cm^2
f) 58 cm^2 g) 90 cm^2 2) 66 cm^2 3) 4 × 12 = 48 m^2, 5 × 6 = 30 m^2............48 + 30 = 78m^2 is the area.
4a) 85 m^2 b) 88 m^2 c) 125 m^2

EXERCISE 9F

1a) 129 m^2 b) 41.5 m^2 c) 51.5 m^2 d) 81 m^2 e) 88 m^2
f) 40.5 m^2

EXERCISE 9G

1a) 12.6 cm^2 b) 78.6 cm^2 c) 95.1 m^2 d) 314.3 cm^2
e) 201.1 km^2 f) 113.1 mm^2
2a) 140 m b) 15386 m^2
3)

Diameter (m)	Radius (m)	Area m^2
	7	154
	21	1386
42		1386
28		616

4a) 14.1 m^2 b) 50.2 m^2 c) 100.5 m^2 d) 379.9 m^2
5a) 218.5 m^2 b) 74.1 m^2

EXERCISE 9H

1a) 13.8 cm^2 b) 109.9 cm^2 2a) 19.63 m^2 2b) 58.88 m^2
2c) 16.13 m^2 3) 21.5% 4) 314 cm^2 5a) 471 cm^2
5b) 66.67% 6) 36.25 m^2 7) 176.7 m^2

CHAPTERS 9 REVIEW SECTIONS

1a) 7 b) 25 2) times 10
3)

Input	9	30	6	15	45	39
Output	15	50	10	25	75	65

4a)

8	3	10
9	7	5
4	11	6

4b)

9	4	11
10	8	6
5	12	7

5) 11.4 cm   6) 37.84 am
7a) 66 cm   b) 264 cm   8) 237 cm²
9a) 15 cm²   b) 24 cm²   c) 27 cm²
 d) 100 cm²   10) 6 cm
11a) 3 cm²   b) 52.5 cm²
c) 92 cm²   d) 420 cm²
12a) 7 m   b) 5 m   c) 4 m   d) 20 m
13a) 142 cm²   b) 72 cm²
14a) 7.1 cm²   b) 314.3 cm²
c) 173.3 cm²   d) 44.2 cm²   e) 285.2 cm²
15a) 294.6 cm²   b) 60%
16) ₦3 142 8 57.14   17) 7626 m²

g) 2, 1, 0   h) 6, 8, 10   i) 5, 0, -5
j) 11, 14, 17   k) 8, 11, 16   l) 10, 13, 18
m) 17, 37, 57   n) 2, 8, 18   o) 2, 11, 26
p) 14, 26, 46   q) 11, 9, 7   r) $1, \frac{1}{4}, \frac{1}{9}$

2a) 40, 44, 48, 52, 56   b) 7, 12, 17, 22, 27
c) 13, 11, 9, 7, 5   d) 2, -1, -4, -7, -10
e) -7, 2, 11, 20, 29
3) 4 and 10
4a) add 5 each time   b) 6   c) 481
5a) 7,   b) 11,   c) 391
6a) 108   b) 972
7) 17, 22, 42
8a) 13, 15, 17   b) -2, -1, 0
c) 1, 4, 9   d) 6, 18, 36

EXERCISE 10A

1a) 8, 9   b) 15, 17   c) 21, 26   d) 49, 97
e) 19, 23   f) -2, -8   g) 15, 20   h) 6.6, 6.5
i) 32, 64   j) 19, 26   k) 81, 243   l) $\frac{5}{162}, \frac{6}{486}$
2a) add 1 each time
b) add 2 each time
c) add 5 each time
3) Could be
i) 1, 8, 15, 22, 29   ii) 5, 12, 19, 26, 33
4a) add 5 each time   b) 4   c) 49
5a) 4, 10, 16, 22   b) 47, 62, 77, 92
c) 2, -5, -12, -19   d) 3, 9, 27, 81
e) 200, 100, 50, 25
6) 88, 5, 88, 616, 1672
7a) i) multiply difference by 2 and add to get
 the next term.   ii) 127
7b) i) Subtract 45   ii) 90, 45
7c) i) Subtract 3   ii) 83
7d) i) Multiply by 10   ii) 3, 3000
7e) i) Difference increases by 1 each then,
then add to the next number   ii) 18
8) 3 086 358 025

EXERCISE 10B

1a) 2, 4, 6   b) 14, 28, 42   c) 4, 6, 8
d) 2, 5, 8   e) 5, 9, 13   f) 1, 4, 9

EXERCISE 10C

1a) 2n + 3   b) 3n + 8   c) n + 3   d) -5n + 35   e) 4n – 3   f) 3n + 6
g) 7n   h) 4n – 4   i) 2n – 3   j) -3n + 20   k) 50n + 450
2)

Term number	4n	Term
1	4	7
2	8	11
3	12	15
4	16	19
70	280	283

3a) 2n + 2   b) -3n + 36   c) 2n   d) 3n – 1   e) n + 2   f) n²
4a) 15 cm
4b)

 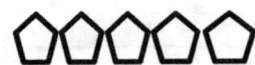
Pattern 4   Pattern 5

4c)

Pattern number	1	2	3	4	5	6
Perimeter (cm)	5	10	15	20	25	30

4d) 5n
4e) 500 cm
4f) 630
5a) 14 rods
5b) 3n - 1
5c) 20
5d)

Pattern number	1	2	3	50
Number of rods	2	5	8	149

EXERCISE 10D

1i) a, b, e ii) c, d, f
iii) c:71 and 97
d: 79 and 105
f: 101 and 123
iv
a) $2n - 1$ b) $-2n + 13$
c) $2n^2 + 4n + 1$ d) $2n^2 + 4n + 9$
e) $10n + 10$ f) $n^2 + 11n + 21$

2a) $n^2 + 1$ b) $n^2 + 5$ c) $2n^2 + 1$
d) $2n^2 - 2$ e) $3n^2 + 3$ f) $2n^2 + n + 1$
g) $2n^2 + 3n - 1$ h) $3n^2 + n - 1$
i) $2n^2 + n$ j) $6n^2$
3) 10005   4) 9, 14, 21
5a) 20 squares b) $n^2 + n$ c) 90300 squares

6a) 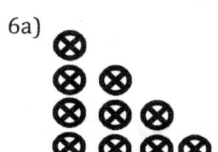 6b) $\frac{n(n+1)}{2}$ c) 5050

EXERCISE 10E
1a) 4, 7, 12 b) 0, 3, 8 c) 11, 14, 19
d) 5, 8, 13 e) 4, 13, 28 f) 5, 28, 87
g) 4, 10, 20 h) -4, 2, 12 i) 5, 12, 23
j) 3, 15, 33
2a) 9809 b) 14 492

CHAPTER 10 REVIEW SECTION

1a) 19 b) $k + 3$ c) $k - 3$
2a) $3n^2 + 3$ b) 120003 3) 40, 121, 364, 1093
4a) i) $3n + 2$ ii) 62 b) i) $-10n + 80$ ii) -120
c) i) $2n^2 - 2$ ii) 798 d) i) $n^2 + 30$ ii) 430
e) i) $\frac{n+2}{2n+3}$ ii) $\frac{22}{43}$ f) i) $n(n+2)$ ii) 440

EXERCISE 11A

1a) 1 : 3 b) 1 : 1 c) 1 : 2
2a) 1 : 2 b) 1 : 2 c) 2 : 1 d) 2 : 1 e) 2 : 1 f) 1 : 8
g) 19 : 30 h) 1 : 3 i) 3 : 2 j) 2 : 3 k) 7 : 11 l) 5:6
m) 1:4:10 n) 4:3:1 o) 1:2:7 p) 1:3:1 q) 2:3:6
r) 6:2:3
3a) 2:3 b: 3:5 c) 1:2 d) 1:3 e) 1:4 f) 1:10
g) 1:4 h) 1:2   4) 6:1   5) 18:7
6) 56 blue and 42 red pens   7) $\frac{16}{33}$
8a) 2:1   8b) 1:5   8c) 1: 2000   d) 20:1

9) 12: 30, 30: 75, 2:5, 42:105, 18:45, 2:5

EXERCISE 11B

1a) £12 and £24 b) £8 and £28 c) £15 and £21
2a) £60 and £360 b) £168 and £252
c) £240 and £180
3a) Mark received £3600 b) £4800 c) 70%
4) £240   5a) 100°, 140°, 40°, 80° b) 100°
6) 1:3   7) 13:22
8) A £20, £70, £40   B £40, £30, £60
C £55, £25, £50   D £39, £91   E £65, £65
F) £52, £78

EXERCISE 11C

1a) £450 b) £5400   2) £45   3) ₦132
4a) 1/5 b) ½ c) 1/5 d) ½
5a) 3/5 b) £2000   6) £400   7a) 1/3 b) 2/3
c) 23 men and 46 women

EXERCISE 11D

1) 5 hours   2a) 40 days b) $\frac{120}{w}$ days   3) 6 hours
4a) 3:2 b) 27 red books

EXERCISE 11E

1) $\frac{150}{10}$ = £15/hour   2a) 10 m/s b) 600 m/min
3) £800/month

EXERCISE 11F

1) 2100 km   2) 2 400 000 m   3) 1200 km
4) 3600 km
5a) 10000 cm × 6000 cm = 100 m × 60 m
5b) Area = 6000 m²   6) 7 cm

EXERCISE 12A

1a) $2x + 6$ b) $3x + 12$ c) $4x + 8$ d) $5x - 15$
e) $7x - 28$ f) $40 - 10x$ g) $24 + 8x$ h) $12x + 60$
i) $8x - 80$ j) $99x + 990$ k) $60 - 12x$ l) $a^2 + 3a$
m) $w^2 + 7w$ n) $3x^2 + 15x$ o) $m^2 + 9m$
p) $42 - 18c$ q) $4d^2 + 8d$ r) $7a - a^2$ s) $2x + 4c - 18$
t) $0.5x^2 + 2x$ u) $3x - 9$ v) $0.18c + 2.7$

EXERCISE 12B

1a) $-2x - 4$ b) $-8x - 24$ c) $-3x - 30$ d) $-5x -10$
e) $-9x - 36$ f) $-x - 2$ g) $-4x - 8$ h) $-7x - 49$
i) $-10x - 60$ j) $-20x - 100$ k) $-42 - 6x$ l) $-3x + 12$
m) $-9x + 9$ n) $-4x + 8$ o) $-8x + 40$ p) $-35 + 7x$
q) $-a^2 - 3a$ r) $-w^2 - 10w$ s) $-36w^3 - 30w^2$
t) $-3v^2 + 2v$ u) $-28w^3 - 24w^2$ v) $-a^2 - 4an + 2a$

EXERCISE 12C

1a) 5x + 9 b) 9x + 27 c) 12x + 52 d) 2x + 8
e) 3x + 11 f) −x − 7 g) x + 4 h) 5x + 6 i) 12y −16
j) 4y + y² k) 30t + 2 l) 6x + 18 m) 40d − 26
n) 11d + 1 o) 2x² − 3x p) -4d² -9d q) 5n² − 15n
r) -98w + 26 s) -6s² + 5s + 59

EXERCISE 12D

1) 12x + 8 2a) 2c + 18 2b) c + 15 2c) 3c + 14
3a) f² − 2f b) f² c) 6f² − 8f d) f³ − 2f²
4a) 3x b) 2x + 6 c) 10x + 36 d) 90x
e) 80x − 36 f) 69x

EXERCISE 12E

1a) x² + 4x + 3 b) x² + 7x + 10 c) x² + 10x + 21
d) n² + 8n + 16 e) n² + 15n + 54 f) x² + x − 2
g) x² − 4x − 21 h) w² − 6w + 5 i) w² − 9w − 10
j) c² − 3c - 108
2a) 12m² + 36m + 15 b) 6x² + 6y² + 20xy
c) x² − 16x + 64 d) a² + 2ab + b²
e) 28c² − 41c + 15 f) 3w³ − 2w² − 16w
3a) 8y + 16 b) 4y² + 16y + 16 c) 16

EXERCISE 12F

1a) 4(x + 4) b) 3(2x + 3) c) 4(3x + 4)
d) 5(a + 2) e) 9(x +10) f) 3(2x − 1)
g) 9(x − 2) h) 5(x − 6) i) 15(2x − 3)
j) 7(m − 7) k) 7(4n + 3) l) 7(x − 2b + 3c)
m) 2n(2 − 3y) n) 2(6x + 7) o) 16(y − 1)
p) bcd(a − 1) q) 5(m + 4n − 1) r) 75(k − 6)

EXERCISE 12G

1a) x(x + 5) b) 4t(t − 4) c) c(bc − d) d) 7p(p + 4)
e) cd(c + d) f) f(f² − 2) g) s(p + q − pr)
h) x(11x + 1) i) df(df − mndf + 1)
j) 3x(x − 3 + 9x²) k) 7y(2 − y + 4y²) l) (c+f)(d+e)
m) (x + 3) (b + c) n) -3(10y² + 3)

EXERCISE 12H

1a) 1, 2, 4, c, 2c, 4c
b) 1, 2, 3, 4, 6, 12, t, 2t, 3t, 4t, 6t, 12t
c) 1, a, c, ac d) 1, 7, x, 7x, y, 7y, 7xy, xy
e) 1, y, y² f) 1, 2, 3, 6, x, 2x, 3x, 6x, y, 2y, 3y, 6y, xy, 2xy, 3xy, 6xy g) 1, 2, 5, 10, x, 2x, 5x, 10x, x², 2x², 5x², 10x² h) 1, 5, 7, 35, e, 5e, 7e, 35e, f, 5f, 7f, 35f, ef, 5ef, 7ef, 35ef

EXERCISE 12I

1a) qr b) n c) 5c² d) 30mn e) b f) 5 g) 6j
h) n i) 5ab j) 6ny

EXERCISE 12J

1a) 12p b) cd c) 15tu d) 7n² e) 8cd² f) 30p²q²
g) 10ab²c h) amn i) 105mn²

EXERCISE 12K

1) 2t 2) $\frac{2x}{y}$ 3) 2/7 4) c/5 5) 2bc 6) $\frac{7a}{2}$ 7) 24y
8) $\frac{df}{eg}$ 9) p/2 10) $\frac{75}{10ps} = \frac{15}{2ps}$ 11) 7x 12) q/5
13) $\frac{b}{9a}$ 14) $\frac{10}{3s}$ 15) x/3 16) 6 17) w²/2y
18) d²/8c 19) 2/ap 20) 4cd/e³ 21) 9w/4y
22) $\frac{g+1}{g-1}$ 23) $\frac{y+2}{3}$ 24) 3x

EXERCISE 12L

1) 4x/7 2) 7n/6 3) 3x/7 4) $\frac{1+n}{y}$ 5) $\frac{5xy}{4w}$
6) $\frac{4c-d}{6}$ 7) $\frac{12x+tu}{4u}$ 8) $\frac{23}{2y}$ 9) $\frac{cx+30}{10c}$ 10) $\frac{15y-7x}{21}$
11) $\frac{7c+3}{10}$ 12) $\frac{8b-15}{15}$ 13) $\frac{28dg-5f}{7g}$ 14) $\frac{79u+v}{9}$
15) k/9 16) $\frac{4x-8}{8} = \frac{1}{2}x - 1$ 17) $\frac{3x+11}{7}$

CHAPTERS 11 & 12 REVIEW SECTIONS

1a) 11 : 10 b) 1 : 7 c) 1 : 14 d) 1 : 4
2) 4 : 1
3) 18:15, 36:30, 48:40, 36:30, 24:20, 42:35
4) Isabel - 15 oranges, Edward – 25 oranges Angela – 35 oranges
5) £45 6a) -4x − 20 6b) -3x + 12
6c) -9s² + 5s + 16 7a) wd(w + d) b) (a+3)(b+c)
c) 4ry(4y + 2r − 1)
8a) HCF = 15wx, LCM = 30w²x
8b) HCF = 5ny, LCM = 25n²y²
9a) 6bc b) 1/2ps
10a) 2x/3 b) 5n/3 c) $\frac{21c-3d}{21}$

EXERCISE 13A

1a) Even chance b) unlikely c) very unlikely
d) Impossible e) certain
2a) Q   b) P   c) R   d) T   e) S
3a) 1/3 b) 3/13 c) 1/13 d) 2/13 e) R and L
f) A
4a) 1 b) 3 c) 19 d) 80
5a) You will live up to 3000 years.
5b) You will die one day
5c) Picking a card greater than 3 from a pack of cards numbered 1 – 9.
6a) R b) Q   c) P d) T e) S

7a) i) ½   ii) ½

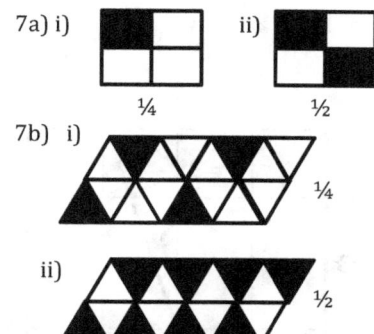

7b) i) ¼   ii) ½

EXERCISE 13B
1a) 0.25 b) 0.5 c) 1/13 d) 1/52
2a) 1/10 b) 3/10 c) 2/5 d) 1/5
3a) 1/8 b) 3/8 c) 3/8 d) ½ e) ¾ f) ¼
g) ½ h) 1/8 i) ½
4a) ½ b) 1/5 c) 1/20 d) 1/10 e) ½ f) 3/20
g) 1/5 h) 2/5
5a) 1/200 b) 1/200

EXERCISE 13C
1) 0.75 2a) NO b) NO c) YES d) NO e) NO
3) 0.3 4) White = 0.25, Green = 0.25
5a) 6/7 b) 1/3 c) 4/5

EXERCISE 13D
1)

	DIE					
COIN		1	2	3	4	5	6
	H	H1	H2	H3	H4	H5	H6
	T	T1	T2	T3	T4	T5	T6

2a)

+	1	2	3	4	5	6
1	2	3	4	5	6	7
2	3	4	5	6	7	8
3	4	5	6	7	8	9
4	5	6	7	8	9	10
5	6	7	8	9	10	11
6	7	8	9	10	11	12

2b) 36 2c) i) 1/12 ii) 0 iii) 13/18 iv) 7/36
v) 1/6

3a)

+	2	3	6	7
1	3	4	7	8
2	4	5	8	9
3	5	6	9	10

3b) i) 1/6 ii) 1/6 iii) 0 iv) 1/3

4a)

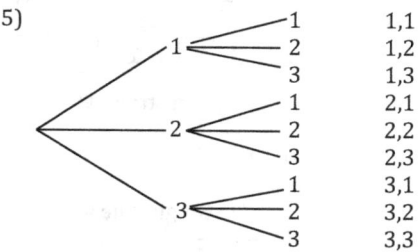

b) i) 2/15 ii) 1/3 iii) 3/5

5)

```
        ┌─1   1,1
     1──┼─2   1,2
     │  └─3   1,3
     │  ┌─1   2,1
 ────2──┼─2   2,2
     │  └─3   2,3
     │  ┌─1   3,1
     3──┼─2   3,2
        └─3   3,3
```

EXERCISE 13E

1) 3/10 2a) i) 0.25 ii) 0.15 iii) 1/16
2b) 32/80 = 2/5 2c) 33/80
2d) Perform more experiments
3a) 18 b) 27 c) 90 4) 26
5a) Blue 0.25 Red 0.1 White 0.3 Pink 0.14
Grey 0.21 5b) 1/5 or 0.2
5c) could be biased as theoretical probability is 0.2
while the relative frequencies are a bit off from
0.2.

CHAPTER 13 REVIEW SECTION

1a) 1/6 b) 0 c) ½ d) 1/6
2a) 4/25 b) 3/5 c) 1/5 d) 4/5
3a) i) Smith 5/8 ii) Deborah 9/14
3b) NO, the umpire is wrong. Alfie was the best
player since he has the biggest fraction of 8/12
(0.6666666..) compared to Smith's 10/16 (0.625)
3c) Stella 3d) Deborah
4a)

+	1	2	3	4	5	6
1	2	3	4	5	6	7
2	3	4	5	6	7	8
3	4	5	6	7	8	9

4b) i) 1/6 ii) 1/9 iii) 0 iv) 2/9

EXERCISE 14
1a) ½ b) ¼ c) 7 d) 9/5 e) – ½ f) – 7/5
g) – 1/11 h) 14/9 i) 7/4 j) 1/12 k) – 1/5
l) -12 m) 11/2 n) -1/4 0) 17/3

2a) – 10/9   b) – 2/3   c) – 4/5   d) – 1/100
e) 7/36   f) 7/26   g) – 2/5   h) – 2/9   i) w   j) 7/n
3) Andrew is WRONG. Reciprocal of 6 is 1/6 = 0.16666.. Reciprocal of ¼ is 4. 4 is greater than 0.166666....
4a) 17/3   b) 3/17   c) – 3/17   d) 1

EXERCISE 15A

1a)
x	1	2	3	4	5
y	4	7	10	13	16
C	1,4	2,7	3,10	4,13	5,16

1b) Graph of y = 3x + 1 drawn as a straight line
c) (0,1)
2a) (-1,2), (0,3), (2,5),(6,9)
2b and c) Students drawing
3a – f) Student's drawings of straight line graphs
4a)
d	0	5	13	20
c	0	15	39	60

4b)
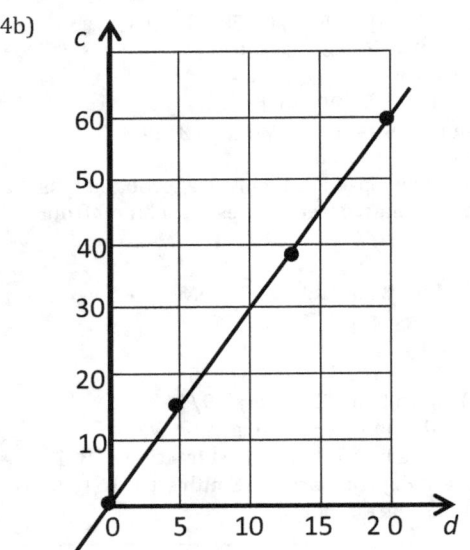

4c)
Diameter (d)	Circumference (c)
4	12
16	48
90	270
1	3

5a) P (2, -2)   Q(-1, 1)

5b)
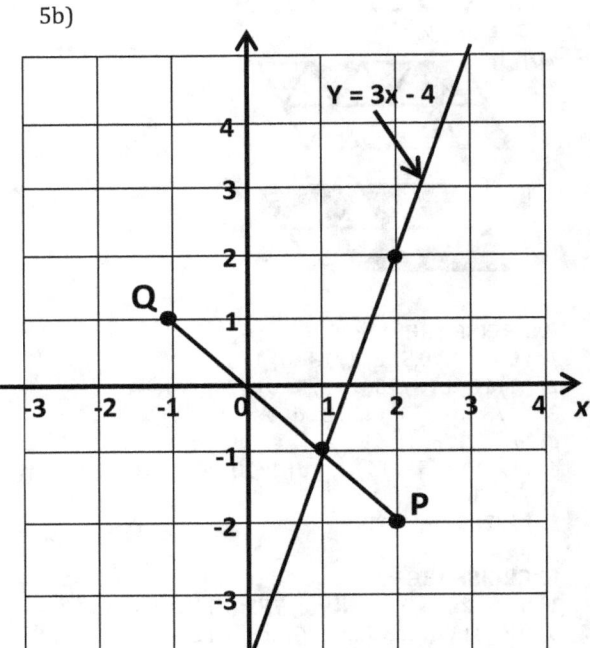

5c) See diagram   5d) (1, -1)

EXERCISE 15B

1) **A** x = -3 **B** x = 2 **C** x = 4 **D** y = 3 **E** y = -3
F x = 0 **G** y = 0

2a)
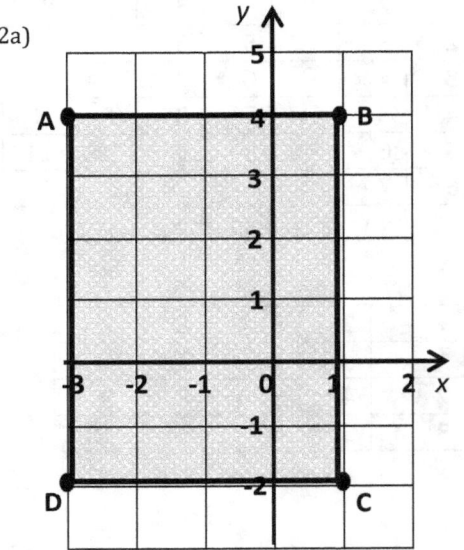

2b) AB y = 4
DC y = -2
AD x = -3
BC x = 1

3a)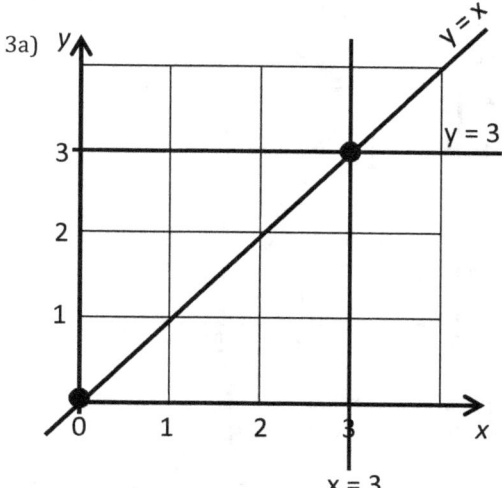

3b) (3, 0) (3, 3) and (0, 0)   3c) 4.5 cm²

4a)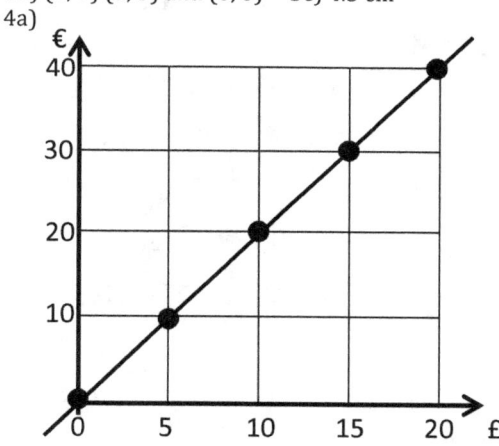

4b) i) €12 ii) €28 iii) €34
4c) i) £2.50 ii) £11 iii) £18.50

EXERCISE 15C

1a) 1 b) 1 c) 6 d) 2 e) 3/2 f) 3 g) -5
h) – 10/7 i) 4 j) -1 k) 1 l) – ¼ m) 4
n) -2 o) 2 p) 2
2a) 1 b) ½ c) – 3/2 d) 2 e) ½ f) – ½ g) – 5/3
3) AB 4   BC – 4/3   DC 5/6   AD – 3/8
4a) 1 b) -2 c)1 d) 7/5

EXERCISE 15D

1a) (4, 6) b) (1, 4) c) (3.5, 4) d) (2, 3),
e) (- 0.5, 0.5), f) (1, -2)
2) A (2, -3) B (3.5, 1) C (4, 5) D (-3, 4)
E (-6, 3.5) F (- 4.5, - 3.5)
3) (-4, 5)
4a)   AB (1, 4)   BC (3, 4)   AC (3, 2) 4b) 8 cm²

EXERCISE 15 E

1a) y = 4x + 4 b) y = -5x + 5 c) y = 4x – 6
d) y = $\frac{2}{3}$x + 4   e) y = -4x – 5   f) -$\frac{5}{6}$x – 4
2a) y = -$\frac{2}{3}$x + 4   b) y = $\frac{4}{5}$x – 2   c) y = 3x - 1

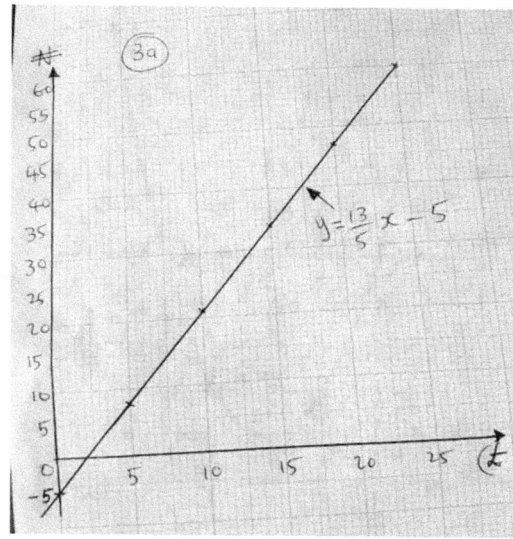

3b) y = $\frac{13}{5}$x – 5
4) A ➡ y = -$\frac{1}{2}$x – 2   B ➡ y = -$\frac{2}{7}$x + 2
C ➡ y = $\frac{1}{3}$x + 3   D ➡ y = x + 5
E ➡ x = - 6   F ➡ y = -x - 5

EXERCISE 15F

1a) Q and S are parallel b) P and R are perpendicular   c) T is neither parallel nor perpendicular

2) Student's drawing and the gradient is 2. Student's drawing of a line perpendicular to y = 2x + 1 with a gradient of -1/2.
3) 5   4) y = x + 4

EXERCISE 15G

1a)

x	-3	-2	-1	0	1	2	3
x^2	9	4	1	0	1	4	9
y	13	8	5	4	5	8	13

1b) Student's own drawing
2a,b,c) Student's own drawing
3a) Student's own drawing
3b) i) x = 1.5   ii) x = - 2.5

EXTENSION WORK
4) Student's drawings
5a) Graph of $y = x^2 + 1$

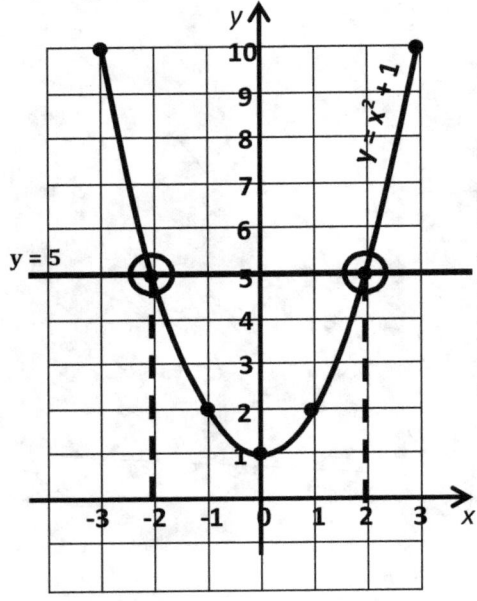

5b) See diagram above.
5c) x = -2 and a = 2

CHAPTERS 14 AND 15 REVIEW SECTIONS

1a) 11/3 b) 3/11 c) – 3/11 d) 1
2a)

x	1	2	3	4	5
y	4	9	14	19	24
C	1,4	2,9	3,14	4,19	5,24

2b) (0, -1)   3a) 1   3b) y = x - 4
4a) (1, 4)   4b) (- 0.5, 0.5)
5a) $y = -\frac{5}{3}x + 7$ b) y = 4x – 1
6) $4 \times -\frac{1}{4} = -1$. Two lines are perpendicular if the product of their gradients is -1.

EXERCISE 16A

1a) Positive correlation   b) Positive correlation
c) Negative correlation   d) Negative correlation
e) Negative correlation   f) No correlation

2a)

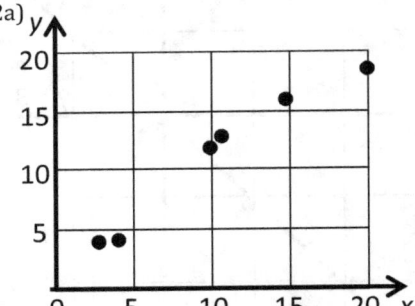

2b) Positive correlation
3a) Negative correlation.   q = 15

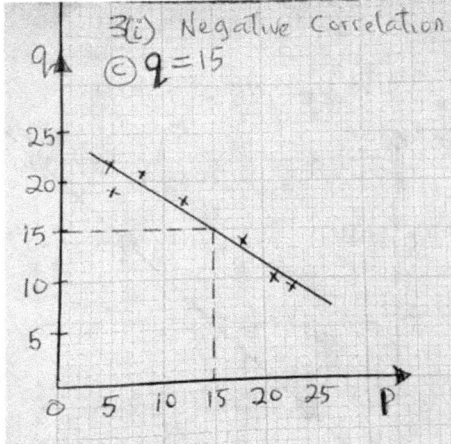

3b)

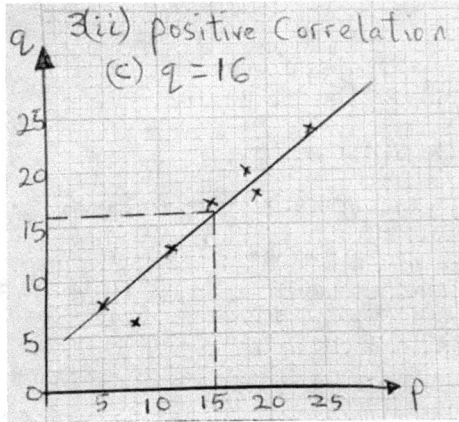

3iii) No correlation.

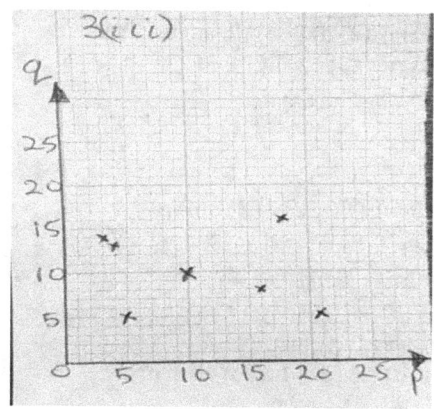

4a) Negative correlation. As the time spent on watching TV increases, the time spent on doing homework decreases.
4b) $1\frac{1}{2}$ hours

EXERCISE 16B

1a)
```
0 | 2   5   9
1 | 2   2   5   7
2 | 2   5   7   9
3 | 0
```
Key: 1/0 represents 10

1b) i) Mode = 12   ii) 16   iii) 28
2a) 11
2b)
```
1 | 50  75  80  90
2 | 25  35
3 | 00  00  33  70
4 | 05
```
Key: 1/50 represents £1.50
2c) £3.00   2d) £2.55   2e) 2.35
3a)
```
1 | 1   2   3   4
2 | 1   2   6
3 | 3   4   4   4
```
Key: 2/6 represents 26 bicycles.
3b) 11 3c) 22 3d) 23 3e) 34
EXTENSION WORK
4) Could be
```
1 | 5   8
2 | 3   5   6   7
3 | 0   8   9
4 | 1   3   5
5 | 0
```
Key: 2/3 represents 23 marks

5a) 35   b) 24   c) 23 to the nearest whole number
d) 23   e) 24

EXERCISE 17C

1a) 1 b) 3 c) 7 and 12 d) 9 e) No mode 2) 8
3) Blue 4) No mode 5a) 13 b) -8 c) 3.5 d) $\frac{2}{5}$
6a) Blue b) π c) Dog and Cat

EXERCISE 17D

1a) 9 b) 12 c) 30 d) 80 2) 2 3) 36 cm
4a) 40 b) 16 c) 5.5 d) 0.4

EXERCISE 17E

1a) 115 b) 110 c) 15 d) 110
2a) 4 b) 5 c) 4 d) 3.6 3a) 1.2 b) 40 matches
4a) 61 b) 56 c) 99 d) 2 e) 6
5a) 0 b) 1.9 c) 5 d) 1

EXERCISE 17F

1a) 17 b) 2 c) 4 d) 2 2) Dog 3a) $\frac{80}{360} = \frac{2}{9}$ 3b) C4
3c) i) 72 students ii) C6 : 20 Students C4: 24 students
Altogether 44 students
4a) 28 b) 28 c) Friday d) 138 apples
5a) Akin b) 10 c) 10.5 d) 20
6a) ₦45 000 b) No mode c) ₦48 000 d) The average salary is ₦48 000 which is less than ₦49 000 stated in the advert. Also, the Carpenter's salary was advertised as ₦40 000.
7) i) No title ii) No label on both axes iii) Bars are of different widths.
8a) **LAGOS**

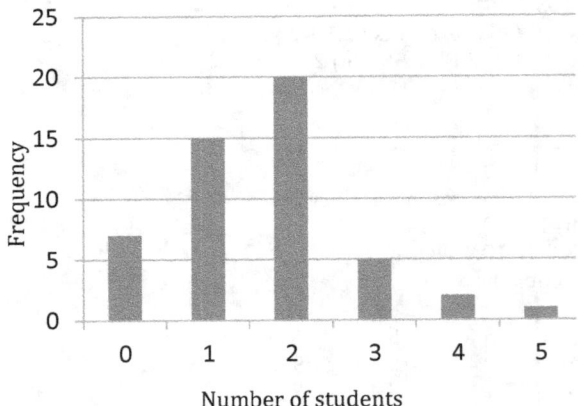

EXERCISE 17A

1a) 4 b) 9 c) 6 d) 4 e) 13.25
2a) ₦3 600 b) ₦240
3a) 1270 g b) 423.33 g
4) 287.5 g 5a) 6 b) $13\frac{1}{3}$ c) $5\frac{3}{4}$
d) 5 e) 6 m f) $1\frac{1}{2}$ 6) 37%
7a) 27 b) 8 8) 7 9a) 5.2 b) 5.8
c) 5.3 d) 2.7

EXERCISE 17B

1a) 6 b) 2.5 c) 4 d) 5 e) 40 f) 4.5
2a) 19 b) 20 c) Chuba's marks were more consistent.
3) 20°C 4) 10 5) 0.9 6a) 5.75 b) 60

ENUGU

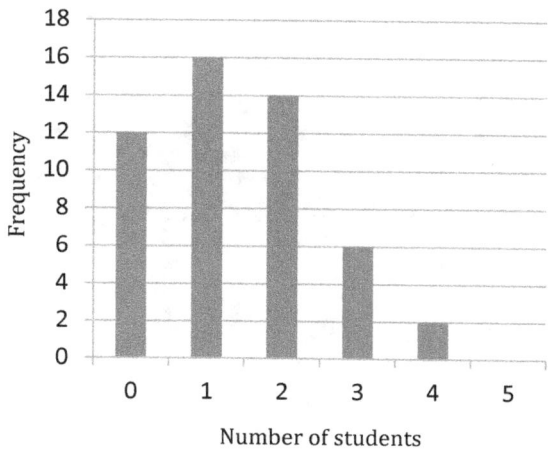

b) For Lagos:
Mean = 1.7 ≈ 2 students
Median = 2 students
Mode = 2 students
Range = 5 students

For Enugu:
Mean = 1.4 ≈ 1 student
Median = 1 student
Mode = 1 student
Range = 4 students

c) From the averages calculated, it shows that fewer students were living in households in Enugu than in Lagos. Also, the spread of students per household is bigger in Lagos than in Enugu (From their ranges).

EXERCISE 17G
1a) 21 – 25 b) 16 – 20 c) 16.7g
2a) $55 \leq w < 60$ b) $60 \leq w < 65$ c) 56.2 kg
d) Midpoints are used instead of actual numbers.
3a) 63.25 mm b) 63.67 mm c) Mean is almost the same 4a) 2.7 – 2.8 b) 2.7 – 2.8

EXERCISE 18A
1a) 10 cm b) 11.40 cm c) 9.22 cm d) 9.43 cm
e) 5.39 cm f) 19.10 cm g) 16.28 cm h) 36.06 cm

EXERCISE 18B

1a) 6 m b) 10.72 m c) 4.58m d) 5.20 m e) 6.71 m
f) 12.69 m g) 4.80 m h) 34.64 m

EXERCISE 18C

1) 7.2 cm 2) 11.3 m 3) 19.1 cm to 1 d.p 4) 8.31 cm
5) 13.9 m 6a) 21.2 cm b) y = 5 cm

EXERCISE 18D

1) P⟶ 4.47 units, Q ⟶2 units, R⟶6.40 units
S ⟶ 7.21 units, T ⟶3 units
2a) 5 units b) 5 units c) $\sqrt{29}$ units d) $\sqrt{18}$ units
e) $\sqrt{61}$ units

EXERCISE 19A

1i) 050° ii) 230° 2i) 280° ii) 100° 3a) 280°
b) 310° c) 250° 4a) i)298° ii) 118° b) i)50°
ii) 230° 5a) 120° b) 300° c) 30° d) 210°

6) Accurate construction.

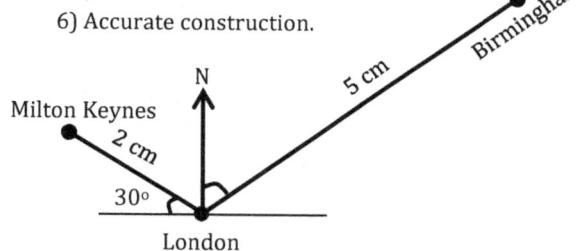

7a) 763.22 km
7b) Accurate diagram drawn by student and should look like this:

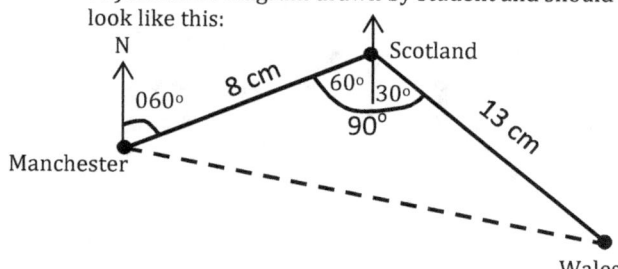

7c) Measure the distance from Wales to Manchester. (Should be around 15.26 cm or 763.22 km)
7d) They are almost the same.

EXERCISE 20A

1) $x = 5$ and $y = 2$
2) $x = 8$ and $y = 5$
3) $x = 20$ and $y = 5$
4) $x = 7$ and $y = -1$
5) $x = 3$ and $y = 1$
6) $x = 5$ and $y = 2$
7) The left side will be zero. The two lines if plotted would be parallel because they have the same gradients (2) and there won't be a point of intersection. Therefore, it is impossible to solve the simultaneous equation.
8) a) $t - h = 37$ and $t + h = 67$
   b) Tom is 52 years and Harry, 15 years.
   c) 780 years

INDEX

A

Adding and subtracting fractions	93
Adding and subtracting mixed numbers	99
Adding and subtracting negative numbers	139
Algebra 1	158
Algebraic fractions	248
Alternate angles	196
Angles and parallel lines	196
Angles at a point	183
Angles in quadrilaterals	192
Area and perimeter	199
Area of 2-d shapes	203
Area of a circle	213
Area of compound shapes	210
Averages and range	285
Averages from diagrams	292
Averages from grouped data	295

B

Bearings	302
Brackets	239

C

Chapter 1 review section	30
Chapter 10 review section	230
Chapter 2 review section	77
Chapter 3 review section	115
Chapter 5 review section	144
Chapter 6 review section	156
Chapter 7 review section	179
Chapter 8 review section	198
Chapter 13 review section	261
Chapter 4 Review section	131
Chapters 11 and 12 review sections	251
Chapters 14 and 15 review sections	279

Circles	182
Circles	194
Circles and perimeter	200
Compound interest	153
Correlation	281
Corresponding angles	197
Cube roots	69
Cubes	68

D

Depreciation	153
Diagrams	280
Direct proportion	235
Directed numbers	133
Dividing by 10, 100, 1000,..	25
Dividing fractions	112
Dividing whole numbers	20
Dividing/sharing ratios	234
Divisibility test	60
Double brackets	243

E

Equations	164
Equations of straight lines	269
Equations with brackets	169
Equations with unknown on both sides	170
Equilateral triangle	185
Equivalent fractions	86
Equivalent ratios	232
Estimation and Approximation	11
Expanding brackets	240
Experimental probability	259

F

Factorisation	245
Factors	45
Fibonacci sequence	221
Forming and solving equations	172
Formula and substitution	163

Fraction of an amount	106
Fractional equations	167
Fractions	82
Fractions to decimals	121
Fractions to percentages	124
Functional maths	29

G

Gradients	270
Graphs and gradients	264
Graphs and sequences	224

H

HCF	54
Horizontal line	268

I

Improper fractions	84
Indices	37
Inequalities	174
Inequalities and regions	177
Intercept	274
Inverse proportion	237
Isosceles triangle	185

L

Laws of Indices	70
LCM	40
LCM of algebraic expressions	247
Length of a line segment	301
Line of best fit	282
Linear sequence	221
Lines and angles	183

M

Map, scale and ratio	238
Mean	286
Median	287
Mid-point of a line segment	273
Mixed numbers	84
Mode	288
Multiples	38
Multiplying and dividing negative numbers	142
Multiplying by 10, 100,…	23
Multiplying fractions	102
Multiplying mixed numbers	109
Multiplying whole numbers	14
Mutually exclusive events	256

N

Negative numbers	134
Nth term	225
Number work	6

O

Order of operations (BIDMAS)	159

P

Parallel lines	276
Parallelogram	190
Percentage change	149
Percentage increase and decrease	148
Percentages 2	145
Percentages of a quantity	146
Percentages to fractions	122
Percentages1	118
Perimeter	199
Perpendicular lines	276
Polygons	182
Polygons	194
Powers, squares and roots	67
Prime factors	48
Prime numbers	47
Probability	252
Probability scale	253
Product of prime factors	50
Product of prime factors for LCM	41
Pythagoras' theorem	297

Q

Quadratic Graphs	277
Quadratic sequences	227
Quadrilaterals	189

R

Range	289
Ratio and proportion	231
Reciprocal and inverse	262
Relative frequency	259
Reverse percentages	154
Rhombus	189

S

Scatter diagram	281
Sequences	220
Shaded areas	212
Significant figures	7
Simple interest	150
Simplifying fractions	91
Simplifying ratios	232
Simultaneous equations	305
Solving inequalities	176
Standard form	72
Stem and leaf diagram	283
Straight line graphs	265
Substitution	160

T

Term numbers	223
Trapezium	190
Tree diagrams	258
Triangle numbers	221
Triangles and properties	185
Two-way tables	257

V

Venn Diagram for HCF	57
Venn Diagram for LCM	42
Venn diagrams	247
Vertical line	268
Vertically opposite angles	183

Y

$Y = MX + C$	274

www.ingramcontent.com/pod-product-compliance
Lightning Source LLC
Chambersburg PA
CBHW082321220526
45470CB00008B/2368